Crust/Mantle Recycling at Convergence Zones

NATO ASI Series

Advanced Science Institutes Series

A Series presenting the results of activities sponsored by the NATO Science Committee, which aims at the dissemination of advanced scientific and technological knowledge, with a view to strengthening links between scientific communities.

The Series is published by an international board of publishers in conjunction with the NATO Scientific Affairs Division

A Life Sciences
B Physics

Plenum Publishing Corporation
London and New York

C Mathematical
and Physical Sciences
D Behavioural and Social Sciences
E Applied Sciences

Kluwer Academic Publishers
Dordrecht, Boston and London

F Computer and Systems Sciences
G Ecological Sciences
H Cell Biology

Springer-Verlag
Berlin, Heidelberg, New York, London,
Paris and Tokyo

Series C: Mathematical and Physical Sciences - Vol. 258

Crust/Mantle Recycling at Convergence Zones

edited by

Stanley R. Hart

and

Levent Gülen

Department of Earth Atmospheric,
and Planetary Sciences, Massachusetts
Institute of Technology, Cambridge, MA, U.S.A.

Kluwer Academic Publishers

Dordrecht / Boston / London

Published in cooperation with NATO Scientific Affairs Division

Proceedings of the NATO Advanced Research Workshop on
Crust/Mantle Recycling at Convergence Zones
Antalya, Turkey
25–29 May 1987

Library of Congress Cataloging in Publication Data

NATO Advanced Research Workshop on 'Crust/Mantle Recycling at
  Convergence Zones' (1987 : Antalya, Turkey)
    Crust/mantle recycling at subduction zones : proceedings of the
  NATO Advanced Research Workshop on 'Crust/Mantle Recycling at
  Convergence Zones', held in Antalya, Turkey, 25-29 May 1987 / edited
  by Stanley R. Hart, Levent Gülen.
      p.   cm. -- (NATO ASI series. Series C, Mathematical and
  physical sciences ; vol. 258)
    Includes index.

    1. Earth--Crust--Congresses. 2. Earth--Mantle--Congresses.
  3. Plate tectonics--Congresses.   I. Hart, Stanley R. (Stanley
  Robert), 1935-   . II. Gülen, Levent, 1953-   . III. Title.
  IV. Series: NATO ASI series. Series C, Mathematical and physical
  sciences ; no. 258.
  QE511.N27 1987
  551.1'3--dc19   88-31488
                                                          CIP

ISBN-13: 978-94-010-6891-8   e-ISBN-13: 978-94-009-0895-6
DOI: 10.1007/978-94-009-0895-6

Published by Kluwer Academic Publishers,
P.O. Box 17, 3300 AA Dordrecht, The Netherlands.

Kluwer Academic Publishers incorporates the publishing programmes of
D. Reidel, Martinus Nijhoff, Dr W. Junk, and MTP Press.

Sold and distributed in the U.S.A. and Canada
by Kluwer Academic Publishers,
101 Philip Drive, Norwell, MA 02061, U.S.A.

In all other countries, sold and distributed
by Kluwer Academic Publishers Group,
P.O. Box 322, 3300 AH Dordrecht, The Netherlands.

This book contains the proceedings of a NATO Advanced Research Workshop held within the programme of activities of the NATO Special Programme on Global Transport Mechanisms in the Geo-Sciences running from 1983 to 1988 as part of the activities of the NATO Science Committee.

Other books previously published as a result of the activities of the Special Programme are

BUAT-MENARD, P. (Ed.) – *The Role of Air-Sea Exchange in Geochemical Cycling* (C185) 1986

CAZENAVE, A. (Ed.) – *Earth Rotation: Solved and Unsolved Problems* (C187) 1986

WILLEBRAND, J. and ANDERSON, D. L. T. (Eds.) – *Large-Scale Transport Processes in Oceans and Atmosphere* (C190) 1986

NICOLIS, C. and NICOLIS, G. (Eds.) – *Irreversible Phenomena and Dynamical Systems Analysis in Geosciences* (C192) 1986

PARSONS, I. (Ed.) – *Origins of Igneous Layering* (C196) 1987

LOPER, E. (Ed.) – *Structure and Dynamics of Partially Solidified Systems* (E125) 1987

VAUGHAN, R. A. (Ed.) – *Remote Sensing Applications in Meteorology and Climatology* (C201) 1987

BERGER, W. H. and LABEYRIE, L. D. (Eds.) – *Abrupt Climatic Change – Evidence and Implications* (C216) 1987

VISCONTI, G. and GARCIA, R. (Eds.) – *Transport Processes in the Middle Atmosphere* (C213) 1987

SIMMERS, I. (Ed.) – *Estimation of Natural Recharge of Groundwater* (C222) 1987

HELGESON, H. C. (Ed.) – *Chemical Transport in Metasomatic Processes* (C218) 1987

CUSTODIO, E., GURGUI, A. and LOBO FERREIRA, J. P. (Eds.) – *Groundwater Flow and Quality Modelling* (C224) 1987

ISAKSEN, I.S.A. (Ed.) – *Tropospheric Ozone* (C227) 1988

SCHLESINGER, M. E. (Ed.) – *Physically-Based Modelling and Simulation of Climate and Climatic Change 2 vols.* (C243) 1988

UNSWORTH, M. H. and FOWLER, D. (Eds.) – *Acid Deposition at High Elevation Sites* (C252) 1988

KISSEL, C. and LAY, C. (Eds.) – *Paleomagnetic Rotations and Continental Deformation* (C254) 1988.

TABLE OF CONTENTS

Dedicated to Prof. Dr. Gürol Ataman

1936-1986

Founder of the Department of Earth Sciences
Hacettepe University, Ankara, Turkey

Front row: Yalcin, Javoy, Turcotte, Kay, Weis, White,
    Nataf, McCulloch
Middle row: Jeanloz, Allegre, Hart, Zindler, Morris,
    Hawkesworth, Gülen, Eyidogan, Ansel, Gülec, Grönvold,
    Vlaar
Back row: Veizer, Jacobsen, Christensen, Jordan, Sacks,
    Eggler, Yilmaz, Canitez, Boztug, Kasapoglu, Davies,
    Göndogdu, Salters, Dupre, Spakman

PREFACE

This book consists of a collection of papers presented at the NATO Advanced Research Workshop (ARW) on "Crust/mantle Recycling at Convergence Zones," held in Antalya, Turkey, between May 25 to 29, 1987. The workshop was attended by 36 earth scientists from ten countries and 28 papers were presented.

Crust/mantle recycling is one of the most fundamental processes in the Earth. The study and understanding of this process requires the consideration of the Earth as a whole system including the atmosphere, the hydrosphere and the core, as well as the crust and the mantle; effective interdisciplinary collaboration is therefore essential to our progress. The Antalya ARW gave us the opportunity to assemble key specialists from relevant branches of the earth sciences and to address our state of knowledge. This ARW proved to be very useful in attaining an interdisciplinary, mutual understanding among specialists from diverse fields such as isotope and trace element geochemistry, mineral physics, theoretical geophysics, seismology, experimental petrology, and structural geology.

Plate tectonics provides a simple model with which to study crust/mantle recycling. Oceanic crust is created from the mantle at mid-ocean ridges (R-process). As a result of seafloor spreading in the course of its transportation to subduction zones, the crust interacts with seawater, and its chemistry is modified due to alteration (A-process). The existing high heat flow, which tends to decay smoothly with increasing age of the oceanic crust, causes metamorphism of the crust (ME-process). Also a sedimentary column, which is ultimately derived from the continental crust, is added on top of the oceanic crust (SE-process). This whole package eventually goes back into the mantle at subduction zones (S-process), thus completing one full cycle, which we may refer to as the RAMESES cycle.

However, crust/mantle recycling cannot be depicted and modeled by the RAMESES cycle only. The accumulated geochemical and geophysical data within the last decade appear to require an additional cycle involving depletion modification and delamination processes in the subcontinental lithosphere. This cycle can be considered as an intra-mantle cycle which eventually mixes non-convecting and convecting parts of the mantle. However, the subduction process may provide an interaction between this cycle and the RAMESES cycle, potentially supplying a "crustal input" to the otherwise purely intra-mantle cycle. Both of these cycles consist of different processes and may require different time scales to be completed. Also the apparent intermittent and possibly partial interactions of the two cycles make the so-called overall crust/mantle recycling quite a complex geodynamical problem.

We think that, for the advancement of our knowledge in the crust/mantle recycling problem, our future efforts should be directed toward the construction of a time dependent, three-dimensional dynamical model of the whole earth which would satisfy all of the observational and experimental facts within the context of the total energy and mass balance of the system.

In reaching for this ultimate goal, geochemists need to better characterize the different terrestrial reservoirs, with a special emphasis given to the subcontinental reservoir. The geochemical data base needs to be augmented with multi-isotope tracers which are sensitive to different processes. The earth and space-based geophysical observations ought to have finer 3-D resolution, better world-wide coverage, and long-term monitoring capabilities. Recent advances in super computers, non-linear dynamics theories, seismic tomography studies, fractal geometry and chaos applications promise a bright and productive future for core and mantle convection studies, and will lead towards a better understanding of the geodynamics of the earth.

We gratefully acknowledge the funds provided for the ARW by the NATO Scientific Affairs Division, Scientific and Technical Research Council of Turkey, Ministry of Energy and Natural Resources of Turkey, Etibank, Mineral Research and Exploration Institute, and Turkish Coal Works.

We also wish to express our appreciation to Profs. Mumin Köksoy, K.E. Kasapoglu, Sefik Ozanözgü, and Niyazi Gündogdu of Hacettepe University for their organizational and logistic support.

Finally, we thank all the participants in this Advanced Research Workshop and the "camera-ready" manuscript contributors who made it possible to put together this end product.

July 11, 1988
Cambridge, MA, USA

Stanley R. Hart
Levent Gülen

LIST OF PARTICIPANTS

Francis Albarede*
Centre de Recherches
Petrographiques et
Geochimiques and Ecole
Nationale Superieure de
Geologie, 54502 Vandoeuvre-
les Nancy Cedex, France

Claude J. Allegre
Institut de Physique du
Globe de Paris, Lab. de
Geochimie et Cosmochimie,
Tour 14, place Jussieu,
75252 Paris, Cedex 05,
France

V. Ansel
Lab. de Geophysique et Geo-
Dynamique Interne,
Universite de Paris-Sud,
91405 Orsay, France

D. Boztug
Dept. of Earth Sciences,
Hacettepe University,
Beytepe, Ankara, Turkey

N. Canitez
Mining Faculty, Istanbul
Technical University,
Tesvikiye, Istanbul, Turkey

U.R. Christensen
Max-Planck Inst. fur
Chemie, Saarstrasse 23,
6500 Mainz, F.R. Germany

Geoffrey F. Davies
Research School of Earth
Sciences, Australian
National University, GPO
Box 4, Canberra, ACT 2601,
Australia

Bernard Dupre
Institut de Physique du
Globe de Paris, Lab. de
Geochimie et Cosmochimie
Tour 14, place Jussieu
75252 Paris Cedex 05,
France

David H. Eggler
Geosciences Department, The
Pennsylvania State
University, University
Park, PA 16802, USA

H. Eyidogan
Istanbul Technical
University, Tesvikiye,
Istanbul, Turkey

K. Grönvold
Nordic Volcanological
Inst., University of
Iceland, Geoscience
Building, 101, Reykjavik,
Iceland

N. Gülec
Dept. of Geology, Middle
East Technical University,
Ankara, Turkey

Levent Gülen
Dept. of Earth, Atmospheric
and Planetary Sciences,
Massachusetts Institute of
Technology, Cambridge, MA
02139, USA

N. Gündogdu
Dept. of Earth Sciences,
Hacettepe University,
Beytepe, Ankara, Turkey

Stanley R. Hart
Dept. of Earth, Atmospheric
and Planetary Sciences,
Massachusetts Institute of
Technology, Cambridge, MA
02139, USA

C.J. Hawkesworth
Dept. of Earth Sciences,
The Open University, Walton
Hall, Milton Keynes, MK7
6AA, United Kingdom

S.B. Jacobsen
Dept. of Earth and
Planetary Sciences, Harvard
University, Cambridge, MA
02138, USA

Marc Javoy
Universite Paris VII, IPGP,
Lab. de Geochimie des
Isotopes Stabiles, 2, place
Jussieu, 75251 Paris, Cedex
05, France

Raymond Jeanloz
Dept. of Geology and
Geophysics, University of
California, Berkeley, CA
94720, USA

Thomas H. Jordan
Dept. of Earth, Atmospheric
and Planetary Sciences,
Massachusetts Institute of
Technology, Cambridge, MA
02139, USA

K. Ercin Kasapoglu
Dept. of Earth Sciences,
Hacettepe University,
Beytepe, Ankara, Turkey

Robert W. Kay
INSTOC, Snee Hall, Dept. of
Geological Sciences,
Cornell University, Ithaca,
NY 14853, USA

M. Köksoy
Dept. of Earth Sciences
Hacettepe University,
Beytepe, Ankara, Turkey

M. T. McCulloch Research
School of Earth Sciences,
The Australian National
University, Canberra, ACT,
2601, Australia

Julie Morris
Carnegie Inst. of
Washington, Dept. of
Terrestrial Magnetism, 5241
Broad Branch Rd., NW,
Washington, DC 20015, USA

H.C. Nataf
Ecole Normale Superieure,
Dept. Geologie, 24, rue
Lhomond, 75231 Paris, France

M. Rabinowicz*
CNES/CRGS, 18, Ave. Edouard
Belin, 31055 Toulouse
Cedex, France

I. Selwyn Sacks
Dept. of Terrestrial
Magnetism, Carnegie Inst.
of Washington, Washington,
DC, 20015, USA

V.J.M. Salters
Dept. of Earth, Atmospheric
and Planetary Sciences,
Massachusetts Institute of
Technology, Cambridge, MA
02139, USA

W. Spakman
Dept. of Geophysics
Institute of Earth
Sciences, PO Box 80.0213508
TA Utrecht, The Netherlands

M. Nafi Toksöz*
Earth Resources Laboratory,
Dept. of Earth, Atmospheric
and Planetary Sciences,
Massachusetts Institute of
Technology, Cambridge, MA
02139, USA

D.L. Turcotte
Dept. of Geological
Sciences, Cornell
University, Ithaca, NY
14853, USA

J. Veizer
Derry Laboratory, Dept. of
Geology, University of
Ottawa, and Ottawa-Carleton
Geoscience Center, Ottawa,
Ontario K1N 6N5, Canada

N.J. Vlaar
Dept. of Geophysics
University of Utrecht, PO
Box 80.021 3508 TA,
Utrecht, The Netherlands

Dominique Weis
Lab. Associes Geologie-
Petrologie-Geochronologie,
Universite Libre de
Bruxelles, Ave. F.D.
Roosevelt, 50B-1050,
Brussels, Belgium

William M. White
Dept. of Geological
Sciences, Cornell
University, Ithaca, NY
14853, USA

H. Yalcin
Dept. of Earth Sciences,
Hacettepe University,
Beytepe, Turkey

Y. Yilmaz
Dept. of Geology, Istanbul
Technical University,
Tesvikiye, Turkey

A. Zindler
Lamont-Doherty Geological
Observatory, Palisades, NY
10964, USA

*was unable to attend, but
contributed to this volume

MANTLE CYCLING: PROCESS AND TIME SCALES

Claude J. ALLEGRE
Laboratoire de Geochimie et Cosmochimie
Institut de Physique du Globe de Paris
4, Place Jussieu, 75252 Paris Cedex 05
France

1. INTRODUCTION

Since the discovery of sea-floor spreading it was clear that the mantle regime was a gigantic cycle from ridge crest, to subduction zone to the convective regime of the mantle.

However, it took a while before we realized that in fact such a cycle was the key process for the understanding of the nature and the structure of the Earth's mantle. R. Armstrong clearly was the first to realize this [1]. However, for a long time we remained mired in concepts like primitive fertile mantle generating basalt or a more or less infinite reservoir for the mantle [2].

The discovery of the depleted character of upper mantle by isotopic studies [3,4,5,6] should have made the idea of recycling obvious, but it did not. Now the idea of recycling is well established and in fact can be substantiated by two steps:

1.1 A simple kinetic calculation has shown that if the convective regime is restricted to upper mantle, no significant volume of pristine mantle exists anywhere [7]. All of the convective upper mantle (with the exception of stagnant subcontinental lithosphere) has passed at least once through the ridge process. With a model of whole mantle convection, the argument is a little less dramatic but not by much (Fig. 1).

1.2 Since the ridge crest produces a cake with an oceanic crust having a basaltic composition and a residual ultramafic lithosphere, such an assemblage has to be mixed and mechanically homogenized to produce basalts with olivine at the liquidus. Ringwood's argument against recycling (residual peridotite, not fertile [2]), and Yoder's argument (olivine on the liquidus in basalts [8]) against a pure eclogitic mantle disappear if we admit that mechanical mixing is a very important process in mantle dynamics. Such mixing produces a marble cake mantle where eclogite and

1

S. R. Hart and L. Gülen (eds.), Crust/Mantle Recycling at Convergence Zones, 1–14.
© 1989 by Kluwer Academic Publishers.

ultramafics are intimately mixed. Within such general
framework several questions have to be addressed:

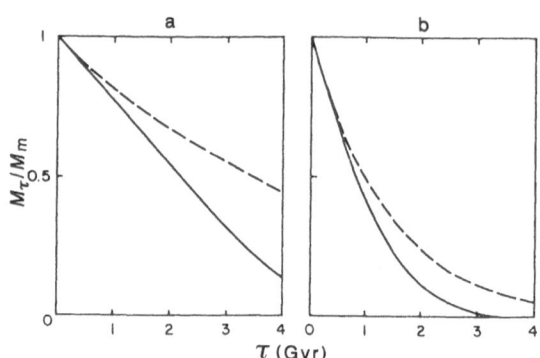

Figure 1: Proportion of virgin mantle which has not
been recycled into the mantle.

- How is the continental extraction coupled with the
sea floor spreading cycle?
- Do we have some process which reinjects into the
mantle not only oceanic crust, but also continental crust?
- Since the oceanic crust is in contact with the
hydrosphere and such hydrosphere is in equilibrium with
atmosphere, do we have reinjection of some atmophile-
hydrophile elements into the mantle?
Instead of giving firm answers to such very important
questions, I will try to insist on the logic of the
geodynamic cycle and ask as many questions as the number of
tentative answers I will give.

2. CONTINENTAL EXTRACTION AND GEODYNAMIC CYCLE

Two a priori possibilities can be considered for the key
process of continental extraction:
(1) The downgoing slab is the source of arc magmas.
Scenario: The melting of the altered downgoing slab
generates arc basalts. By magmatic differentiation, the arc
basalts produce andesites and then, eventually, continental
crust. In this model the oceanic crust is stripped of its
"sialic" elements. This slab is mixed into the mantle
creating the depleted character of the mantle. This
depleted character is then "automatically" the complement of
the continental crust (Fig.2).

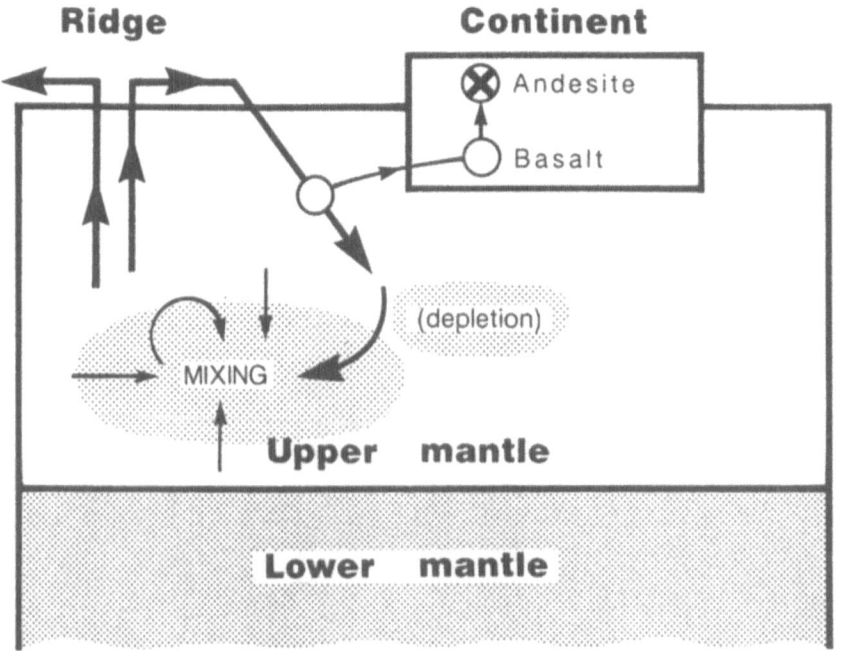

Figure 2: Scenario in which the downgoing slab is melted, generating the arc magma. The residual slab goes into the upper mantle and mixes with the surrounding material creating the depleted upper mantle.

(2) The melting, which creates arc magmas, occurs only at subduction wedge.

In this case, the residual material of the continental extraction stays on the wedge.

(a) If the residual wedge is entrained with the subduction slab and mixed together into the mantle, then the whole process has a very similar result as the one above, in respect to the depletion mechanism (Fig. 3a).

(b) If after melting, the residual wedge follows a different path than the subduction, then the residual mantle created by continental extraction is not spacially related to subduction.

Different scenari can be built up in this case. For example:

- The slab penetrates the lower mantle, and the residual wedge is mixed with the upper mantle (Fig. 3b). This scenario will accommodate Creager and Jordan's interpretation of seismic paths at subduction zones (9), as well as isotopic constraints on MORB genesis.

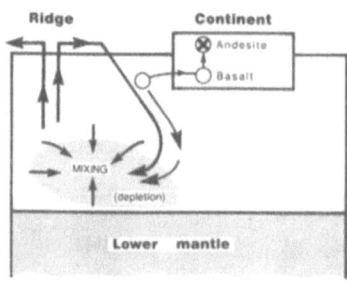

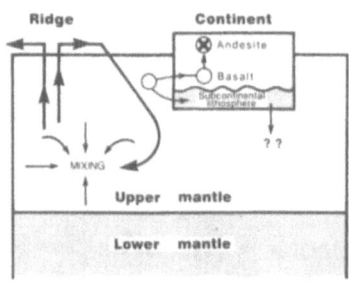

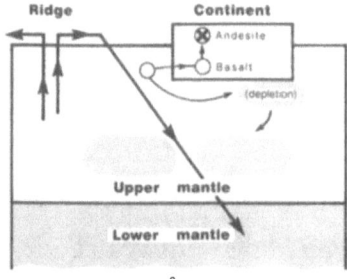

Figure 3: In this scenario, the melting which creates arc magmas occurs at the wedge. But in (a) the residual wedge is entrained with the slab and mixed with the upper mantle creating the depletion. In (b) the residual wedge is incorporated as continental lithosphere. The depletion of the upper mantle occurs because of delamination. In (c) the residual wedge is mixed in the upper mantle independently of the subduction kinematics.

- The residual wedge ultimately becomes the subcontinental lithosphere. In this scheme the residual counterpart (depleted) of the continental crust would be the subcontinental lithosphere (Fig. 3c). Isotopic composition of the MORB does not support such a model because in such a case, upper convecting mantle would be more or less pristine unless the continental lithosphere were continuously re-mixed with the upper mantle.

However these two extreme models are not the only possibilities and <u>partial</u> decoupling of continental extraction and the sea floor spreading petrogenetic cycle can occur, creating large regional heterogeneities within the convecting upper mantle. This third type of model will decouple the sea floor spreading cycle from continental growth regarding the mass transfer processes.

(3) The slab-triggered wedge melting model (like dehydration of the slab-inducing melting on the wedge) is a more moderate middle-ground model than the previous models and is certainly one of the most popular today. The question of what happens to the residual wedge and the subducted slab is crucial. The important point of this scenario is the fact that the geodynamic cycle can be <u>different for different lithophile elements</u>. An element soluble in water like Rb (or Sr) will have a more important contribution to the mantle wedge than an insoluble element (Nd, Sm). This kind of behavior may explain the partial decoupling between different elements. Geodynamic cycles will explain the diversity of geochemical cycles (Fig. 4).

3.   RECYCLING OF SURFACE MATERIAL INTO THE MANTLE

The recycling of oceanic crust is not a problem. It is mandatory. The only question is the mechanical path of this recycling. On the other hand, there are four geodynamic units for which mantle recycling is a question: continental crust, oceanic water, sediments and continental lithosphere.

3.1   Continental Crust

There are two mechanisms in theory: mantle erosion during subduction; and erosion and sediment. The first mechanism is conjectured. I don't know any observational evidence for it. Recent results on ^{10}Be and lead isotope in the Antilles show clearly that sediments are involved in subduction (10,11). On a global scale, the isotopic <u>evolution budget</u> shows that <u>recycling</u> of part of the <u>continental crust</u> during geological time is mandatory (12).

3.2   Sediments

The problem seems to be qualitatively, but not quantitatively, solved (1). Of the 3 to 6 km of sediment produced annually, how much goes to the mantle? How much is recycled through the continents? In such a dichotomy do all types of sediments behave uniformly? Should we have to recalculate the Garrels''s cycle (13) in such a perspective?

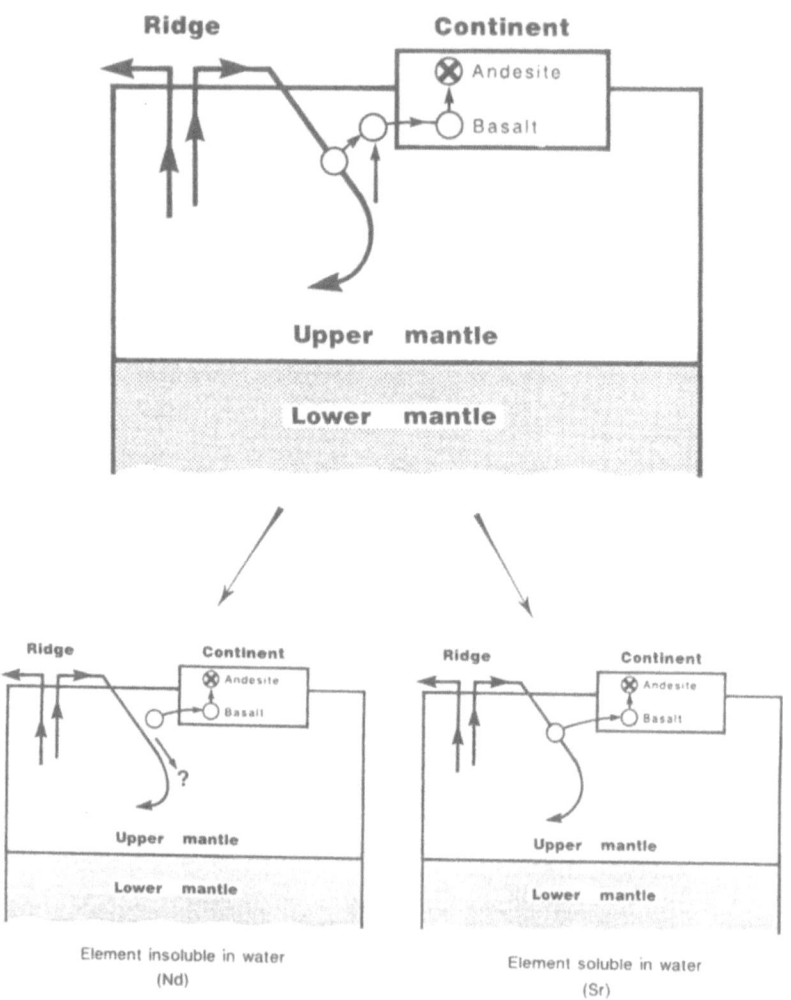

Figure 4: In this scenario a double mechanism is involved at the subduction zone. One dehydrates the slab; the other melts the wedge. The first triggers the second. Depending on the element, the mechanism for transfer from mantle to crust is determined by one process or the other.

Another important question is related to carbonate recycling. Is such recycling volumetrically important? Is the CO_2 of the upper mantle completely recycled (14)? If carbonates are recycled, and since Sr is enriched in these sediments, should we consider a preferential recycling of Sr rather than Nd or Pb?

3.3 Sea Water

The oceanic crust is altered by interaction with sea water
(15, 16, 17). Two processes are important, one at high and
one at low temperature (17).
      Such interactions result in an enrichment of oceanic
crust in uranium, rubidium, strontium and some adsorption of
dissolved gases and, of course, an increase of sea water
content. What happens to this material at subduction zones?
Is this material completely released by dehydration and/or
melting? Does some altered material pass this barrier and
go into the mantle (18)? There are no clear, obvious
answers to such fundamental questions, yet.

3.4 Subcontinental lithosphere

To explain the continental plateau, and also the presence of
100 km lithosphere under a crustal doubling in Tibet,
geophysicists have proposed the concept of delaminated
continental lithosphere (19,20). This concept was used by
McKenzie and O'Nions (21) to explain some features observed
on oceanic islands. This hypothesis is indeed very
attractive, especially for the Indian Ocean anomaly in OIB,
but better documentation is needed (22).

4. PROCESS AND WITNESS

The different materials of the Earth's surface (minerals,
rocks, continents, mantle units, etc.) can be considered as
distributions of elements (multidimensional vectors with
space and time coordinates). The different geological
phenomena can be considered as operators which transform the
distributions into each other. These operations can be
separated into two types: differentiation and mixing (23).
      Differentiation is a chemical operation which separates
one distribution into two, three or more distributions,
which can then be considered as a transformation of one
distribution (initial) to another (final).
      Mixing is an operation which combines several
distributions to form a single one.
      An adequate mathematical formula is available to
describe the different operations for concentrations,
isotopic ratios or trace element ratios when their physical
nature is determined. However, the physics of many of these
mechanisms are still subject to debate. They fall into the
general class of problems named process identification which
is a fundamental question in geochemistry.
      Keeping the formula of operators and distributions, we
can try to describe the geodynamic cycles with this formula.
Such an approach (without going into the mathematical

formula) will emphasize what is known and what is unknown or debatable.

The geodynamic cycle can be described in the following way:

(1) At ridge crest, a differentiation process, the ridge process, occurs (R-process). Starting with mantle materials, it differentiates oceanic crust (OC) from residual ultramafic lithosphere (OL). It creates a distribution:

Symbolically we can write:

$$R \{Mantle\} = \begin{bmatrix} OC \\ OL \end{bmatrix}$$

(2) At ridge flank, and during spreading, an hydrothermal interaction between ocean and oceanic crust occurs. Such an operation is a mixing process which creates a distribution:

Oceanic crust ± oceanic input = Altered oceanic crust = AOC

Then, the $\begin{bmatrix} AOC \\ OL \end{bmatrix}$ arrives at the subduction zone.

(3) At the subduction zone a second differentiation occurs, a subduction process named (S-process). Such a process extracts material and transfers it through the surface. Some of this material is andesitic in composition. Where such a process occurs, at slab (Ss) or at wedge (Sw), is a question of debate but a fundamental one

Andesitic material = S {altered oceanic crust + wedge}

Therefore, residual altered oceanic crust is $S^{-1}(AOC)$.

(4) After the S-process the modified oceanic crust and residual oceanic lithosphere mix with the mantle and also with the surrounding average mantle. Such mixing creates the marble cake mantle:

$$Marble\ cake = Mixing\ \begin{bmatrix} S^{-1}(AOC) + Average\ Mantle \\ OL \end{bmatrix}$$

(5) Linked or not with this cycle the <u>external differentiation of continental crust</u> is a process which manufactures a piece of continental crust from mantle material. How such cooking is done, nobody knows exactly. It seems to have been done by magmatic process but the

majority of continental rocks are sedimentary and
metamorphic. How can we reconcile these two observations?

Continental Crust = S {Altered Oceanic Crust}?
or Continental Crust = S {wedge upper mantle}
or a linear combination of both.

(6) The genesis of oceanic islands (OIB) is not
obviously part of the normal geodynamic system. Initially,
oceanic islands were considered to have been created in the
pristine lower mantle. This hypothesis is not supported
anymore. An alternative hypothesis for the generation of
OIB has been proposed, i.e., they are created in a transient
boundary layer situated at the bottom of the upper mantle.
This boundary layer is alimented by a subduction slab,
perhaps containing thin sediments, and also by average
mantle. After a certain period of stagnancy, this boundary
layer is mixed normally with the upper mantle. The time
scale of this path is certainly variable depending on the
spreading rate and the mechanisms of the boundary layer
instabilities. Since (OIB) show large heterogeneities, the
question is what kind of materials contribute to their
genesis (sediments, continental lithosphere, etc.) since it
is known that freshly injected slab + normal upper mantle
are normal components of the OIB source.

Oceanic island = Melting ({Marble cake} + {"external
addition"})

given that the external addition is continental lithosphere,
or altered oceanic crust. Then the question is--what is the
external addition?
    To test all these hypotheses, we use two types of
isotopic tracer materials: a) volcanic rocks, where only
isotopic ratios are properly conserved but which sample
large volumes of mantle; and b) ultramafics where isotopic
ratios and chemistry are conserved but scarce and sample a
limited part of the mantle. In the past, mainly basalts
have been used; in the future, we should combine both
approaches to get an answer.

5.   TECHNICAL COMMENTS: FROM AVERAGE VALUES TO STATISTICS

In another point of view the depletion of the upper mantle
can be measured for isotopic ratios (or element contents or
ratios) by a distribution with an average value, a
dispersion and a skewness. If the coupling between
continental extraction and geodynamic cycle is good, the
dispersion can be expected to have random variation around
the mean value (gaussian histogram). If the coupling is

poor it is a two-parameter distribution and will have far
more complex histograms of values.

5.1 TIME SCALES

Isotope variations are linked with a certain time scale. In
a simple closed system the link is quite obvious and
isotopic ratios, parent/daughter ratios and time are simply
related. In a more complex system like a convective mantle,
the relationship is far more complex than just the simple
"mantle isochron" concept.

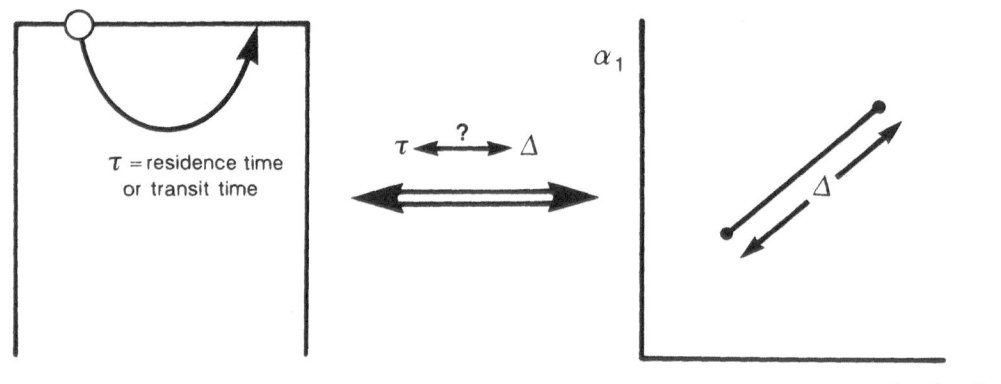

Figure 5: We can define a transfer time for material
from the surface to the mantle. We can measure the
difference in isotopic ratios. How can we link one to
the other?

As shown previously (24), the slope of a Pb-Pb or Rb-Sr
line in a medium convecting for 4.5 b.y. is not related to
any particular age of differentiation but to a mixing
parameter (M) defined in the mean-field approximation
approach. When M=1 (perfect mixing) all experimental points
plot as the same point. When M=0 (no mixing), the apparent
age is $3.3 \ 10^9$y.
This interpretation in contradiction to the so-called
mantle isochron concept is conforted by the different
apparent ages for different isotopic systems (like Pb-Pb and
Sm-Nd as modelled by a continuous convecting model and in
contradiction to the isochron concept).
A recent survey on various MORB by White et al. (25)
has shown variations in slope of Pb-Pb corresponding in a
continuous mixing model to variations in mixing parameters.
On the other hand, box model modelling with mean-age
concept (26) shows that the MORB source is depleted with a
mean age of $2 \pm 0.3 \ 10^9$y. Such a mean age is easily

computed by Rb-Sr systems because Rb/Sr in MORB is almost equal to zero.

5.2   Oceanic island heterogeneities

The fact that OIB are more radiogenic in Pb and Sr than "normal" MORB is clear. The question (in each case) is how are such heterogeneities created and how can they be quantitatively explained?

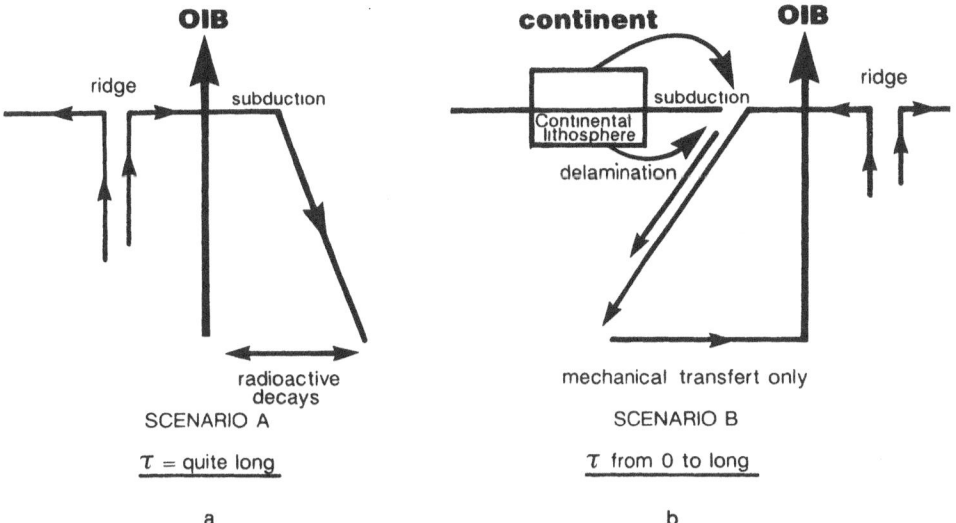

Figure 6:   Two scenari for injection: a) pure reinjection of oceanic crust. Therefore, the isotopic difference is created in situ into the mantle by readioactive decay;   b) reinjection with material from the continent. The isotopic difference is created at least partially before injection.

a)   The heterogeneities are extrinsic from the mantle, which means the heterogeneity has been created at continental crust of continental lithosphere and then transferred to the mantle. Therefore, the importance of heterogeneities is that they do not measure in any way the time scale taken in the mantle to generate (OIB). The OIB can be produced instantly after the mixing.
b) The heterogeneities are intrinsic from the mantle, which means the isotopic heterogeneities result from a chemical heterogeneity within the mantle which has been transformed with time to isotopic heterogeneities. Therefore, the time scale of the production of OIB should be compatible with isotopic observations. With extreme cases like St. Helena we need 10^9y. isolation in the mantle. With

a moderate case like Canary, we need 500 m.y. if we inject
layered oceanic crust.

We have not reached a consensus of such a dichotomy so
far. A few important time constraints should be taken into
account.

a) The discovery of Indian Ocean anomalies is a
significant signature for the structure of the OIB source.
Weiss and Demaiffe (27) have shown that this signature
lasted for at least 300 years and perhaps as long as 1 b.y.
Such a time scale will support the idea that the same
process operating in the mantle lasts for the same period of
time.

b) The discovery by Hamelin et al. (28) that the Indian
Ocean signature is present also in MORB, and by Gopel et al.
(29) that such an effect was also seen in 100 m.y.-old
ophiolites generated at the Indian Ocean, also support the
previous idea, because the OIB source should have enough
time to mix with the upper reservoir. However, at the
moment, these constraints have not been used in a definite
way to generate answers about the extrinsic-intrinsic
dichotomy.

REFERENCES

[1] R.L. Armstrong, 'A model for the evolution of strontium
     and lead isotopes in a dynamic earth,' Rev. Geophysics,
     6, 175-199, 1968.
[2] A.E. Ringwood, "Phase transformations and
     differentiation in subducted lithosphere: implications
     for mantle dynamics, basalt petrogenesis and crustal
     evolution,' J. Geol., 90, 611-643, 1982.
[3] P.W. Gast, G.R. Tilton and C. Hedge, 'Isotopic
     composition of lead and strontium from Ascension and
     Gough Islands,' Science, 145, 1181, 1964.
[4] M. Tatsumoto, 'Genetic relations of oceanic basalts as
     indicated by lead isotopes,' Science, 150, 886-888,
     1965.
[5] P. Richard, N. Shimizu and C.J. Allegre, '$^{143}Nd/^{146}Nd$ a
     natural tracer: an application to oceanic basalts,',
     Earth Planet. Sci. Lett., 31, 269-278, 1976.
[6] D.J. De Paolo and G.J. Wasserburg, 'Nd isotopic
     variation and petrogenetic model,' Geophys. Res. Lett.,
     3, 249-252, 1976.
[7] C.J. Allegre and D. Turcotte, 'Implications of a two-
     component marble-cake mantle' Nature, 323, 123-127,
     1986.
[8] H.S. Yoder, 'Generation of basaltic magma,' Nat. Acad.
     Sci. Washington, 1976.

[9] K.C. Creager and T.H. Jordan, 'Slab penetration into the lower mantle beneath the Mariana and other island arcs of the Northwest Pacific. 'J. Geophys. Res., 91 (B3), 3573-3589, 1979.

[10] J. Morris, F. Tera, I.S. Sacks, L. Brown, J. Klein and R. Middleton, 'Sediment recycling at convergent margins: constraints from the cosmogenic isotope ^{10}Be.' in: "Crust/Mantle recycling at convergence zones" Abstr. Vol. NATO Adv. Res. Workshop, Antalya, Turkey, 1987.

[11] W. White and B. Dupre, 'Sediment subduction and magma genesis in the lesser Antilles: isotopic and trace element constraints,' J. Geophys. Res., 91, 5927-5941, 1986.

[12] E. Lewin and C.J. Allegre, 'Kinematic evolution of the continent-mantle system studied by inversion of historical isotopic data,' in prep.

[13] R.M. Garrels and F.T. MacKenzie, 'Sedimentary rocks' in: Evolution of Sedimentary Rocks, W.W. Norton & Co. Inc., New York, 1971.

[14] M. Javoy, F. Pineau and C.J. Allegre, 'Carbon geodynamic cycle,' Nature, 300, 171-173, 1982.

[15] J.M. Edmond, C. Measures, R.E. McDuff, L.H. Chan, R. Collier, et al., 'Ridge crest hydrothermal activity and the balances of the major and minor elements in the ocean: the Galapagos data,' Earth Planet. Sci. Lett. 46, 1-18, 1979.

[16] F. Albarede, A. Michard, J.F. Minster, and G. Michard, '^{87}Sr/^{86}Sr ratio in hydrothermal water and deposits from the East Pacific Rise at 21°N.' Earth Planet. Sci. Lett. 55, 229-236, 1981.

[17] S.R. Hart and H. Staudigel, 'The control of alkalies and uranium in seawater by ocean crust alteration,' Earth Planet. Sci. Lett., 88, 202-212, 1982.

[18] T. Staudacher and C.J. Allegre, 'Recycling of oceanic crust and sediments: the noble gas subduction barrier,' Earth Planet. Sci. Lett., in press.

[19] P. Bird, 'Continental delamination and the Colorado Plateau,' J. Geophys. Res., 84, 7561-7571, 1979.

[20] C.J. Allegre, V. Courtillot, P. Tapponnier et al., 'Structure and evolution of the Himalaya-Tibet orogenic belt,' Nature, 307, 17-22, 1984.

[21] D. McKenzie and R.K. O'Nions, 'Mantle reservoirs and ocean island basalts,' Nature, 301, 229-231, 1983.

[22] C.J. Allegre, B. Hamelin, A. Provost and B. Dupre, 'Topology in isotopic multispace and origin of mantle chemical heterogeneities,' Earth Planet. Sci. Lett., 81, 319-337, 1986.

[23] C.J. Allegre, 'Isotope Geodynamics,' Earth Planet. Sci. Lett., in press, 1987.

[24] C.J. Allegre, O. Brevart, B. Dupre and J.F. Minster, 'Isotopic and chemical effects produced in a continuously differentiating convecting Earth mantle,' Phil. Trans. R. Soc. Lond. A., 297, 447-477, 1980.

[25] W.M. White, A.W. Hofmann and H. Puchelt, 'Isotopic geochemistry of Pacific Mid-Ocean Ridge Basalt,' J. Geophys. Res., 92, 4881-4893, 1987.

[26] C.J. Allegre, S.R. Hart and J.F. Minster, 'The chemical structure of the mantle determined by inversion of isotopic data, Part I: Theoretical methods, Part II: Numerical experiments and discussion,' Earth Planet. Sci. Lett. 66, 190-213, 1983.

[27] D. Weiss and D. Demaiffe, 'A depleted mantle source for kimberlites from Zaire, Nd-Sr and Pb isotopic evidence,' Earth Planet. Sci. Lett., 73, 269-277, 1985.

[28] B. Hamelin, B. Dupre and C.J. Allegre, 'Pb-Sr-Nd isotopic data of Indian Ocean ridges: new evidence of large scale mapping of mantle heterogeneities,' Earth Planet. Sci. Lett., 76, 288-298, 1985.

[29] C. Gopel, C.J. Allegre and R.H. Xu, 'Lead isotopic study of the Xigaze ophiolite (Tibet) the problem of the relationship between magmatites (gabbros, dolerites, lavas) and tectonites (harzburgites),' Earth Planet. Sci. Lett., 69, 301-310, 1984.

ISOTOPIC CHARACTERIZATION AND IDENTIFICATION OF RECYCLED COMPONENTS

S.R. Hart
Center for Geoalchemy
Department of Earth, Atmospheric and Planetary Sciences
Massachusetts Institute of Technology
Cambridge, MA 02139

H. Staudigel
Scripps Institution of Oceanography
University of California
San Diego, California

It has been 23 years since the oceanic mantle was first shown to be isotopically heterogeneous (Gast, Tilton & Hedge, 1964). The richness and diversity of this heterogeneous character has been expanded over the intervening years with the application of new isotopic tracer systems (Nd, Hf, He, Xe) to volcanic rocks from most of the oceanic islands. Currently, this isotopic heterogeneity can be embraced as mixtures between four or five fairly distinct isotopic components (DMM: depleted MORB mantle; HIMU: a high U/Pb component; and two enriched components, EMI and EMII; Zindler and Hart, 1986). The DMM component comprises the upper mantle, and has been depleted over geologic time by processes associated with extraction of continental crust. This is the only mantle reservoir for which a location and evolutionary process are well established. Chase (1981) and Hofmann & White (1982) proposed that subducted oceanic crust ± sediment would acquire significant isotopic heterogeneity while aging in the mantle, and that recycling of this material could provide some of the isotopic diversity observed in oceanic basalts. This model has been embraced by many geochemists, but has never been quantitatively scrutinized. Our objective here is to provide some of this scrutiny; we conclude that this model is only viable if significant trace element fractionations are imprinted on the oceanic crust during its passage through the subduction zone.

We will focus first on the geochemical nature of the oceanic crust, not in its pristine form, but in the altered form it acquires prior to subduction. We have prepared a

S. R. Hart and L. Gülen (eds.), Crust/Mantle Recycling at Convergence Zones, 15–28.
© 1989 by Kluwer Academic Publishers.

variety of composites from the upper 500 meters of crust cored at DSDP site 417/418. This crust is sufficiently old (≈120 m.y.) to have run the full gamut of high and low temperature alteration processes. These composites were carefully constructed to contain all of the crustal units encountered, in representative proportions (in contrast to most geochemical studies of oceanic crust which emphasize the "fresher" units). For further details regarding these composites, see Staudigel et al., 1988. The concentrations of various key elements in this average upper crustal section are given in the table below, in comparison with the concentrations in unaltered crust from the same site.

TABLE I. Trace element & isotopic data for upper ocean crust composite

Element	Unaltered[+] Crust (Site 418A)	Average Upper Crust Composite
H_2O*	--	2.68%
CO_2*	--	2.98%
K	430	4805 ppm
Rb	0.37	9.04
Cs	0.0086	0.173
Sr	101	117.5
Ba	7.3	16.85
Th	0.063	0.0723
U	0.035	0.321
Pb	0.28	0.69**
Sm	2.41	2.552
Nd	6.60	6.76
87/86Sr	--	0.704636 ± 25
143/144Nd	--	0.513077 ± 19
206/204Pb	--	19.30

    *From Staudigel et al., 1988; all other data by isotope dilution.

    **Many of the sub-composites contained samples which had suffered catastrophic Pb contamination during drilling and ship-board handling; this Pb value therefore does not reflect the same "sample suite" as the other data in this column.

    [+]Derived from data for basalt glasses and least-altered whole rocks from site 418A (Staudigel et al., 1979; Staudigel & Hart, 1983; Staudigel et al., 1981); U and Th concentrations were assigned so as to give a typical MORB K/U ratio of 13, 700 and a Th/U appropriate for this U (Jochum et al., 1983); Pb was derived from U/Pb relationships in MORB glasses (Cohen et al., 1980; Cohen & O'Nions, 1982 a,b).

The success of this upper crust composite in reflecting the average upper crust at these sites was favorably assessed by Staudigel et al., 1988. It can also be noted that independent estimates for K, Rb and Cs concentrations in the upper crust at these sites were made previously by Hart and Staudigel (1982) and the agreement (3990 ppm, 9.3 ppm and 0.18 ppm respectively) with the composite results in the above Table is quite satisfactory.

Table I shows that average upper oceanic crust is strongly enriched in the alkalies and U, whereas Sr, Th and the REE are relatively unchanged (relative to primary unaltered MORB). The Nd isotope ratio is MORB-like, while the Sr isotope ratio is significantly higher than unaltered glass at this site ($\approx$.7030). The 206/204 Pb ratio is high for normal Atlantic MORB, but most of this can be accounted for by 120 m.y. of radiogenic growth, given the rather high 238U/204Pb ratio for this composite ($\approx$30). The initial Sr isotope ratio at 120 m.y. is 0.70426, still much higher than MORB and reflecting the interaction of the oceanic crust with seawater (87/86Sr $\approx$0.7092)

In a qualitative sense, all of these alteration effects are normal, and have been demonstrated previously in many studies (Hart, 1969; Staudigel et al., 1979; Hart et al., 1974; Philpotts et al., 1969; Mitchell & Aumento, 1977; MacDougall, 1977). The importance of the above data lies in the quantitative coupling of the various parent/daughter ratios for a given well-averaged upper crustal section: Rb/Sr and U/Pb ratios are markedly increased during alteration, the Th/U ratio is markedly decreased, and Sm/Nd is relatively unchanged.

While the above data applies only to the upper 500 m of crust, the nature of deeper crustal changes can be assessed from the geochemistry of the axial high temperature hot springs (Albarede et al., 1981; Chen et al., 1986; Michard et al., 1983; Von Damm et al., 1985). The hot spring data shows that Sr, Th and the REE are not strongly mobilized in the deeper crust, whereas Rb is strongly extracted, Pb is mildly extracted and U is strongly added. Thus, the deeper crust will be similar to the upper crust in having high U/Pb, low Th/U, and unchanged Sm/Nd. While the Rb/Sr ratio of the deeper crust is decreased, the absolute crustal budget of both Rb and Sr is still dominated by the upper crust.

The upper crust composite data of Table I has been used to model the Sr, Nd and Pb isotopic evolution of oceanic crust subducted at various times in the past, to see if any of the isotopic diversity of oceanic basalts can be ascribed to subduction, aging and remelting of oceanic crust. The model is constructed as follows; from 4.55 to 4.0 billion years, the mantle is allowed to isotopically evolve with bulk earth parent/daughter element ratios. At 4.0 b.y., the

upper mantle is depleted, isolated, and allowed to evolve so
as to have a present day isotopic character equivalent to
the "extreme" MORB mantle of Allegre et al. (1983) - 87/86
Sr = 0.7022; 143/144 Nd = .5133; 206/204 Pb = 17.30 and
208/204 Pb = 37.20. It would make no sensible difference to
the modeling if an "average" present day MORB mantle were
assumed, or if the depletion age were allowed to be younger
than 4.0 b.y., or indeed if a continuous evolution model was
used for the MORB mantle depletion, instead of the two-stage
model used here.   We simply need to start with some
reasonable MORB mantle evolution curve.

At various times in the past, oceanic crust is derived
from this MORB mantle and altered by interaction with sea
water so as to have trace element abundances as given in
Table I.   The Nd and Pb isotope ratios are assumed to be
unchanged by this alteration, as Nd and Pb abundance levels
in seawater are very low. The Sr isotope ratio is strongly
affected by interaction with sea water, as is obvious from
the data in Table I (0.70464 in the composite versus ≤0.7030
in fresh MORB glass), and numerous published studies.   To
model this process throughout geologic time, we will assume
that the sea water Sr/MORB Sr mixing ratio is constant and
equal to the 28% value which can be calculated from Table I.
The Sr isotopic ratio of sea water is assumed to vary with
time according to the lower bound of the curve given by
Veizer and Compston (1976, Figure 2).

The Sm and Nd data were modified slightly, since it can
be seen that the Sm/Nd ratio given in Table I (0.228 atomic)
is in fact slightly higher than the second-stage model MORB
mantle value of 0.222.   The oceanic crust should have a
Sm/Nd ratio less than the MORB source mantle, both because
the melting and differentiation process decreases Sm/Nd, and
because the alteration process will tend to enhance the
light rare earths relative to the heavy rare earths, if only
slightly (see discussion in Staudigel et al., 1979).   To
account for this, we have arbitrarily fixed the altered
crust Sm/Nd ratio at a value 10% below the model mantle
ratio; we view this as a minimum possible value. A summary
of the adopted parent/daughter element ratios is given in
Table II. We have included some sediment values also, to be
discussed below.

The results of the model calculations are shown in
various isotope correlation plots, Figures 1-3, in
comparison with the field of modern oceanic basalts for
comparison. The model results are shown as locii of oceanic
crust which has been altered (and subducted) at various
times in the past; note that these locii are not growth
curves in the conventional sense. On the Nd-Sr isotope plot

Table II.   Recycling Model Parameters

	87Rb 86Sr	147Sm 144Nd	238U 204Pb	232Th 204Pb	232Th 238U
Bulk Earth (4.55 by-4.0 by)	0.0884	0.1967	8.04	29.02	3.61
MORB Mantle (4.0 by-present)	0.0456	0.2217	7.75	30.8	3.98
Sediment Source Reservoir (4.0 b.y. -present)	0.33	0.11	9.51	38.3	4.03
Altered Oceanic Crust	0.223	0.1995	30.0	6.96	0.232
Sediment	1.038	0.1177	6.6	40.3	6.1

(Fig. 1), the locus has a negative slope mildly similar to the oceanic mantle array, but displaced well above it. This position is dictated by the use of an extreme present day MORB Nd isotope ratio (0.5133); if the lowest reasonable Nd isotope ratio were used instead *(0.5130), the locus (dashed line) moves closer to the oceanic array, but is still not a reasonable explanation for any of the oceanic data.   In addition to supposing that the altered crust may recycle in a pure form (i.e., lie on the locus), we may also suppose that altered crust (of various ages) is mixed with the depleted MORB mantle to varying extents. For recent mixing, because the Nd/Sr ratio is essentially unchanged by the alteration process, mixing will generate linear lines connecting the DMM point with various points on the altered crust locus.   While this process serves to fill in the "gap" between DMM and the "zero age" end of the altered crust locus, it still does not produce either positions or trends which would explain any of the oceanic mantle field. Altered oceanic crust could, when mixed with components such as HIMU or EMI (as defined by Zindler and Hart, 1986), produce mantle sources lying within the oceanic field.

These mixing possibilities, however, are more severely constrained by the Pb isotope data, shown in Fig. 2. Because the altered crust U/Pb ratio is higher, and the Th/Pb ratio is lower, than the MORB reservoir (DMM) itself, 206/204Pb is accelerated and 208/204 Pb is retarded, relative to DMM, producing a locus of negative slope. As this particular calculation is sensitive to the Pb concentration utilized for altered crust, and because the Pb value in Table I does not pertain to the whole composite,

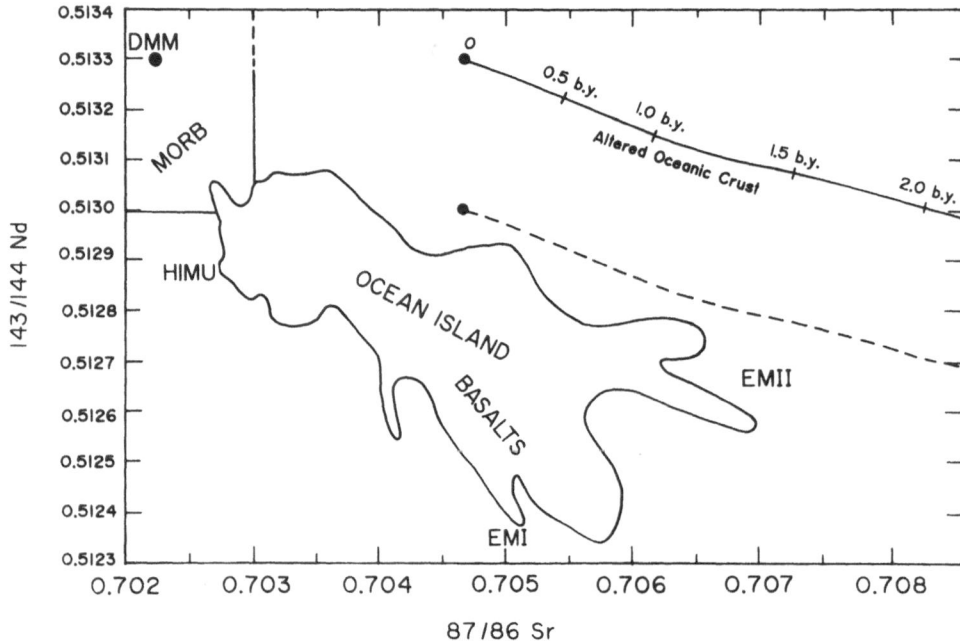

Fig. 1   Nd-Sr isotope correlation plot showing the locus of altered subducted oceanic crust as a function of the age of alteration. Solid line represents crust derived from the most depleted model mantle, with Nd isotope ratio unchanged during alteration and Sr isotope ratio increased by sea water alteration, in proportions given by the 418A upper crust composite (Table I). Dashed line is similar model but starting from a less depleted MORB mantle. Limits to present day MORB and oceanic island basalt fields shown for reference (from Hart, 1988).   The four mantle components defined by Zindler & Hart (1986) are also shown (DMM, EMI, EMII, HIMU).

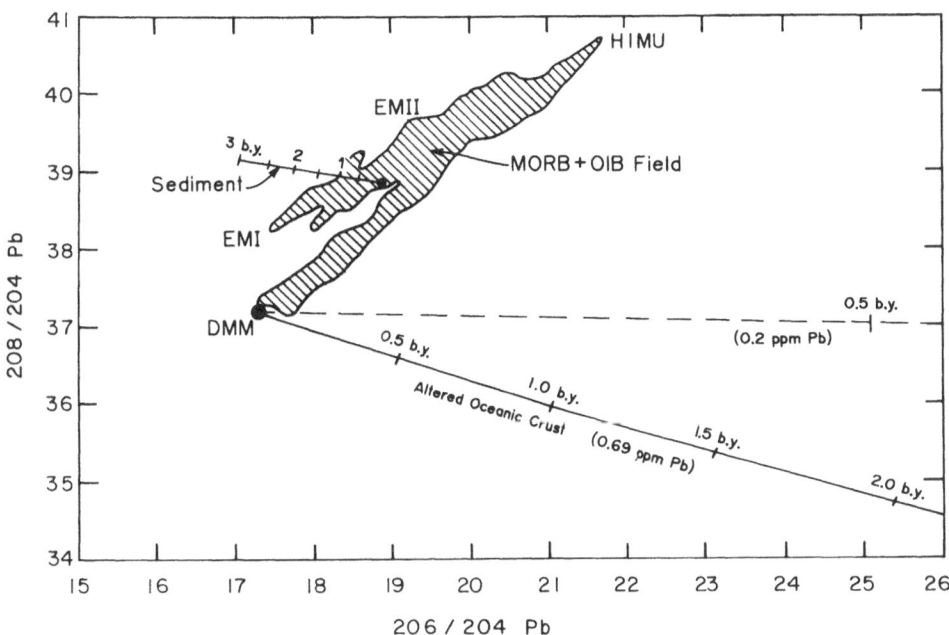

Fig. 2   Pb-Pb isotope correlation plot showing locus
of altered subducted oceanic crust as a function of the
age of alteration.   The solid line represents a model
which uses the altered crust Pb concentration given in
Table I (0.69ppm); the dashed line shows an "upper
bound" locus which uses a probable lower limit to the
Pb concentration (0.2 ppm).   The field bounding all
available oceanic basalt data (both MORB and OIB) is
shown for reference, along with the four mantle
components of Zindler & Hart (1986).   The solid curve
marked "sediment" represents the locus of sediment
subducted at various times in the past (see text for
details of this model).

but only a sample subset, we also calculated a model for a crustal Pb concentration of 0.2 ppm, assuming this to be a firm lower limit. The locus is now almost horizontal, but still clearly bearing no similarity whatsoever to the oceanic array (MORB + OIB). Note that mixing arrays on this figure are linear, and that mixtures of altered crust (from the locus) with any of the possible mantle components (DMM, EMI, EMII, or HIMU) will typically generate negative-sloped arrays, in contrast to the almost universally positive sloped arrays defined by individual oceanic islands.

Fig. 3, a Sr-Pb plot, simply further reinforces the inadequacy of old recycled oceanic crust as an explanation for the oceanic basalt data field. While young altered crust falls close to EMI on this plot, it clearly does not mimic EMI on either Fig. 1 or Fig. 2. While a very old (> 3 b.y.) crust might show some EMII tendency on Fig. 1 (off to the right on the dashed line), the 206/204 Pb ratio would be much too high (on Fig. 3, altered crust of 0.5 b.y. has about the right 206/204Pb to be EMII).

In summary, altered oceanic crust does not seem even remotely adequate for explaining any of the arrays or components needed to delineate the oceanic basalt isotope fields. While recycled sediment is not the principal focus of this study, it is a component in the original ocean crust recycling model of Hofmann & White (1982), and is also argued for by Cohen and O'Nions (1982b). We have carried out some simple model calculations to assess whether or not an altered ocean crust plus sediment package would provide a more reasonable explanation of the oceanic data.

For trace element concentrations in oceanic sediments, we have used the pelagic/terrigennous sediment data of White et al. (1985); relevant parent/daughter element ratios are given in Table II. Note that, in comparison with altered oceanic crust, sediments are also enriched in Rb/Sr and Nd/Sm, though to an even greater extent. However, the sediments are marked by low U/Pb and high Th/Pb, in contrast to altered oceanic crust which has high U/Pb and low Th/Pb. For the sedimentary protolith reservoir, it is obviously not correct to use the depleted MORB mantle. Instead, for Pb, we have utilized the Pb isotope ratios for average modern pelagic sediments (18.814, 15.651, 38.839; 52 samples, various literature sources), and assumed that a protolith reservoir (basically the continental crust) evolved from bulk earth values at 4.0 b.y. to those present day sediment Pb values (we did not use the Pb isotopic ratios which accompany the sediment data of White et al. (1985) because we feel these are biased toward older continental crust).

The Pb isotope locus of sediment subducted into the mantle at various times in the past is shown in Fig. 2; it also has a mild negative slope, like the altered oceanic crust locus, but trends in the opposite direction (because

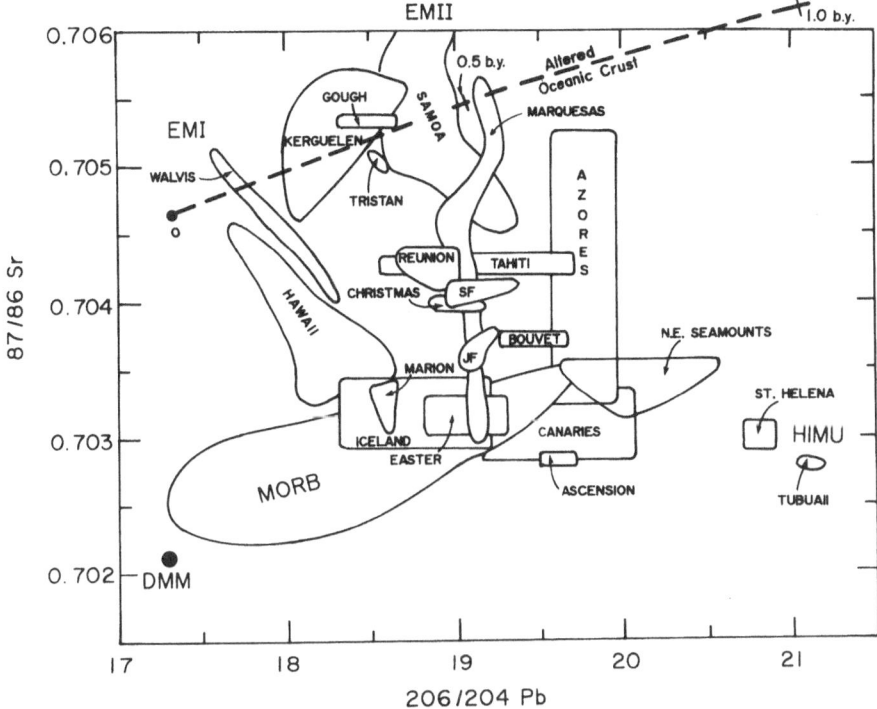

Fig. 3   Sr-Pb isotope correlation plot showing the locus (dashed line) of altered subducted oceanic crust as a function of the age of alteration. Also shown are fields for MORB and various oceanic islands (figure adopted from Hart, 1988), as well as the four mantle components of Zindler & Hart (1986).

of the low U/Pb of sediment, it's 206/204Pb growth is
retarded and it falls progressively further behind with
time). The sediment locus is not shown on Figures 1 and 3
because all of it falls well beyond the limits of the plots.
In Nd/Sr space (Fig. 1), the locus has a slope similar to
the mantle array, but displaced well to the southeast from
it. Mixing of sediment with DMM could mimic many of the
arrays comprising the oceanic field. On Fig. 3, the
sediment locus is nearly vertical, with its lower end well
north of the plot at ≈0.719 and 18.81; again mixing of
sediment with DMM or HIMU could produce the EMII arrays such
as Samoa and Marquesas (on the reasonable assumption that
both the DMM and HIMU sources have Sr/Pb ratios which are
much higher than the sediment). While recycled old sediment
would appear to be a viable candidate for the EMII component
in terms of Nd-Sr and Sr-Pb relationships, it's
appropriateness on Fig. 2 is less evident, as sediment-DMM
mixing would not embrace EMII. No reasonable perturbations
of U/Pb and Th/U ratios in sediments will remove this
discrepancy; the discrepancy can be eliminated only by
moving the array to the northeast (that is, using a sediment
protolith reservoir with 206/204 Pb > 20; this would require
an older and higher μ protolith than is reflected in any
common oceanic sediments at the present time).

Since sediments and altered oceanic crust would be
subducted as a package, it might be more realistic to test
mixtures of these two components as possible mantle sources
for oceanic basalts. We have not attempted to model this in
detail, but the qualitative aspects can be noted by
reference to Figures 1-3. Fig. 2 is the most
straightforward case, because all mixing lines are linear;
the locus of mixtures thus becomes lines joining equivalent
age positions on the sediment and ocean crust locii. While
such mixtures could embrace the left end of the oceanic
field (DMM, EMI), it would fall far short of explaining
HIMU. On Fig. 3, since all such mixtures would lie north of
the altered crust locus, most of the oceanic field would not
be embraced.

In summary then, neither altered ocean crust nor
typical ocean sediment, nor mixtures of the two, subducted
at any time in the past, can be reasonably identified with
the oceanic basalt components EMI and HIMU, or with arrays
between these and DMM. By stretching the sediment case, a
component with an EMII "flavor" at least can be generated.
We know ocean crust and sediment do get subducted--does this
discussion preclude any recycling of these components?
Clearly not, because the modeling done here makes the
fundamental assumption that ocean crust (± sediment) can be
subducted into the mantle <u>as a closed system</u>. In other
words, the model ignores the myriad of possible processes
operating in the subduction zone itself, which may short-

circuit the long-term recycling process, and which may do so on a very element-specific basis. The subducted slab clearly goes through a complex de-volatilization process, and it is possible that this process can selectively strip out some trace elements relative to others. The final subducted "product" then, which undergoes aging and possible eventual recycling into oceanic magmas, may have parent/daughter element characteristics quite different from those used in the present modeling (Table II). Even a perfect knowledge of the composition of subducted oceanic crust and sediment will not help this situation, and we must therefore turn to other types of studies for further illumination--for example, study of arc lavas may help identify relative element fractionations associated with slab de-watering and melting. Direct experimental measurement of element mobility and partitioning during basalt de-volatilization, such as the study reported at this meeting by Zindler & Watson (1987), can be expected to greatly enhance our understanding of what's removed, and what's left, during the passage of oceanic lithosphere through the upper mantle. As a corollary to this, the materials removed from the lithosphere during subduction are largely trapped in the mantle wedge (only a small fraction can be removed in arc magmas); this sub-arc mantle, as well as the final subducted lithosphere, are both mantle reservoirs which can age and become ultimately available for recycling into oceanic basalts. The oceanic lithosphere is thus, in a sense, split into two complementary mantle reservoirs; the quantitative details of how this splitting takes place remain as a formidable challenge to our science.

Acknowledgements

The site 417/418 samples were provided by the Ocean Drilling Project. We are grateful for the analytical support provided by L. Gulen, and the mass spectrometry maintenance and wizardry provided by K. Burrhus. And thanks to D. Watson-Mitchell, who kept track of the financial support afforded by NSF grants OCE 8400794 and EAR 8708372 as well as handling manuscript preparation.

REFERENCES

Albarede, F., Michard, A., Minster J.F. and Michard G. $^{87}Sr/^{86}Sr$ ratios in hydrothermal waters and deposits from the East Pacific Rise at 21°N.' Earth Planet. Sci. Lett. 55: 229-236, 1981.

Chase C.G., 'Ocean island Pb: two-stage histories and mantle evolution.' Earth Planet. Sci. Lett. 52: 277-84, 1981.

Chen, J.H., Wasserburg, G.J., von Damm, K.L., and Edmond, J.M. 'The U-Th-Pb systematics in hot springs on the East Pacific Rise at 21°N and Guaymas Basin.' Geochim. Cosmochim. Acta 50: 2467-2479, 1986.

Cohen, R.S. and O'Nions, R.K. 'Identification of recycled continental material in the mantle from Sr, Nd and Pb isotope investigations.' Earth Planet. Sci. Lett. 61: 73-84, 1982b.

Cohen, R.S. and O'Nions, R.K., 'The lead, neodymium and strontium isotopic structure of ocean ridge basalts.' J. Petrol. 23: 299-324, 1982a.

Cohen, R.S., Evensen, N.M., Hamilton, P.J., O!Nions, R.K. 'U-Pb, Sm-Nd and Rb-Sr systematics of mid-ocean ridge basalt glasses.' Nature 283: 149-53, 1980.

Gast, P.W., Tilton, G.R. and Hedge, C., 'Isotopic composition of and strontium from Ascension and Gough Islands,' Science 145: 1181, 1964.

Hart, S.R. 'K, Rb, Cs contents and K/Rb, K/Cs ratios of fresh and altered submarine basalts.' Earth Planet. Sci. Lett. 6: 295-303, 1969.

Hart, S.R. 'Heterogeneous mantle domains: signatures, genesis and mixing chronologies.' Earth Planet. Sci. Lett. in press, 1988.

Hart, S.R. and Staudigel, H. 'The control of alkalies and uranium in seawater by ocean crust alteration'. Earth Planet. Sci. Lett. 58: 202-212, 1982.

Hart, S.R., Erlank, A.J., and Kable, J.D. 'Sea floor basalt alteration: some chemical and Sr isotopic effect.' Contr. Mineral. and Petrol. 44: 219-230, 1974.

Hofmann, A.W., and White, W.M., 'Mantle plumes from ancient oceanic crust.' Earth Planet. Sci. Lett. 57: 421-36, 1982.

Ito, E., White, W.M., and Gopel, C. 'The O, Sr, Nd and Pb isotope geochemistry of MORB.' Chem. Geol. 62: 157-176, 1987.

Jochum, K.P., Hofmann, A.W., Ito, E., Seufert, H.M., and White, W.M. 'K, U, and Th in mid-ocean ridge basalt glasses. The terrestrial K/U and K/Rb ratios and heat production in the depleted mantle.' Nature 306, 431-436, 1983.

MacDougall, J.D. 'Uranium in marine basalts: concentration, distribution and implications.' Earth Planet. Sci. Lett. 35: 65-70, 1977.

Michard, A., Albarede, F., Michard, G., Minster, J.F., Charlou, J.L. 'Rare-earth elements and uranium in high-temperature solutions from East Pacific Rise hydrothermal vent field (13°N).' Nature 303: 795-797, 1983.

Mitchell, W.S. and Aumento, F., 'The distribution of uranium in oceanic rocks from the Mid-Atlantic Ridge at 36°N.' In: Initial Reports of the Deep Sea Drilling Project, 37: 547-560, 1977.

Philpotts, J.A., Schnetzler, C.C., and Hart, S.R. 'Submarine basalts: some K, Rb, Sr, Ba Rare-Earth, H_2O, and CO_2 data bearing on their alteration, modification by plagioclase, and possible source materials.' Earth Planet. Sci. Lett. 7: 293-299, 1969.

Staudigel, H., and Hart, S.R. 'Alteration of basaltic glass: mechanisms and significance for the oceanic crust-seawater budget.' Geochim. Cosmochim. Acta 47: 337-350, 1983.

Staudigel, H., Frey, F.A., and Hart, S.R. 'Incompatible trace-element geochemistry and $^{87/86}Sr$ in basalts and corresponding glasses and palagonites.' In: Initial Reports of the Deep Sea Drilling Project, LI, LII, LIII, , 1137-1144, 1979.

Staudigel, H., Hart, S., Smith, B.M., and Schmincke, H.-U. 'The CO_2 budget of ocean crust formation: a sink or a source?' Submitted to Earth Planet. Sci. Lett., 1988.

Staudigel, H., Hart, S.R., and Richardson, S.H. 'Alteration of the oceanic crust: processes and timing.' Earth Planet. Sci. Lett. 52: 311-327, 1981.

Veizer, J. and Compston, W., '$^{87}Sr/^{86}Sr$ in Precambrian carbonates as an index of crustal evolution'. Geochim. Cosmochim. Acta 40: 905-914, 1976.

Von Damm, K.L., Edmond, J.M., Grant, B., Measures, C. I., Walden, B., and Weiss, R.F. 'Chemistry of submarine hydrothermal solutions at 21°N, East Pacific Rise.' Geochim. Cosmochim. Acta 49: 2197-2220, 1985.

28

White, W.M., Dupre, B., and Vidal, P. 'Isotope and trace element geochemistry of sediments from the barbados Ridge-Demerara Plain region, Atlantic Ocean.' Geochim. Cosmochim. Acta 49: 1875-1886, 1985.

White, W.M., Dupre, B., Vidal, P. 'Isotope and trace element geochemistry of sediments from the Barbados Ridge-Demerara Plain region, Atlantic Ocean.' Geochim. Cosmochim. Acta 49: 1875-86, 1985.

Zindler, A., Hart, S.R., 'Chemical geodynamics,' Ann. Rev. Earth Planet. Sci. 14: 493-571, 1986.

HYDROTHERMAL ALTERATION OF THE OCEANIC CRUST

Francis Albarede and Annie Michard,
Centre de Recherches Petrographiques
et Geochimiques and Ecole National e
Superieure de Geologie
54501 Vandoeuvre-les Nancy Cedex, France

Recycling of the oceanic crust is potentially a dominant process in the formation of OIB mantle sources (Hofmann et al. 1978; 1986) and is a significant contribution to the flux of material between the continent and the depleted mantle (Albarede and Michard, 1986). The subducted oceanic crust is expected to comprise widely different end-members: continent-derived sediments of which the abyssal clays are an important fraction, metalliferous ridge-flank sediments, and the magmatic layers (basalts, dolerites, gabbros,...). First, these magmatic layers interact at high temperature with the overlying seawater in the hydrothermal systems recently discovered along most ridge segments around the world (black smokers). SiO_2 geothermobarometry of the vent solutions suggests that the last major chemical exchanges between hydrothermal solutions and the plutonic part of the oceanic crust take place 0.5-3.5 km below the seafloor, a depth that recent seismic data correlate with the top of an axial magma chamber. In addition, seawater reacts at low temperature with the uppermost basaltic flows and dikes, giving rise to the alteration most commonly studied on dredge samples and DSDP cores.

The purpose of this work is to review some of the chemical and isotopic changes induced in the oceanic crust by the hydrothermal activity at ridge crest and to assess some criteria which could help identify the contribution of the altered oceanic components in the formation of mantle-derived rocks.

COMPUTATION OF ELEMENTAL FLUXES

Computation of hydrothermal fluxes at ridge crest using the ^3He/heat constraints usually suffers an intrinsic uncertainty: Sleep and Wolery (1978) pointed out that a large fraction of the hydrothermal activity may escape detection due to warm porous flow, conductive cooling and

S. R. Hart and L. Gülen (eds.), Crust/Mantle Recycling at Convergence Zones, 29–36.
© 1989 by Kluwer Academic Publishers.

extensive dilution by seawater even at ridge crest.   Edmond
et al. (1979) and Hart and Staudigel (1982) realized that
there is not enough potassium in the oceanic crust to
account for the hydrothermal flow calculated from vent data.
Michard and Albarede (1985) arrived at the same conclusion
for lead. Sleep et al. (1985) recently assessed the
fraction of heat loss through black smokers to 10-20 percent
of the total hydrothermal heat loss. It is likely that low
temperature reaction with wall rocks, deposition of various
minerals and clogging of the conduits strip the hydrothermal
solutions from alkali, alkaline-earth or transition elements
which are thereby retained in the oceanic crust.
Examination of stockwork mineral assemblages confirms the
importance of these secondary processes (Anderson et al.,
1983). Therefore, fluxes inferred for elements which are
removed from the crust must be taken as maximum values.
      In contrast, the uptake of elements from seawater
during water-rock hydrothermal interaction is very likely
independent of the cooling and venting stages: the 9 m.y.
time which Edmond et al. (1979) and Michard et al. (1983)
suggest for complete renewal of seawater Mg, SO_4 and U
respectively by hydrothermal removal at ridge crest depends
only on assumptions on the global hydrothermal heat and ^{3}He
flux estimate and not on the removal mechanism itself.   Mg,
SO_4, O, and U and, to a lesser extent, radiogenic ^{87}Sr
deserve special attention for they are the only species
quantitatively stripped off from seawater during the
hydrothermal process.

THE IMPORTANCE OF THE HYDROTHERMAL ALTERATION ON ISOTOPIC
SYSTEMS.

The chemical and isotopic systematics of the altered oceanic
crust are presumably strongly modified relative to what can
be expected as pristine magmatic characters. The extent to
which these modifications may be estimated will be briefly
reviewed for those isotopic systems with the most prominent
effects (S, U-Pb) but also for some others which are widely
used in mantle geochemistry (Rb-Sr, Sm-Nd).

a) Sulfur

Iron from the hydrothermally altered rocks observed in the
oceanic crust in general and in the deep part of DSDP hole
504B in particular (Alt and Emmermann, 1985) is oxidized to
20-40 percent. Such an extensive oxidation is demonstrated
by the occurrence of epidote, hematite or magnetite in those
rocks and must be balanced with an oxygen addition from
seawater. Water reduction is certainly an inappropriate
reaction as the maximum hydrogen (+ methane) flux calculated
for the EPR 21°N vent solutions from Welhan and Craig data

(1983) is two orders of magnitude smaller than required by iron oxidation. Anhydrite precipitated at low temperature is likely to be reduced upon reaction with ferrous iron from basaltic rocks, sulfide minerals, or hydrothermal solutions. Although $\delta^{34}S$ values as high as +9.6 were known from sulfides in altered mid-oceanic basalts (Grinenko et al., 1975), this process has also been recently advocated to account for rather high $\delta^{34}S$ values in Kuroko ores (Ohmoto et al., 1983) and Juan de Fuca Ridge hydrothermal deposits (Shanks and Seyfried, 1985). If the 4×10^{12} moles/year seawater sulfate entering the system are stripped off from the solution and not returned to seawater, this flux provides enough oxygen to oxidize about 40 percent of the iron held by a 5 km thick basaltic/gabbroic layer with 10 percent FeO. The agreement between the flux of sulfate and the demand of oxygen to oxidize iron to the extent which is observed in deep oceanic crust suggests that a large fraction of precipitated anhydrite is reduced within the hydrothermal system prior to cooling of the crust, which therefore behaves as a sink for seawater sulfate. The strength of this sink cannot be far from Edmond et al.'s (1979) estimate which, probably not coincidentally, equals the sulfate removal from the continental crust by runoff.

The hydrothermal flux derived by Edmond et al. (1979) may be combined with an estimate of the amount of sulfur sequestered by continental crust to derive a time constant of continental sulfur removal through hydrothermal processes of about 100 Ma. For any acceptable initial concentration of sulfur in the continent crust, this element should have been quantitatively transferred from the continental crust to the mantle. This major imbalance requires that a hidden return flow of sulfur from the oceanic crust to the continental crust exists with a fast turnover rate. Arc volcanics are known to be sulfur-rich and $\delta^{34}S$ measured in anhydrite microphenocrysts from El Chichon may be as high as +9.2 (Rye et al., 1984). Up to now, the preferred interpretation of those high $\delta^{34}S$ values favored incorporation of sedimentary sulfur (evaporites) at shallow levels in the continental crust. However, Ueda and Sakai (1984) reported a $\delta34S$ value of +9.8 for a basalt dredged in the intraoceanic Mariana arc. Harmon (pers. comm.) has measured the sulfur isotope composition of mafic volcanics from continental (Aegean Sea) or intraoceanic arcs (Lesser Antilles, Marianas) and found $\delta34S$ value as high as +15 to +20. Such high values in arcs which may not be suspected to contain significant amounts of evaporates at depth (in particular for the Mariana arc) have the implication that sulfur added to the oceanic crust through hydrothermal alteration is outgassed from the oceanic crust by the process which generates arc magmas. Arc magmas are known to be often much more oxidized than ridge tholeiites (Gill,

1981, p. 109). The higher oxygen fugacity in the terrestrial mantle than in extra-terrestrial is probably one of the best indications that oxidized oceanic crust and its sediments have been subducted throughout the Earth's history.

b) Rubidium and Strontium

The evolution of $^{87}Sr/^{86}Sr$ in the ocean throughout geological time may be quantified thanks to an estimate of the Sr cycling through the ridge hydrothermal system (1 x 10^{10} moles/y) (Albarede et al., 1981). The $^{87}Sr/^{86}Sr$ ratio of 0.710 deduced from this calculation for the continental runoff (Albarede et al., 1981) is in strong agreement with an independent estimate based on continental geothermal systems and major drainage basins. Goldstein (1987 and pers. comm.) pointed out that the ridge system acts as a drain for the radiogenic Sr produced in the continental crust and prevents the continental Sr from becoming as radiogenic as the average upper crustal Rb/Sr would suggest.
In contrast with the $^{87}Sr/^{86}Sr$ ratio, the average Rb/Sr of the altered oceanic crust is poorly known. Vent solution and altered oceanic rocks suggest that Sr concentrations probably undergo only minor changes in the alteration process. However, the extent to which the low temperature Rb enrichment of the uppermost layers of the oceanic crust balances the Rb loss by high temperature seawater/rock interaction merits further document (Hart and Staudigel, 1982). The removal of Rb from the basalt is probably nearly quantitative (Edmond et al., 1979; Michard et al.; 1984) but, again, the fraction of Rb which is redeposited in the upper layers through low-temperature venting is uncertain.

c) Samarium and Neodymium

Michard et al. (1983, 1986) have shown that the hydrothermal solutions spouting out from the East Pacific Rise at 13 and 21°N are enriched in light REE and EU relative to seawater. However, the low water/rock ratios prevailing during the hydrothermal reaction (1-10) makes the fraction of the initial REE subtracted from the oceanic crust quite negligible. Therefore, the Sm/Nd isotope systematics of the altered oceanic crust are not expected to differ from the unaltered material.

d) Uranium, thorium, and lead

Michard et al. (1983), Michard and Albarede (1985), and Chen et al. (1986) measured U concentrations in vent solutions from the EPR at 13 and 21 °N and the Juan de Fuca Ridge and found that the high temperature components (>320 °C) are essentially free of uranium. Hence, 3.3 ppb of U are quantitatively stripped off from seawater upon transfer through the 350 °C isotherm, probably as a consequence of reduction during the hydrothermal process. Abundance of ferrous iron and low pH should convert massively uranyle to quadrivalent uranium with a rather low solubility very similar to Th and REE. The alkalinity drop should also enhance massive decomplexation of uranyle species by carbonate ions.

The U content of the oceanic crust increases by about 20 percent solely through the high temperature hydrothermal activity at mid-ocean ridges (Michard et al., 1983). An alternative form of uranium uptake from seawater is low temperature alteration of basalts and sediments (Aumento, 1971). The strength of this sink is highly uncertain and estimates vary from 0.1 to 5×10^8 moles/y (Bloch, 1980; Hart and Staudigel, 1982). The depth at which oceanic crust undergoes U gain by low temperature alteration is presently not known: neither is the proportion of this flux which actually represents uranium stripped at low temperature from seawater in the recharge zone of the axial ridge systems. Global U fluxes are therefore probably not known to be more than an order of magnitude.

The Th behavior is probably very similar to that of REE since Chen et al. (1986) found very little of this element in vent fluids. Accordingly, the mean Th/U ratio of the altered oceanic crust is expected to be at least 20 percent lower than that of the upper mantle which gave rise to the ridge magmas.

Lead is extracted very efficiently from the basalt during the hydrothermal processes at ridge crest. Michard and Albarede (1985) and Chen et al. (1986) found up to 100 ppb of Pb in the vent solutions, i.e., more than the crust could produce if all the hydrothermal heat escapes through black smokers. This result shows the importance of low-temperature venting with extensive redeposition of Pb, but also suggests that the magmatic layers of the altered oceanic crust lose a large fraction of their Pb in the process. The U/Pb and hence the μ value of the altered mafic rocks should increase dramatically above the magmatic value. Again quantifying this increase is a difficult task.

THE RECYCLING OF OCEANIC CRUST AND BASALT GENESIS.

Hydrothermal alteration is not expected to produce significant changes in the Sm–Nd evolution pattern of the oceanic crust. The $^{87}Sr/^{86}Sr$ ratio of the basaltic layer of the oceanic crust is probably somewhat higher (0.704-0.705) than the ratio of the depleted upper-mantle (0.7025), but, because of our ignorance of the Rb balance, little is known of the $^{87}Sr/^{86}Sr$ evolution in the subducted material. It is the present author's prejudice that the Rb low-temperature enrichment of the uppermost basaltic layer has been overemphasized and that the $^{87}Sr/^{86}Sr$ ratio of the altered crust may increase not much faster than the ratio of the depleted mantle itself.

Hydrothermal alteration has potentially major relevance to the U-Th-Pb systematics because the altered oceanic basalts will be subducted with an excess of U and a large deficit of lead (Michard and Albarede, 1985). A significant fraction of lead contained in the oceanic crust will be removed at ridge crest, even if redeposition of sulfide upon conductive cooling of vent solutions limits the efficiency of the process. As shown by lead isotope data from the East Pacific Rise (Dasch, 1981), lead extracted by hydrothermal alteration is transferred to the ridge flank metalliferous sediments. The altered oceanic crust presents, therefore, two layers of contrasting μ: the high μ basaltic layer overlain by a veneer of low μ and high Th/U hydrothermal + abyssal sediments. Following the original suggestion of Hofmann and White (1980), Michard and Albarede (1985) have calculated that altered oceanic basalts stored in the mantle for a period of one Ga or so may reproduce the lead isotopic properties of the source of some ocean island basalts. Michard et al. (1986) have attributed the existence of a component with a very low $^{206}Pb/^{204}Pb$ and high $^{208}Pb/^{204}Pb$ ratios in basalts from the Rodrigues Triple Junction (Indian Ocean) to the presence of hydrothermal and abyssal sediments in their mantle source, whereas Alibert et al. (submitted) make this recycled component a distinctive constituent of the lamporite mantle source (Australia, Smoky Butte, Arkansas).

An important unsolved problem in understanding the importance of oceanic crust recycling in the generation of magmas from the mantle is to estimate the fraction of the useful tracers (S,U/Pb, Rb/Sr) which escape incorporation into the continental crust at subduction zones through calc-alkaline magmas. We have emphasized the fact that a large fraction of sulfur must go from the altered oceanic crust back to the surface quite efficiently, otherwise runoff would wipe this element out of the continental crust in a few hundred Ma (Albarede and Michard, 1986). Identification of the oceanic crust residue through this element will be therefore more difficult. The same might be true for Pb and the process responsible for the Pb/Ce fractionation between the continental crust and oceanic basalts observed by Hofmann et al. (1986) may correspond to the melting of the subducting crust strongly zoned in U, Th, and Pb. The next step is therefore to understand what the subducted oceanic crust looks like when it is irreversibly incorporated into the mantle material.

REFERENCES

Albarede F. et al. (1981) Earth Planet. Sci. Lett. <u>55</u>, 229-236.

Albarede F. and Michard A. (1986) Chem. Geol. <u>57</u>, 1-15.

Alt, J.C. and Emmermann, R. (1985) Init. Rept. DSDP <u>83</u>, 249-262

Anderson R.N. et al. (1983) Nature <u>300</u>, 589-594.

Aumento F., (1971) Earth Planet. Sci. Lett. <u>11</u>, 90-94.

Bloch, S. (1980) Geochim. Cosmochim. Acta <u>44</u>, 373-377.

Chen, J.H. et al. (1986) Geochim. Cosmochim. Acta <u>50</u>P, 2467-2479.

Dasch, J.E. (1981) Geol. Soc. Amer. Mem. <u>154</u>, 199-208.

Edmond et al. (1979) Earth Planet. Sci. Lett. <u>46</u>, 1-18.

Gill, J. (1981) <u>Orogenic Andesites and Plate Tectonics,</u> Springer

Grinenko, V.A. Et al. (1975) Geochem. Intern. <u>12</u>, 132-137.

Hart, S.R. and Staudigel, H. (1982) Earth Planet. Sci. Lett. <u>58</u>, 202-212.

Hofmann, A.W. et al. (1978) Carnegie Inst. Wash. Yb. <u>79</u>, 477-483.

Hofmann, A.W. et al. (1986) Earth Planet. Sci. Lett. <u>79</u>, 33-45.

Michard, A., et al. (1983) Nature <u>303</u>, 795-797.

Michard A. and Albarede F. (1985) Chem. Geol. <u>55</u>, 51-60.

Michard A., et al. (1986) Earth Planet. Sci. Lett. <u>78</u>, 104-114.

Ohmoto H. et al. (1983) Econ. Geol. Mem. <u>5</u>, 570-604.

Rye, R.O. et al. (1984) J. Volc. Geotherm. Res. <u>23</u>, 109-123.

Shanks, W.C. and Seyfried W.E. (1985) EOS <u>66</u>, 928-929.

Sleep, N.H. and Wolery T.J. (1978) J. Geophys. Res. <u>83</u>,
    5913-5922.

Sleep, N.H., et al. (1985) EOS <u>66</u>, 920.

Ueda, A. and Sakai, H. (1984) Geochim. Cosmochim. Acta <u>48</u>,
    1837-1848.

Welhan J.A. and Craig, H. (1983) in <u>Hydrothermal Processes
    at Seafloor Spreading Center</u>, Rona P. et al. eds,
    Plenum, NY 391-409.

USE OF MULTISPACE TECHNIQUE TO DETERMINE THE CONTRIBUTION OF REINJECTED SEDIMENTS IN THE MANTLE (E.G. LESSER ANTILLES)

B. DUPRE AND C.J. ALLEGRE
Institut de Physique du Globe de Paris, Laboratoire de Géochimie et
Cosmochimie
Tour 14 - 3è étage, 4, place Jussieu 75252 Paris Cedex 05

ABSTRACT . Based on the technique described by Allègre et al (1987) we demonstrate in this paper that sediments having isotopic characteristics similar to the sediments located near the island arc do not contribute to the genesis of Lesser Antilles volcanics. Two solutions are possible to explain the isotope results :
1) very peculiar sediments exist in the source of the basalts or,
2) an old lithosphere having isotope characteristics similar to the source of tholeiites from Parana and eastern USA exists beneath the Lesser Antilles.

The second conclusion is that this technique is clearly more sensitive than the two classical diagrams ($^{207}Pb/^{204}Pb$ - $^{206}Pb/^{204}Pb$, $^{208}Pb/^{204}Pb$ - $^{206}Pb/^{204}Pb$).

INTRODUCTION

Recycling of sediments in the mantle is a very important process but it has not been completely solved.
Island arcs are places where sediments most likely perform as very important source components for volcanics. (Amstrong and Cooper, 1971). We have chosen the Lesser Antilles arc as an example for the multispace approach. In the Lesser Antilles, many different types of samples (sediments, old oceanic crust and island arc basalts) have been analysed. (White and Dupré, 1986 ; White et al, 1985 ; Davidson, 1983 ; Vidal et al, en préparation). The technique described by Allègre is able to analyse the structure of a cluster of points and to compare two different populations. We use this method (with 3D program Mac Spin ; D^2 Software adapted to microordinator) with three Pb isotopic ratios ($^{206}Pb/^{204}Pb$ - $^{207}Pb/^{204}Pb$ - $^{208}Pb/^{204}Pb$). In this case, variations of isotopic compositions derived by mixing between two sources and between three sources make respectively a straight line and a plane. For discussion, we use two diagrams to describe the isotopic structure : a "plane view" and "thickness view". The plane view corresponds to vectors L_1 and L_2 and thickness view corresponds to vectors L_1 and L_3. Vectors, L_1, L_2 and L_3 represent $L_1 > L_2 > L_3$ in the ellipsoid axis.

37

S. R. Hart and L. Gülen (eds.), Crust/Mantle Recycling at Convergence Zones, 37–42.
© 1989 by Kluwer Academic Publishers.

A) Isotopic structure of basalts from the Lesser Antilles

The structure is largely elongated with L_1. This implies that the L_1 is much greater than L_2 and L_3 (Fig. 1). On the other hand, L_3 is 2.4 times smaller than L_2 and nearly forms a plane structure (Fig. 1).

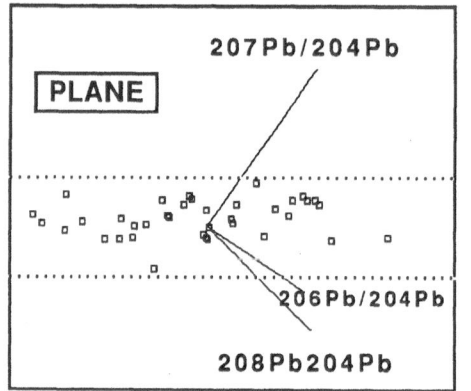
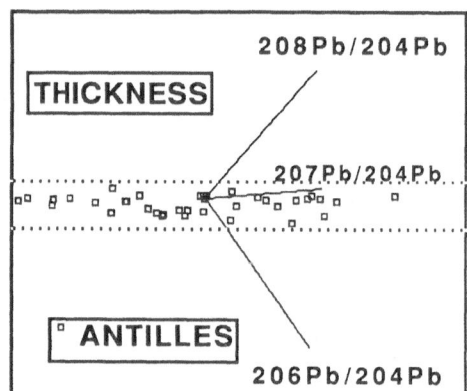

We have checked that this plane is not due to mass discrimination. Therefore the structure of the Lesser Antilles data can be compared with data from other samples and reservoirs.

B) Comparison between Lesser Antilles volcanics and other populations

1) MORB. In the thickness view the MARB (MID ATLANTIC RIDGE BASALTS) vector (including old tholeiites) appears in the Lesser Antilles plane, whereas an angle between the main elongation of MARB and the Lesser Antilles samples is seen in the plane view (Fig. 2). Old tholeiites (leg 78, White et al, 1985) and recent tholeiites (North Atlantic Ridge, Dupré et al, 1980) are taken as a representative oceanic crust.

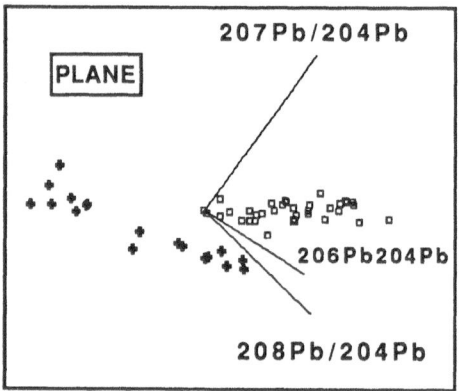
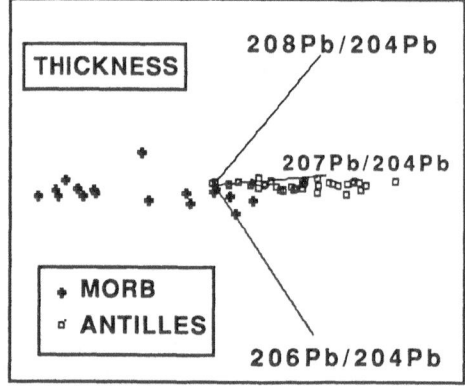

2) <u>SEDIMENTS.</u> For this comparison we use mainly surface sediments and deep sediments (leg 78) described by White et al (1985). In the thickness view, the sediments clearly plot away from the Lesser Antilles plane. However the sample RC 1647 exceptionally belongs to the Lesser Antilles plane (Fig. 3).

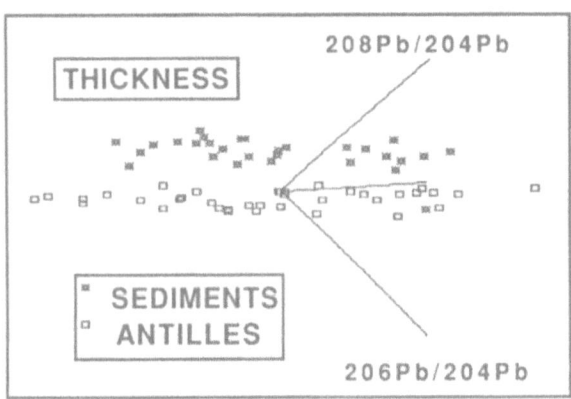

Figure 4 is a schematic diagram that portrays the plane for the Lesser Antilles volcanics and MORB and the plane for sediments. We have also plotted the OIB plate for comparison. <u>This observation strongly suggests that neither surface nor deep sediments are required to explain the isotopic structure of the Lesser Antilles volcanics.</u> If sediments are involved in the source of Lesser Antilles volcanics, the isotopic signature must be different that the observed sediments data. The angle between sediments plate and IAB Antilles plate is exagerated in the shematic diagram.

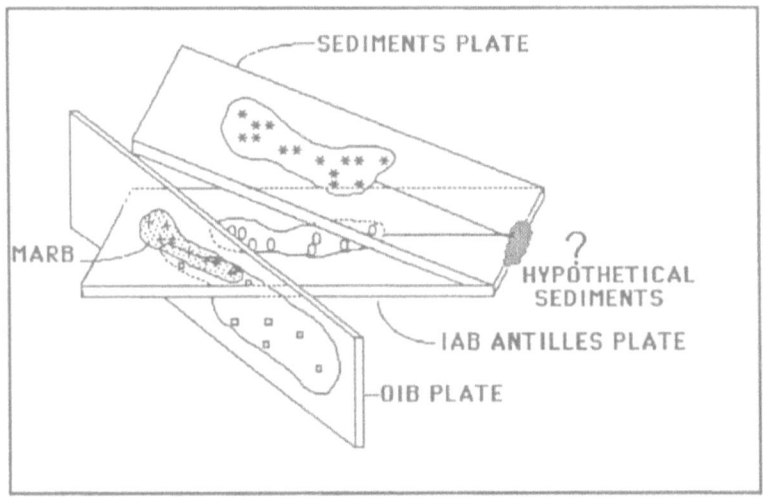

OTHER EXPLANATIONS

A) First, we examine a model proposed by Morris and Hart (1983). In this model the source of IAV is more or less similar to the OIB source. In order to evaluate this model, the OIB structure is compared with the Lesser Antilles volcanics. In the thickness view, the Lesser Antilles volcanics plot clearly away from the "OIB plate". Therefore, the involvement of the OIB source is unlikely to explain the isotopic structure of the Lesser Antilles volcanics(Fig.5).

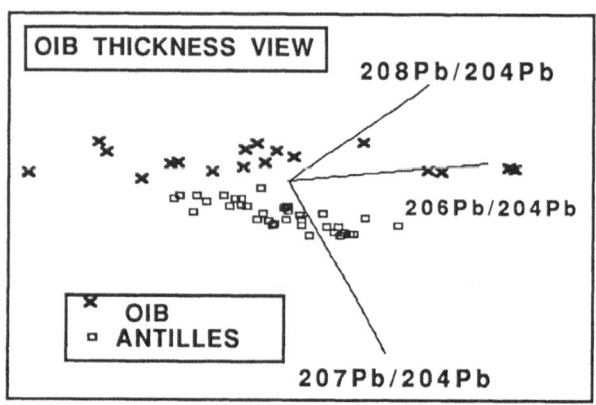

B) The second interpretation comes from the schematic diagram of Fig. 4 . The fact that a small angle exists between the planes for the sediments and the Lesser Antilles volcanics suggests that the two data sets could share a common component. The geographic location of each sediment sample discussed by White et al (1985) is correlated to its location on the sediment plane in that the most Pb-Radiogenic samples are located in the south, closer to the South American Landmass ; further, these investigators noted that deeper sediments tended to be more Pb radiogenic than shallower and surface sediments. Conceivably, sediments that are more Pb-Radiogenic than those analysed by White et al (1985) could exist in this region and these hypothetical sediments could have been a source component for the volcanics. If so, however, these sediments would have Pb isotopic compositions which are remarkably homogeneous and which are much more radiogenic than any previously reported from ocean basin sediments. In fact Golstein (unpublished data) have analysed the Orenoque sediments and the values plotted exactly in the plane defined by IAV Antilles.

C) Finally, we compare the isotopic structure of CFB and the Lesser Antilles. Using the same techniques we introduce CFB (including Columbia River, westernt U.S.A., eastern U.S.A. and Parana). (CFB data come from W. Pegram (1986), C. Hawkesworth et al (in press), R. Carlson, (1984), W. Hart (1985)).

CFB from Columbia River and western U.S.A. are very similar but the location of these data are completly isolated from the Lesser Antilles plane (not shown). CFB from eastern U.S.A. and Parana have an interesting isotopic structure which is approximately identical to the structure of IAB (Fig.6).

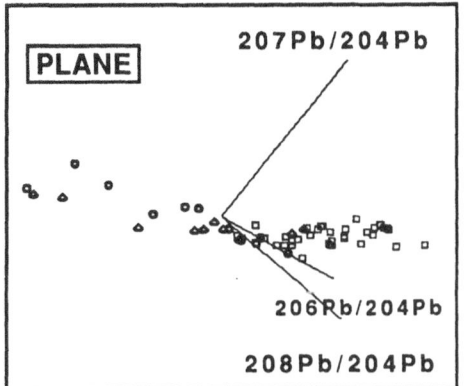 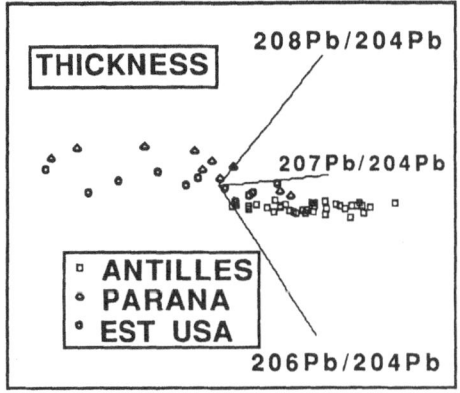

The isotopic characteristics of the CFB from the eastern U.S.A. and Parana can be explained by the existence of very old, evolved subcontinental lithosphere as suggested by Pegram (1986) and Hawkesworth (1987). These observations would suggest that this unique lithosphere exists beneath the Lesser Antilles Islands as well but in this case we don't explain the very radiogenic Pb found in the Lesser Antilles.

CONCLUSIONS

The use of multispace techniques is well adapted to the understanding of the genesis of the Lesser Antilles volcanics. On the basis of the above discussion, we can apparently eliminate as significant source components : 1) sediments found in front of the trench and 2) the OIB source. We propose two possible solutions to account for the data structure described above : 1) the Pb isotopic composition of subducted sediments involved in the genesis of the volcanics is unsusually radiogenic, much more so than those analysed by White et al (1985) or 2) the subarc mantle involves in the Lesser Antilles contains and ancient component similar to that dominating the subcontinental mantle of the eastern U.S. and Parana. We prefer this second interpretation because the isotopic value of Orenoque sediments is on the IAV Antilles plane. In this case, the plane structure which is obtained in the Antilles is created by the mixing between one sedimentary pole and a heterogeneous mantle pole. This explains why MARB are contained in the plane defined by the IAV Antilles.

Acknowlegments : We thank E. Lewin for various discussions, E. Nakamura and W. Pegram for English and M. Sennegon for her assistance in preparing the manuscript. The I.P.G.P contribution number is 973.

REFERENCES

Allègre C.J., Hamelin B., Provost A., Dupré B., 'Topology in isotopic multispace and origin of mantle chemical heterogeneities', *Earth Planet. Sci. Lett.*, **81**, 319-337, (1987)

Amstrong R.L. and Cooper J.A., 'Lead isotopes in island arcs', *Bull. Volcanol.*, **35**, 27-37, (1971)

Dupré B. and Allègre C.J., 'Pb-Sr-Nd isotopic correlation and the chemistry of the North Atlantic mantle', *Nature*, **286**, 17-22, (1980)

Carlson R., 'Isotopic constraints on Columbia River Flood Basalt genesis and the nature of the subcontinental mantle'. *Geochim. Cosmochim. Acta,* **48**, 2357-2372, (1984)

Davidson J.P., 'Lesser Antilles isotopic evidence of the role of subducted sediments in island arc magma genesis', *Nature*, **306**, 253-256, (1983)

Hart W., 'Chemical and isotopic evidence for mixing between depleted and enriched mantle, northwestern U.S.A.', *Geochim. Cosmochim. Acta*, **49**, 131-144, (1985)

Hawkesworth C.J., Mantovani M.S.M., Taylor P.N. and Palacz, 'Couple crust mantle systems : evidence from the Parana of South Brazil', *Nature, 322*, 356, (1986)

Morris J.D. and Hart S.R., 'Isotopic and incompatible element constraints on the genesis of island arc volcanics from Cold Bay and Amak Island, Aleutians and implications for mantle structure',*Geochim. Cosmochim. Acta*, **47**, 2015-2030, (1983)

Pegram W., 'The isotope, trace element and major element geochemistry of the Mesozoic Appalachian Tholeiite Province', *Thesis of M.I.T.*, (1986)

White W. and Dupré B., 'Sediment subduction and magma genesis in the Lesser Antilles : Isotopic and trace elements constraints'. *Journal of Geophysical Research*, **91**, 5927-5941, (1986)

White W.M. , Dupré B., Vidal P., 'Isotope and trace element geochemistry of sediments from Barbados Ridge, Demerara Plain region, Atlantic Ocean', *Geochim. Cosmochim. Acta*, **49**, 1875-1886, (1985)

GEOCHEMICAL EVIDENCE FOR CRUST-TO-MANTLE RECYCLING IN SUBDUCTION ZONES

William M. White
Dept. of Geological Sciences
Cornell University
Ithaca, NY 14853 USA

Return of oceanic crust to the mantle is implicit in plate tectonic theory. During its life, oceanic crust acquires a veneer of sediment, which for the most part is derived from the continental crust. It also reacts with sea water and hydrothermal fluids produced from sea water, changing its composition. To the degree this sediment is subducted along with the basaltic portion of the oceanic crust, continental crust is recycled into the mantle. That this process might have profound effects on the chemical evolution of both the continental crust and mantle was first recognized by Armstrong (1968). Since then a number of papers appeared appealing to just such a process to explain the geochemistry of the mantle (e.g., Hawkesworth, et al., 1979; Chase, 1981; Hofmann and White, 1982; White and Hofmann, 1982; DePaolo, 1983; Dupre and Allegre, 1983; Weaver, et al., 1986), but it is by no means clear that this process is in fact responsible for all that it is called upon to account for.

The composition of subducted oceanic crust, with or without its sedimentary veneer, should differ in a number of ways from ordinary mantle, in part because of the effects of magmagenesis, in part because of the reaction with sea water, and in part because of the presence of sediment. These compositional differences can provide the basis for testing the recycling hypothesis: any mantle reservoir alleged to contain recycled oceanic crust should share the compositional characteristics of oceanic crust. This paper attempts to trace the chemical and isotopic signature of oceanic crust as it is subducted into the mantle. The evidence suggests oceanic crust and sediment is often present in the sources of at least some island arc magmas and also in the sources of at least some back-arc or marginal basin magmas. The evidence that mantle plumes and oceanic island magmas contain a recycled component is, however, contradictory; the case against, particularly for sediment recycling, being at least as strong as the case for. A resolution of this question will require a much

S. R. Hart and L. Gülen (eds.), Crust/Mantle Recycling at Convergence Zones, 43–58.
© *1989 by Kluwer Academic Publishers.*

better understanding of the net compositional change experienced by oceanic crust through interaction with sea water and of the chemical processes and elemental fluxes in subduction zones.

ISLAND ARCS

Magmas produced in island arcs differ consistently in their chemistry from magmas in most other tectonic environments, and the association of island arc volcanism with subduction in both space and time has been well established for some time (e.g., Gill, 1981) and is clearly not coincidental. It seems reasonable, therefore, to assume that subduction is in some way responsible for the distinctive geochemistry of these magmas. This idea can be traced to Coats (1962) and predates the plate tectonic concept. At present, there is wide agreement that subduction influences the compositions of island arc magmas, though there is less agreement on the nature of this influence. There is evidence that material derived from the subducting oceanic crust and sediment can be divided into three types: petrological, trace element, and isotopic. These three lines of evidence will be briefly considered.

The petrological evidence for a subducted component in island arc magmas comes principally from the observation that such magmas are almost invariably richer in water than most magmas from other tectonic environments. Not only do island arc magmas have high concentrations of H_2O, many of their petrological features indicate melting and crystallization under relatively high water pressures: e.g., their siliceous composition, early appearance of oxides and hydrous phases. Since there is no reason to believe the subarc mantle should be more water-rich than the mantle elsewhere, it is reasonable to conclude this water must come from the subducting oceanic crust, present in both hydrous clay minerals of sediment and hydrous alteration minerals in the basaltic portion of the crust. These minerals become unstable at the pressures and temperatures encountered beneath arcs and all thermal models of the slab predict dehydration of the slab at depths of 30-100 km. These predictions together with the water-rich nature of arc magma are convincing evidence that the subducting slab contributes water to the magmas.

Water released from the slab is supercritical and quite likely carries a number of elements in solution. Island-arc volcanics (IAV) have higher $^{87}Sr/^{86}Sr$ ratios than mid-ocean ridge basalts (MORB), though their ratios are rather similar to oceanic island basalts (OIB). The higher $^{87}Sr/^{86}Sr$ ratios in IAV compared to MORB have been interpreted as resulting from the presence of subducted altered oceanic crust in the IAV source (e.g., Meijer, 1976) or the presence of subducted sediment. But since $^{87}Sr/^{86}Sr$ ratios in IAV and OIB are similar, IAV could be derived from OIB-type sources without involvement of oceanic crust or sediment (e.g., Stern, 1979). Nd isotope data combined with the Sr data were interpreted in favor of subducted crust involvement, as IAV appeared to have higher $^{87}Sr/^{86}Sr$ for a given $^{143}Nd/^{144}Nd$ than OIB (e.g., DePaolo and Wasserburg, 1976). This interpretation was based on the observation that alteration of oceanic crust should increase $^{87}Sr/^{86}Sr$ but not $^{143}Nd/^{144}Nd$. However, both the OIB and IAV fields in $^{143}Nd/^{144}Nd-^{87}Sr/^{86}Sr$ space have expanded with the result that they overlap considerably, and this argument loses the force it once had; the same is true of $^{176}Hf/^{177}Hf-^{87}Sr/^{86}Sr$ and $^{176}Hf/^{177}Hf-^{143}Nd/^{144}Nd$ fields (White and Patchett, 1984). Although $^{176}Hf/^{177}Hf$ and $^{143}Nd/^{144}Nd$ ratios in IAV are lower than in MORB, suggesting the presence of subducted sediment in IAV sources (White and Patchett, 1984), the Hf-Sr-Nd data alone can still be explained by melting of an OIB-type source (e.g., Stern, 1979; Morris and Hart, 1983).

Pb isotope data provide much stronger evidence for subducted sediment in IAV sources. Armstrong (1971) was the first to recognize this on the basis of the similarity of Pb isotopic compositions in Lesser Antilles and New Zealand lavas to those in marine sediments. Subsequent Pb isotopic data have generally confirmed the presence of sediment in IAV lavas (e.g., Kay et al., 1978; Barreiro, 1983, White and Dupre, 1986). Meijer's (1976) work in the Marianas is an example where the opposite conclusion was reached; however, a more recent and extensive Pb isotope study of the Marianas by Woodhead and Fraser (1985) found evidence for a component of subducted sediment in Marianas lavas.

The features of the Pb isotope data that suggest IAV sources contain subducted sediment are: (1) Though there is often considerable overlap between IAV and MORB or OIB fields on Pb-Pb plots, in at least some cases, such as the Lesser Antilles, the IAV data extends to points entirely outside the oceanic basalt field. (2) The typically steeper slopes of IAV arrays compared to MORB and OIB. Since sediment has generally higher $^{207}Pb/^{204}Pb$ but similar $^{206}Pb/^{204}Pb$ compared to oceanic basalts, these steep arrays suggest mixing of sediment and mantle to form the source of IAV. (3) There is a consistent relationship between Pb isotope ratios in sediments in front of arcs and Pb isotope ratios in arcs themselves. This is illustrated in Figure 1. Furthermore, the local sediment generally falls at the high $^{207}Pb/^{204}Pb$ end of the IAV arrays, suggesting it is the end-member which mixes with mantle (of more or less MORB composition).

The other isotope data set which is convincing and virtually unequivocal evidence of a sediment component in arc lavas is ^{10}Be. ^{10}Be has a half-life of 1.6 Ma and is

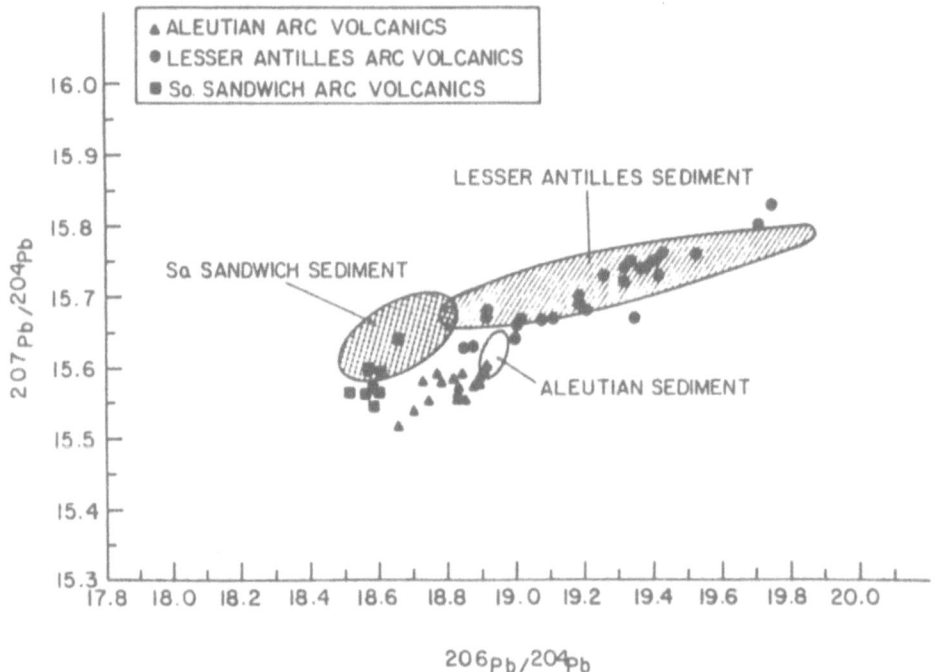

FIGURE 1: Pb isotopic compositions of 3 selected island arcs compared with Pb isotope ratios of sediment potentially begin subducted beneath those arcs.

continually produced by cosmic ray spallation reactions in the atmosphere. This cosmogenic [10]Be is incorporated into oceanic sediment. Since it does not exist in the Earth's interior (this theoretical expectation is confirmed by its absence in all but island-arc magmas), its presence in IAV is strong evidence that oceanic sediment is present in the source of such lavas (Tera et al., 1986). Indeed, the only alternative explanation for the Pb and Be isotope data is assimilation of sediment as magmas ascend through the arc crust. While this process may occasionally occur, even proponents of assimilation agree it is highly unlikely assimilation can explain all the isotope data (e.g., Davidson, 1986). [10]Be appears to be absent in many arcs, but as only the youngest, uppermost sediment is expected to contain [10]Be (sediment isolated from the atmosphere by burial for as little as 10 Ma will not have [10]Be above background levels), this does not mean sediment is not subducted beneath these arcs (Tera, et al., 1986). Indeed, there appears to be no [10]Be in the Lesser Antilles (L. Brown, pers. comm., 1984), where the Pb isotope evidence for subducted sediment is the strongest. This apparent conflict is explained by geological and geophysical studies of the Lesser Antilles forearc, which indicate at least the uppermost 200 m of sediment (^{10}Be would be limited to the uppermost 100 m or less) is being scraped off.

The isotope data on the whole are consistent with the involvement of both oceanic crust and sediment in IAV genesis. On the other hand, they do limit the amount of sediment involved to a rather small proportion of the source. Putting a number on this limit is difficult because the arc source appears to be a three-component mixture (mantle--oceanic crust--sediment), with two of the components being isotopically similar, but this limit appears to lie between 1% and 10% sediment, depending on whether mantle or subducted oceanic crust is the dominant 'depleted' component (e.g., White and Patchett, 1984).

The trace element geochemistry of IAV is characterized by relatively high concentrations of the alkali and alkaline earth elements (K, Rb, Cs, Sr, Ba) and low concentrations of high field strength elements such as Ta, Nb, Hf and Zr (e.g., Perfit et al., 1982). Rare earths patterns vary from light-rare-earth depleted to light-rare-earth enriched. Overall, the trace element geochemistry of IAV bears some marked similarities to that of sediments. Kay (1980) and Hole et al. (1984) successfully modeled the trace element geochemistry of Aleutian and Marianas lavas as mixtures of mantle, sediment and subducted oceanic crust. Hole et al., noted, however, that their model predicted much higher ^{87}Sr/^{86}Sr in Marianas lavas than actually observed. Similarly, White and Dupre (1986) concluded from attempts to model both the isotopic and trace element geochemistry of Lesser Antilles magmas that only a small part of the alkali-alkaline earth enrichment could be ascribed to subducted sediment. They concluded most of this enrichment must arise from partitioning of these elements into fluids produced by dehydration of the oceanic crust.

Morris and Hart (1983) argued IAV are derived from OIB type sources and the apparent enrichment in alkali and alkaline earth elements reflected depletion of the rare earths (due to their retention in a trace phase stable beneath arcs but not elsewhere) rather than actual enrichment of the alkali and alkaline earths. They did concede, however, that high Cs/Rb ratios in arcs were difficult to account for other than by the presence of subducted sediment in arc magma sources.

Trace element data require that conditions and processes of magma generation be different beneath arcs than elsewhere. It seems likely that the essential difference is the presence of water derived from the subducting oceanic crust and lithosphere. Thus recycling of oceanic crust clearly affects the subarc mantle. But the trace element data are not sufficient proof of this. The same may be said of Sr and Nd isotope ratios. Very convincing evidence for subduction of sediment to the arc magma generation zone comes from Pb and Be isotope data. However, it is not yet clear that subducted sediment is present beneath every arc. It follows that sediment can be, and often is, subducted and recycled into the mantle to depths of at least 100 km.

BACK-ARC BASINS

Are the effects of subduction and recycling of oceanic crust and sediment limited to the sub-arc mantle, or are wider areas of the mantle affected? Hart et al. (1972) pointed out that basalts from the Mariana trough had both similarities to and differences from typical MORB. Subsequent data has borne out this observation: most back-arc basin volcanics are intermediate between those erupted in island arcs and those erupted at mid-ocean ridges, though there is overlap with both. They show an enrichment in alkalies and alkaline earth, slightly higher $^{87}Sr/^{86}Sr$ and lower $^{143}Nd/^{144}Nd$ than MORB (e.g., Hart et al., 1972; Saunders and Tarney, 1979). They may share the high-field strength element depletion of IAV, but the data are limited. Pb and Be data are available only for the Valu Fa Ridge (Jenner et al., 1987; Vallier et al., 1987). The Valu Fa samples do seem to have slightly elevated $^{207}Pb/^{204}Pb$, but ^{10}Be is at background levels.

Saunders and Tarney (1979) concluded fluids produced by continued dehydration of the subducting slab also affect the source of back-arc basin basalts, although to a lesser degree than IAV sources. This accounts for the intermediate nature of back-arc basin basalt geochemistry. If this interpretation is correct, the affects of recycling are not limited to the subarc mantle. It would also demonstrate that the incompatible element and volatile content of the slab are not exhausted in the production of island arc magmas. These two conclusions would have important implications for mantle evolution. However, Jenner et al. (1987) point out that it is difficult to distinguish this hypothesis from one which calls for simple mixing of subarc mantle ('slab-metasomatized' mantle wedge) and normal depleted asthenosphere. This mixing might occur as the subarc mantle is invaded by asthenosphere in response to spreading.

In summary, basalts and their derivative liquids erupted in back-arc or "marginal" basins appear to share many of the geochemical characteristics of island arc magmas, though the data on such rocks are very limited. This may suggest the subducting oceanic crust (± sediment) is affecting the composition of a broader region of mantle than the subarc wedge. This, however, remains to be demonstrated. Back-arc basin volcanics are certainly important keys to understanding recycling and its affects. They deserve more intensive study than they have thus far received.

SUBOCEANIC MANTLE

Recycling of oceanic crust (± sediment) was invoked principally to explain some of the features of the isotope geochemistry of oceanic island basalt sources, including higher $^{87}Sr/^{86}Sr$ and lower $^{143}Nd/^{144}Nd$ ratios in OIB compared to MORB (e.g., Hofmann and White, 1982), $^{207}Pb/^{204}Pb-^{206}Pb/^{204}Pb$ correlations or mantle isochrons (e.g., Chase et al., 1981), and deviations from the linear $^{87}Sr/^{86}Sr-^{143}Nd/^{144}Nd$ correlation or mantle array (e.g, Hawkesworth et al., 1979). Since these papers were written, a somewhat clearer picture of the isotope geochemistry of sediments and of the effects of hydrothermal alteration of the oceanic crust has emerged. This allows more quantitative testing of the recycling hypothesis by comparing the predicted composition of recycled ancient sediment and oceanic crust with the observed isotope systematics of OIB sources.

It is now reasonably clear that sediments have low but rather constant $^{147}Sm/^{144}Nd$; $^{143}Nd/^{144}Nd$ varies, of course, but the variation is relatively small compared with $^{87}Sr/^{86}Sr$. Wide variations in Rb/Sr in marine sediments lead in time to widely varying $^{87}Sr/^{86}Sr$, as well as variable Sr/Nd ratios. Mixing of ancient oceanic sediment with depleted mantle would thus produce just the pattern observed in oceanic basalts: a wedge-shaped field on a plot of $^{87}Sr/^{86}Sr-^{143}Nd/^{144}Nd$, the narrow region located at low $^{87}Sr/^{86}Sr$ and high $^{143}Nd/^{144}Nd$ and the wide region at high $^{87}Sr/^{86}Sr$ and low $^{143}Nd/^{144}Nd$ values. Predicting the composition of 'mature' (i.e., altered) oceanic crust, and hence the isotopic composition of recycled oceanic crust, is still problematic. Staudigel and Hart (1983) calculated a bulk $^{87}Sr/^{86}Sr$ ratio of about 0.705. The Rb/Sr ratio should increase, but the magnitude of the increase is hard to quantify since Rb is lost from the crust during hydrothermal alteration but gained during low temperature alteration. Nd systematics are not expected to be affected by sea water interaction, thus oceanic crust differs from depleted mantle only in having a slightly higher $^{147}Sm/^{144}Nd$, leading ultimately to slightly higher $^{143}Nd/^{144}Nd$. To the degree we can predict the effects of recycling on Sr and Nd isotope evolution of mantle reservoirs, the observed evolution of at least some reservoirs is qualitatively consistent with the predictions.

Pb isotope systematics of sediment and mature oceanic crust should provide a better test of recycling. As noted above, sediments have high $^{207}Pb/^{204}Pb$ for a given $^{206}Pb/^{204}Pb$ compared to the mantle, although $^{206}Pb/^{204}Pb$ ratios are comparable. They also have higher $^{208}Pb/^{204}Pb$ ratios. Sediments, however, have rather low $^{238}U/^{204}Pb$ and high $^{232}Th/^{238}U$ ratios. As a result, ancient sediment, including anciently recycled sediment, would have relatively low present day $^{206}Pb/^{204}Pb$ ratios. Because of the short half-life of ^{235}U, the $^{207}Pb/^{204}Pb$ ratio was largely determined early in Earth's history and there has not been much change in $^{207}Pb/^{204}Pb$ late in Earth's history. Ancient sediment would therefore have high $^{207}Pb/^{204}Pb$. High $^{232}Th/^{238}U$ in ancient sediments would lead to high present day $^{208}Pb/^{204}Pb$ in relation to $^{206}Pb/^{204}Pb$. Thus, anciently recycled sediment should be characterized by high $^{208}Pb/^{204}Pb$ and $^{207}Pb/^{204}Pb$ and low $^{206}Pb/^{204}Pb$ compared to mantle values. Only a few oceanic islands such as Kerguelen, Tristan da Cunha and Samoa, and Indian Ocean MORB have such Pb isotopic compositions. Recycling of sediment is a viable hypothesis for such islands from the viewpoint of Pb isotope systematics, but it apparently is not for other islands, which is to say for most islands.

One must be a little cautious here, because this argument makes the uniformitarian assumption that ancient sediments are chemically similar to modern ones. However, U is apparently extracted from sea water by absorption on organic material in high productivity areas, most of which occur over continental shelves. Though there are few data of high quality, the low $^{238}U/^{204}Pb$ ratios of pelagic sediment are probably balanced by high $^{238}U/^{204}Pb$ ratios in organic-rich shelf sediments. The involvement of biological processes in this fractionation suggests caution in adopting a uniformitarian view. Also, the marine geochemistry of U would have been much different if the oceans were ever reducing.

It is now clear that hydrothermal activity at mid-ocean ridges removes Pb from the oceanic crust and adds U (Chen et al., 1986; Michard and Albarede, 1985). Thus, mature oceanic crust will have higher $^{238}U/^{204}Pb$ than fresh MORB and the mantle. This would lead to high $^{206}Pb/^{204}Pb$ in recycled oceanic crust. The magnitude of this effect is not yet known. Newsom et al. (1986) noted that if the effect is large, this hydrothermal transport of U and Pb could be the cause of the so-called 'Pb paradox', i.e., the increase in the mantle $^{238}U/^{204}Pb$ ratio, when a decrease associated with mantle depletion is expected.

Depending on the magnitude of the change in $^{238}U/^{204}Pb$ of the oceanic crust, recycling of oceanic crust could account for the high $^{206}Pb/^{204}Pb$ ratios in some OIB such as St. Helena basalts (Palacz and Saunders, 1986). However, it is not clear whether the Sr-Nd systematics of St Helena and related reservoirs are consistent with this hypothesis.

Hf and Nd isotope systematics also provide a test of the sediment recycling hypothesis. Hf and Nd isotopes are well correlated in oceanic basalts (Patchett, 1983), but somewhat less well correlated in sediments (White et al., 1986). More importantly, while Sm/Nd ratios in sediments are relatively constant, Lu/Hf ratios vary widely (Patchett et al., 1984). This would lead in time to widely varying $^{176}Hf/^{177}Hf$ ratios but relatively constant $^{143}Nd/^{144}Nd$ ratios in ancient sediment. Recycling and mixing of such sediment with depleted mantle should produce a wedge-shaped array (i.e., poor correlation at the low $^{176}Hf/^{177}Hf$, low $^{143}Nd/^{144}Nd$ end of the array), much like the $^{87}Sr/^{86}Sr$-$^{143}Nd/^{144}Nd$ array. Since this is not observed, Patchett et al. conclude that if sediment recycling is important, recycled sediment must always consist of a homogeneous mixture of constant proportions of terrigenous and pelagic sediment. This seems geologically unlikely, implying that sediment recycling is not important in mantle evolution.

Recycling has also been invoked to explain the trace element geochemistry of some OIB sources. Weaver et al. (1986) found La/Nb, Ba/Nb and Th/Nb ratios in Tristan, Gough and Walvis Ridge lavas differed from those of MORB, primitive mantle and other OIB. Noting a similarity of these ratios in sediment to those in these island lavas, they suggested the sources of Tristan, Gough and Walvis attained their peculiar geochemistry by the addition of a small amount (about 1%) of sediment to a more typical OIB source. Palacz and Saunders (1986) suggested low Th/U and Pb/Nb ratios in the sources of islands such as Mangaia and St. Helena reflect recycling of hydrothermally altered oceanic crust.

There is also trace element evidence that recycling of sediment, and possibly oceanic crust, has been very limited. The evidence is based on the constancy of some trace element ratios in oceanic basalts: Cs/Rb, Rb/Ba, Pb/Ce, U/Nb (Hofmann and White, 1983; Newsom et al., 1986; Hofmann et al., 1986). While ratios of these elements are uniform in all high-precision analyses of fresh oceanic basalts, these ratios are variable, and generally different than the oceanic basalt values, in continental crust and sediment. Thus sediment recycling should lead to variable ratios in the mantle. The uniformity of Pb/Ce ratios, together with the high concentrations of Pb in sediment and low Pb concentrations in the mantle are a particularly strong constraint on sedimentary recycling. As indicated in Figure 2, even additions of less than 1% sediment to the mantle would lead to Pb/Ce ratios in the mixture that are higher than any observed in fresh oceanic basalts.

Alteration of the oceanic crust would certainly lead to variations in all the above ratios, but the magnitude of the effects are unknown. If the effects are large, the constancy of these ratios in various mantle reservoirs would argue against extensive recycling of oceanic crust. On the other hand, concentrations of elements like Cs and Pb are much lower in altered oceanic crust than in sediment, so recycling of oceanic crust leads to much smaller variations in these ratios than does sediment recycling.

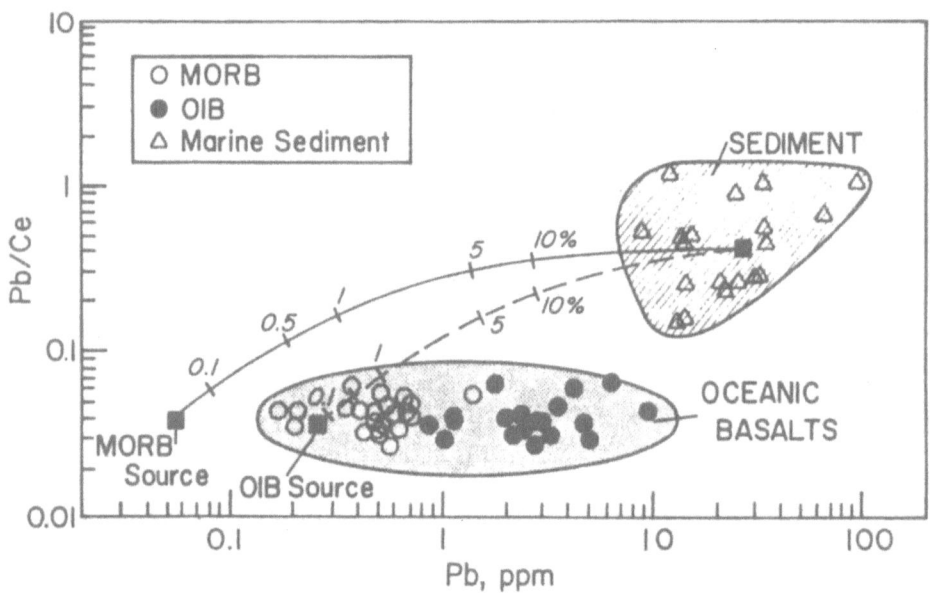

FIGURE 2: Pb concentrations and Pb/Ce ratios in oceanic basalts and modern marine sediments. Solid line sediment-MORB source mixing curve. Numbered ticks show percentage of sediment in the mixture. MORB source is assumed to have Pb and Ce concentrations 10 times lower than average MORB. Dashed line illustrates the effect of mixing sediment with a hypothetical OIB source having Pb and Ce concentrations 5 times higher than the MORB source. Addition of less than 1% sediment increases Pb/Ce ratios beyond the range observed.

The evidence for recycling to the suboceanic mantle is thus conflicting. There are clearly some trace element and isotopic data suggesting this process is important in the evolution of magma sources, but other evidence, particularly the $^{176}Hf/^{177}Hf-^{143}Nd/^{144}Nd$ correlation and the uniformity of Pb/Ce ratios suggests recycling of sediment is very limited.

SUBCONTINENTAL MANTLE

It has also been suggested that subducted oceanic crust and sediment influences the composition of the subcontinental mantle (e.g., Carson, 1984). Oceanic lithosphere is often subducted beneath continents. Just as subduction appears to influence the composition of the mantle beneath back-arc basins, it would influence the composition of the subcontinental mantle. It has been argued that island arc volcanism and accretion of island arcs is an important mechanism of continental growth (e.g., Taylor, 1977). If so, the subcontinental lithosphere, whose evolution is presumably coupled to that of the continent above it, may in large measure be similar to mantle beneath arcs and back-arc basins. The influence of recycling in this case, as in the subarc mantle, would arise indirectly from metasomatism of the mantle by fluids or melts derived from the subducting slab.

Subducted oceanic crust may also be stored in the subcontinental lithosphere. MacGregor and Manton (1986) have made a strong case from stable and radiogenic isotopes and trace elements that Roberts Victor Group II eclogites represent ancient subducted oceanic crust. Extremely high $^{143}Nd/^{144}Nd$ ratios (Jagoutz et al., 1983) are consistent with this hypothesis if the crust experienced light-rare-earth depletion through partial melting or dehydration in the subduction zone. $^{143}Nd/^{144}Nd$ ratios also indicate the crust has been stored in the subcontinental mantle since at least the late Archean (Jagoutz et al., 1983). Nelson et al. (submitted) have suggested carbonatites are derived from recycled oceanic crust. Perhaps the strongest argument in favor of this hypothesis is the extreme light-rare-earth enrichment, which suggests an eclogitic or garnet-rich residual mineralogy.

SUMMARY

Return of oceanic crust to the mantle through subduction is implicit in plate tectonics. Beyond that, relatively little regarding the role of subducted, recycled oceanic crust and sediment can be regarded as established. It does seem clear that subducted material inevitably influences island-arc volcanism. In some arcs, such as the Lesser Antilles and the Aleutians, Pb and Be isotope evidence strongly suggest sediment is also subducted and this subducted sediment finds its way into island arc magmas. There is also evidence, although it is sparser and weaker, that subduction influences the composition of the mantle beneath back-arc or marginal basins. Whether this occurs through metasomatism by melts or fluids derived directly from the subducting slab or by mixing depleted asthenosphere with 'metasomatized' subarc mantle is not clear. The back-arc basalt data nevertheless suggest the effects of recycling are not limited to island arc volcanism.

There is evidence both for and against recycling influencing the composition of various suboceanic mantle reservoirs. At present, it would seem the weight of the evidence, particularly Pb/Ce ratios and the Hf-Nd isotope correlation, argues against significant recycling of sediment to the suboceanic mantle. Identification of recycled material in sources of continental volcanics is hampered to some degree by potential contamination of the volcanics as they ascend through the continental crust, the geochemical effects of which would be similar to recycling.

Resolving the question of the importance of recycling will depend on making rather specific geochemical predictions and tests. Making such predictions will depend critically on advances made in two areas.

The first is understanding the chemical effect of seawater-basalt interaction on the overall composition of the oceanic crust. After all, what is subducted is not MORB, but MORB after processing by low-temperature and hydrothermal alteration. In many cases, we now know only the direction of the same chemical effects associated with these processes (e.g., decrease in Pb concentration, increase in U concentration). We do not in general know the magnitude of the effects. In other cases, low temperature and high temperature alteration produce opposing effects (e.g., the alkalies) and we cannot even be sure of the direction of the net change.

The second process we must learn more about is the geochemical processing of the subducting slab in the island arc magma generation zone. If island arc magmas contain components of recycled material, there must be a corresponding depletion occurring in the subducting slab. Island arc magmas cannot be accounted for by simple mixing of mantle, basaltic crust, and sediment: some elements appear to be extracted from the slab more efficiently than others. If for example, Pb is extracted more efficiently than Ce, the Pb/Ce ratio of the slab would be lowered and high Pb/Ce ratios in sediments need not be an objection to recycling. To understand the processing occurring in subduction zones requires intensive studies of island arc volcanos as well as a detailed knowledge of the composition of material, altered oceanic crust and sediment, being subducted beneath the arc.

REFERENCES

Armstrong, R.L., 'A model for Sr and Pb isotope evolution in a dynamic Earth.' Rev. Geophys. 6: 175-199, 1968.

Armstrong, R.L., 'Isotopic and chemical constraints on models of magma genesis in volcanic arcs.' Earth. Planet. Sci. Lett. 12: 137-143, 1971.

Barreiro, B., 'Lead isotopic compositions of South Sandwich Island volcanic rocks and their bearing on magmagenesis in intra-oceanic island arcs.' Geochim. Cosmochim. Acta 47: 817-822,. 1983.

Carlson, R.W., 'Isotopic constraints on Columbia River flood basalt genesis and the nature of the subcontinental mantle.' Geochim. Cosmochim. Acta 48: 2357-2372, 1984.

Chase, C.G., 'Oceanic island Pb: two stage histories and mantle evolution,' Earth Planet. Sci. Lett. 52: 277-284, 1981.

Chen, J.H., G.J. Wasserburg, K.L. von Damm, and J.M. Edmond, 'The U-Th-Pb systematics in hot springs on the East Pacific Rise at 21°N and Guaymas Basin.' Geochim. Cosmochim. Acta 50: 2467-2479, 1986.

Coats, R.R., 'Magma type and crustal structure in the Aleutian Arc,' in The Crust of the Pacific Basin, edited by G.A. Macdonald and H. Kuno, pp. 92-109, American Geophysical Union, Washington, 1962.

Davidson, J.P. 'Isotopic and trace element constraints on the petrogenesis of subduction-related lavas from Martinique, Lesser Antilles,' J. Geophys. Res. 91: 5943-5962, 1986.

DePaolo, D.J., 'The mean life of continents: estimates of continental recycling rates from Nd and Hf isotope data and implications for mantle structure.' Geophys. Res. Lett. 10: 705-708, 1983.

DePaolo, D.J. and G.J. Wasserburg, 'The sources of island arcs as indicated by Nd and Sr isotopic studies.' Geophys. Res. Lett. 4: 465-468, 1977.

Dupre, B. and C.J. Allegre, 'Pb-Sr isotope variations in Indian Ocean basalts and mixing phenomena,' Nature, 303: 143-146, 1983.

Gill, J., Orogenic Andesites and Plate Tectonics, 390 pp., Springer Verlag, Berlin, 1981.

Hart, S.R., W.E. Glassley, and D.E. Karig, 'Basalts and sea floor spreading behind the Mariana Island arc.' Earth Planet. Sci. Lett., 15: 12-18, 1972.

Hawkesworth, C.J., M.J. Norry, J.C. Roddick and R. Vollmer, $^{143}Nd/^{144}Nd$ and $^{87}Sr/^{86}Sr$ ratios from the Azores and their significance in LIL-element enriched mantle.' Nature, 280: 28-31, 1979.

Hofmann, A.W., K.P. Jochum, M. Seufert, and W.M. White, 'Nb and Pb in oceanic basalts: new constraints on mantle evolution.' Earth. Planet. Sci. Lett., 79: 33-45, 1986.

Hofmann, A.W. and W.M. White, 'Ba, Rb, and Cs in the Earth's mantle.' Z. Naturforsch, 38: 256-266, 1983.

Hofmann, A.W. and W.M. White, 'Mantle Plumes from ancient oceanic crust.' Earth Planet. Sci. Lett. 57: 421-436, 1982.

Hole, M.J., A.D. Saunders, G.F. Marriner, and J. Tarney, 'Subduction of pelagic sediments: implications for the origin of Ce-anomalous basalts from the Mariana Islands.' J. Geol. Soc. London, 141: 453-472, 1984.

Jagoutz, E., J.B. Dawson, B. Spettel, and H. Wanke, 'Identification of early differentiation processes on the earth,' paper presented at Annual Meeting, Meteoritical Society, Mainz, 1983.

Jenner, G.A., P.A. Cawood, M. Rautenschlein, and W.M. White, 'Composition of back-arc basin volcanics, Valu Fa Ridge, Lau Basin: evidence for a slab-derived component in their mantle source,' J. Volcanol. Geotherm. Res., in press, 1987.

Kay, R.W., S.-S. Sun, and C.-N. Lee-Hu, 'Pb and Sr isotopes in volcanic rocks from the Aleutian Islands and Pribilof Islands, Alaska,' Geochim. Cosmochim. Acta, 42: 263-274, 1978.

Kay, R.W., 'Volcanic arc magmas: implications of a melting-mixing model for element recycling in the crust-upper mantle system,' J. Geol., 88: 497-522, 1980.

MacGregor, I.D. and W.I. Manton, 'Robert Victor Eclogites: ancient oceanic crust.' J. Geophys. Res. 91: 14063-14079, 1986.

Meijer, A., 'Pb and Sr isotopic data bearing on the origin of volcanic rocks from the Mariana island-arc system.' Geol. Soc. Am. Bull., 87: 1358-1369, 1976.

Michard, A. and F. Albarede, 'Hydrothermal uranium uptake at ridge crests,' Nature, 317: 244-246, 1985.

Morris, J.D. and S.R. Hart, 'Isotopic and incompatible element constraints on the genesis of island arc volcanics: Cold Bay and Amak Island, Aleutians.' Geochim. Cosmochim. Acta, 47: 2015-2030, 1983.

Nelson, D.R., A.R. Chivas, B.W. Chappell, and M.T. McCulloch, 'Geochemical and isotopic evidence for a subducted oceanic lithospheric origin for carbonatites,' Geochim. Cosmochim. Acta, submitted, 1987.

Newsom, H.E., W.M., White, K.P. Jochum, and A.W. Hofmann, 'Siderophile and chalcophile element abundances in oceanic basalts, Pb isotope evolution and growth of the Earth's core,' Earth Planet. Sci. Lett. 80: 299-313, 1986.

Palacz, Z.A. and A.D. Saunders, 'Coupled trace element and isotope enrichment in the Cook-Austral-Samoa Islands, southwest Pacific,' Earth Planet. Sci. Lett. 79: 270-280, 1986.

Patchett, P.J., 'Importance of the Lu-Hf isotope system in studies of planetary chronology and chemical evolution.' Geochim. Cosmochim. Acta, 47: 81-91, 1983.

Patchett, P.J., W.M. White, H. Feldmann, and S. Kielinczuk, A.W. Hofmann, 'Hafnium/rare earth element fractionation in the sedimentary system and crustal recycling into the Earth's mantle.' Earth Planet. Sci. Lett. 69: 365-378, 1984.

Perfit, M.R., D.A. Gust, A.E. Bence, and R.J. Arculus, S.R. Taylor, 'Chemical characteristics of island-arc basalts: implications for mantle sources,' Chemical Geology, 30: 227-256, 1980.

Saunders, A.D. and J. Tarney, 'The geochemistry of basalts from a back-arc spreading center in the East Scotia Sea.' Geochim. Cosmochim. Acta 43: 555-572, 1979.

Staudigel, H. and S.R. Hart, 'Alteration of basaltic glass: mechanism and significance for the oceanic crust-seawater budget.' Geochim. Cosmochim. Acta 47: 337-350, 1983.

58

Stern, R.J., 'On the origin of andesite in the northern Mariana Island arc: implications from Agrigan.' Contrib. Mineral. Petrol., 68: 207-219, 1979.

Taylor, S.R., 'Island arc models and the composition of the continental crust,' in Island Arcs, Deep Sea Trenches and Back-Arc Basins, edited by M. Talwani and W.C. Pitman III, pp. 325-336, American Geophysical Union, Washington, 1977.

Tera, F., L. Brown, J. Morris, and I.S. Sacks, J. Klein, R. Middleton, 'Sediment incorporation in island-arc magmas: inferences from [10]Be.' Geochim. Cosmochim. Acta 50: 535-550, 1986.

Vallier, T.L., G. A. Jenner, F.A. Frey, J. Gill, J.W. Hawkins, J.D. Morris, P. Cawood, J. Morton, D.W. Scholl, W. White, M. Rautenschlein, R.W. Williams, A.M. Volpe, A.I. Stevenson and L.D. White, 'Subalkaline andesites from Valu Fa Ridge, a back-arc spreading center in Lau Basin; chemical constraints on magma genesis and tectonic implications.' Geol. Soc. Amer. Bull., to be submitted.

Weaver, B.L., D.A. Wood, J.A. Tarney, and J. L. Joron, 'Role of subducted sediment in the genesis of ocean-island basalts: geochemical evidence form South Atlantic Islands,' Geology, 14: 275-278, 1986.

White, W.M. and B. Dupre, 'Sediment subduction and magma genesis in the Lesser Antilles: isotopic and trace element constraints,' J. Geophys. Res., 91: 5927-5941, 1986.

White, W.M. and A.W. Hofmann, 'Sr and Nd isotope geochemistry of oceanic basalts and mantle geochemistry,' Nature, 296: 821-825, 1982.

White, W.M. and P.J. Patchett, 'Hf-Nd-Sr and incompatible-element abundances in island arcs: implications from magma origins and crust-mantle evolution,' Earth Planet. Sci. Lett., 67: 167-185, 1984.

White, W.M., P.J. Patchett, and D. BenOthman, 'Hf isotope ratios of marine sediments and Mn nodules: evidence for a mantle source of Hf in seawater,' Earth Planet. Sci. Lett., 79: 46-54, 1986.

Woodhead, J.D. and D.G. Fraser, 'Pb, Sr and [10]Be isotopic studies of volcanic rocks from the Northern Mariana Islands: implications for magma genesis and crustal recycling in the Western Pacific.' Geochim. Cosmochim. Acta, 49: 1925-1930, 1985.

KERGUELEN ARCHIPELAGO: GEOCHEMICAL EVIDENCE FOR RECYCLED MATERIAL

D.Weis - J-F.Beaux - I.Gautier - A.Giret and P.Vidal.
Laboratoires Associés Géologie-Pétrologie-Géochronologie
Université Libre de Bruxelles
Avenue F.D.Roosevelt, 50
B-1050 Brussels
Belgium

The most recent review of chemical geodynamics led to the definition of at least four different types of mantle sources (DMM, HIMU, EMI, EMII; (1,2)) from which oceanic basalts may be produced by variable mixing relationships, assuming that oceanic basalt isotope systematics reflect mixing. The existence of two additional components, BSE and PREMA, is also compatible with the observed isotopic heterogeneities in the mantle. The combinations of end-member components for generating individual compositions by mixing are nearly endless. Much more work, especially with a more quantitative approach, will be necessary before we fully understand the geochemical characteristics of the mantle.

Preliminary isotope studies on the Kerguelen Archipelago (3, 4) have indicated the enriched characteristics of these rocks, i.e. $^{87}Sr/^{86}Sr$ > 0.7045, ΣNd < 0 and relatively high $^{207}Pb/^{204}Pb$ and $^{208}Pb/^{204}Pb$ ratios for a given $^{206}Pb/^{204}Pb$ ratio in comparison with other oceanic basalts. These features, together with those observed for the Walvis Ridge, Samoa, Tristan and Gough Islands require the existence of enriched mantle components which appear to be located in the Southern Hemisphere (Dupal anomaly, (5, 6)).

In order to obtain better constraints on the geochemical definition of these enriched components, their geographical and temporal location, and their mode of formation, we have undertaken a much more systematic isotopic and geochemical study of the Kerguelen rocks, both volcanic and plutonic.

The Kerguelen Archipelago is located in the southern Indian Ocean. It consists mainly of transitional or alkaline flood basalts (~85%) which are intruded by alkaline volcano-plutonic complexes belonging to two distinct series:

1. a silica-saturated series, with rocks ranging in composition from gabbro to nordmarkite (Rallier du Baty peninsula, Ile de l'Ouest and Iles Nuageuses complexes, in the western part of the archipelago); and

2. a silica-undersaturated series, with a range from gabbro to ne pheline-bearing syenite (Société de Géographie peninsula, Monts Ballons and Montagnes Vertes)

S. R. Hart and L. Gülen (eds.), Crust/Mantle Recycling at Convergence Zones, 59–63.
© *1989 by Kluwer Academic Publishers.*

60

Gabbros from both plutonic series appear similar but the most differentiated members of each series are clearly distinguished by their alkali to silica ratio: moderately alkaline in the oversaturated series and highly alkaline in the undersaturated series. The evolution of each individual plutonic series can be explained by a fractional crystallization process.

The transitional and alkaline lava series of Kerguelen have clearly distinct geochemical characteristics and reflect evolution by fractional crystallization involving early and dominant fractionation of plagioclase in the transitional series, and of olivine in the alkaline series. Further differentiation, marked by clinopyroxene crystallization, produces similar evolutionary trends for both series except in the case of the hawaiites where kaersutite fractionation induces a TiO_2 decrease and an increase of Th/Ta, Th/U and Th/Tb ratios in the alkaline magmas only. Both series are relatively enriched in incompatible elements (especially Th) and impoverished in MgO compared to other OIB. The alkaline series is more enriched in alkalis (essentially K_2O), P_2O_5, incompatible and LRE elements while the transitional series is enriched in CaO, FeO and HREE. The alkaline series itself evolves into two separate differentiation trends, one weakly alkaline and the other strongly alkaline.

Figure 1: Nd-Sr correlation diagrams.

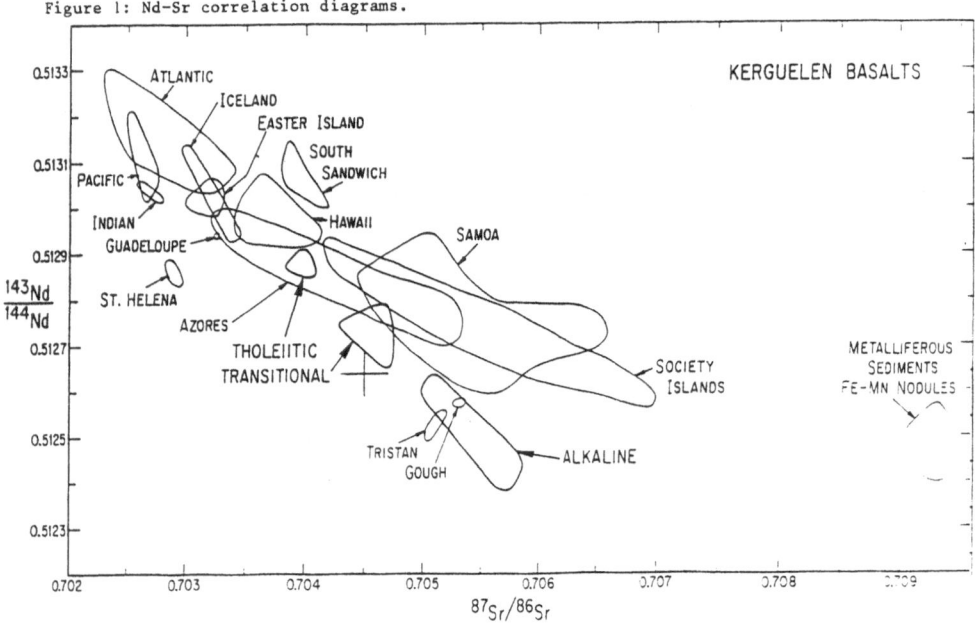

Figure 1a: $^{143}Nd/^{144}Nd$ vs $^{87}Sr/^{86}Sr$ for oceanic basalts (data from the literature, for sources see 1, 2 and 8).
Comparison with the three groups, tholeiitic, transitional and alkaline, of the Kerguelen basaltic series.

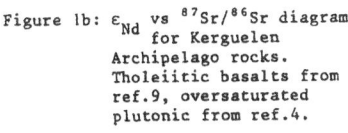

Figure 1b: ε_{Nd} vs $^{87}Sr/^{86}Sr$ diagram for Kerguelen Archipelago rocks. Tholeiitic basalts from ref.9, oversaturated plutonic from ref.4.

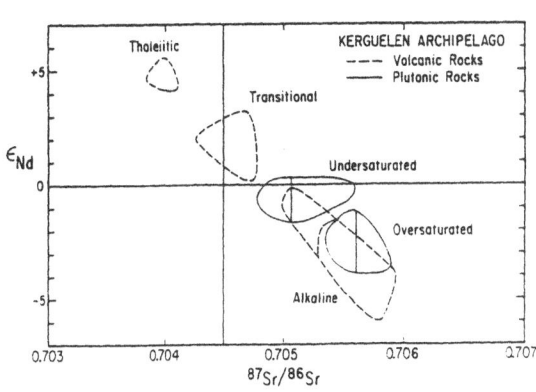

In the undersaturated complex of the Société de Géographie Peninsula, the measured $^{87}Sr/^{86}Sr$ ratios vary between 0.70510 and 0.78546 for $^{87}Rb/^{86}Sr$ ratios between 0.22 and 783. On the Rb-Sr evolution diagram, with the exception of some of the most evolved samples with Rb/Sr > 50, these data define an isochron corresponding to 14.5 + 0.3 Ma with an initial $^{87}Sr/^{86}Sr$ ratio of 0.70506 + 1, i.e. similar but slightly lower than the one measured (0.7057) for the oversaturated rocks of the complex of Rallier du Baty (3). Pb isotopic compositions ($^{206}Pb/^{204}Pb$ = 18.3 to 18.4; $^{207}Pb/^{204}Pb$ = 15.56, $^{208}Pb/^{204}Pb$ = 38.8 to 39.0) fall typically within the range observed for other alkaline rocks of Kerguelen, both volcanic or plutonic (2, 3). ΣNd values show a small range (0 to -1.4) that is slightly higher than the values observed at Rallier (-1.8 to -2.7, (4)). There is no direct evidence from these data on the undersaturated rocks for any continental signature and they confirm the "enriched" characteristics of the Kerguelen mantle sources.

For the volcanic rocks, the $^{87}Sr/^{86}Sr$ ratios vary from 0.70430 to 0.70564 and the $^{143}Nd/^{144}Nd$ ratios from 0.51280 to 0.51250 (ΣNd from +3.0 to -2.8) and are correlated falling along the so-called 'mantle-array' (fig 1a). These data cover the whole range of values observed for the Kerguelen Islands (4, 7, fig 1b). The Pb isotopic compositions show very little scatter and are entirely comparable to those observed in the plutonic rocks (fig 2; $^{206}Pb/^{204}Pb$ = 18.2 to 18.5, $^{207}Pb/^{204}Pb$ = 15.55 to 15.56, $^{208}Pb/^{204}Pb$ = 38.4 to 39.0).

There is a clear distinction in the ΣNd versus $^{87}Sr/^{86}Sr$ diagram (fig.1b) between the transitional and alkaline basaltic series, which have only slightly different major element contents. The former have $^{87}Sr/^{86}Sr$ ratios lower than 0.7050 and ΣNd values above 0.0 while the latter have $^{87}Sr/^{86}Sr$ ratios above 0.7050 and slightly negative ΣNd values. The field is extended to higher $^{87}Sr/^{86}Sr$ and lower ΣNd values by the most alkaline lavas which are also more LILE enriched. On the other hand, there is no systematic difference for

the Pb ratios (fig.2) which are almost the same within analytical
errors.

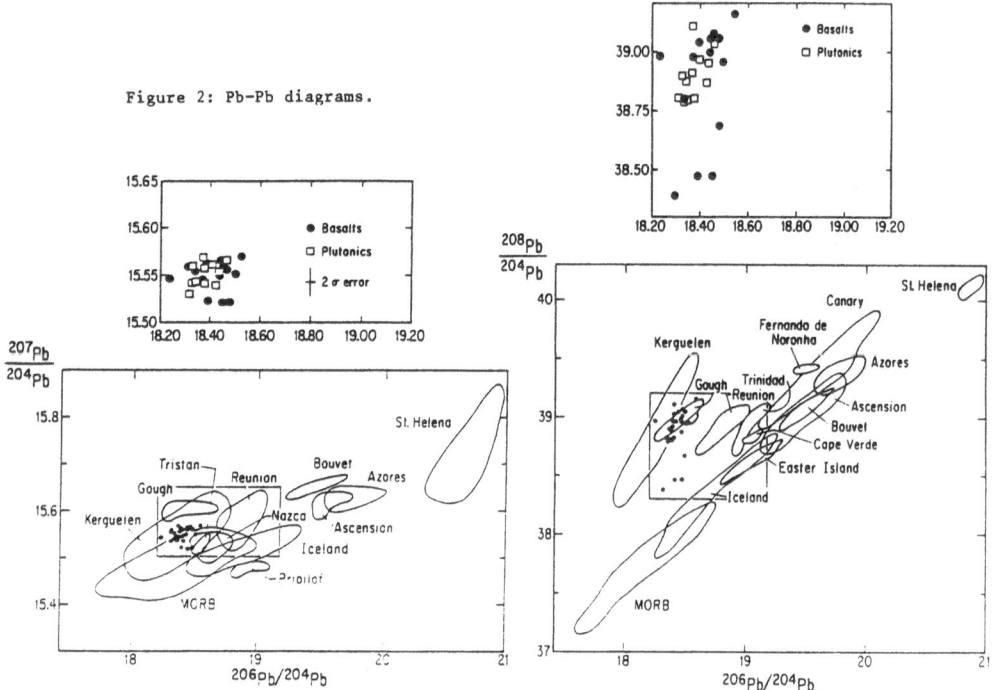

Figure 2: Pb-Pb diagrams.

Kerguelen basalts and plutonics. Comparison with other oceanic basalts (for sources, see 1, 2, 8, 9
and 10). New data (black dots and small enlarged diagrams above) for Kerguelen from Weis
(7, this paper) and White (unpublished).

These isotopic characteristics belong to the so-called Dupal
anomaly, typical of the Southern Hemisphere. They seem to indicate
two different parental magmas, both with enriched features for the
two basaltic series of Kerguelen. They also indicate a similar
source for the silica undersaturated alkaline plutonic series and
the alkali basalt series of Kerguelen. The same observation is valid
for the oversaturated series and the alkali-enriched basalt series.
The similar Pb data for all these plutonic and volcanic rocks
indicate that the Sr and Nd isotopic variations between the
different series reflect younger process(es), probably heterogeneity
or mixing in the source-region or perhaps even later during the
differentiation, on the scale of the magma chamber. This last
conclusion is especially significant for the definition of the
mantle characteristics under Kerguelen Archipelago.
On the basis of these isotopic data, it appears that mixing
processes between at least three different components have been
involved in the genesis of the igneous rocks on Kerguelen as
indicated by the relatively higher $^{207}Pb/^{204}Pb$ and $^{208}Pb/^{204}Pb$ ratios.

In addition, the mixing occurred on at least two different time-scales:
1) incorporation in the mantle source-region of continentally derived sediment or crust responsible for the enriched features of Kerguelen.
2) varying degrees of mixing between a very enriched OIB-type component and a depleted MORB-type component.
The characteristic times for these two processes are approximately 1.5 b.y. and < 50 m.y. respectively.

(1) Zindler A. and Hart S.R. (1986), 'Chemical Geodynamics', *Ann. Rev. Earth Planet. Sci.* **14**, 493-571.
(2) White W.M. (1985), 'Sources of oceanic basalts: radiogenic isotopic evidence', *Geology* **13**, 115-8.
(3) Dosso L., Vidal P., Cantagrel J.-M., Marot A. and Zimine S. (1979), '"Kerguelen: continental fragment or oceanic island?": petrology and isotopic geochemistry evidence', *Earth Planet. Sci. Lett.* **43**, 46-60.
(4) Dosso L. and Murthy V.R. (1980), 'A Nd isotopic study of the Kerguelen Islands: inferences on enriched oceanic mantle sources', *Earth Planet. Sci. Lett.* **48**, 268-276.
(5) Dupré B. and Allègre C.J. (1983), 'Pb-Sr isotopic variations in Indian Ocean basalts and mixing phenomena', *Nature* **303**, 142-6.
(6) Hart S.R. (1984), 'A large-scale isotopic anomaly in the Southern Hemisphere mantle', *Nature* **309**, 753-7.
(7) Weis D., Beaux J.-F. and Giret A. (1986), 'Sr, Pb and Nd isotopic compositions in alkaline undersaturated rocks from the Société de Géographie Peninsula (Kerguelen Archipelago)', *EOS* **67**, 1272.
(8) Zindler A., Jagoutz E. and Goldstein S. (1982), 'Nd, Sr and Pb isotopic systematics in a three-component mantle: a new perspective', *Nature* **258**, 519-23.
(9). White W.M. and Hofmann A.W. (1982), 'Sr and Nd isotope geochemistry of oceanic basalts and mantle evolution', *Nature* **296**, 821-5.
(10) Sun S.S. (1980), 'Lead isotopic study of young volcanic rocks from mid-ocean ridges, ocean islands and island arcs', *Phil. Trans. R. Soc. London Ser. A* **297**, 409-45.

SUBDUCTION OF YOUNG LITHOSPHERE: LITHOSPHERIC DOUBLING, A POSSIBLE
SCENARIO

N.J.Vlaar
Department of Geophysics
University of Utrecht
P.O.Box 80.021
3508 TA Utrecht
The Netherlands

ABSTRACT. The behaviour of oceanic lithosphere when subducted depends
strongly on its temperature, hence its age. The parameters affecting
this behaviour are stratification, rheology, and buoyancy. It is proposed that young oceanic lithosphere (< 20 Ma) and oceanic ridges,
when being subducted upon being overridden by cool and strong (continental) lithosphere resides subhorizontally in the shallow upper
mantle underneath a cooler upper lithospheric plate. This situation,
termed "lithospheric doubling", gives rise to upper mantle thermal
anomalies. The thermally anomalous layering leads to a number of geophysical, geological, and petrological consequences.

1. Introduction

Vlaar (1975) proposed that between creation at an oceanic spreading
center of young, and destruction at a subduction zone of old oceanic
lithosphere, a transition should take place from gravitationally
stable to unstable stratification of the lithosphere with respect to
the upper mantle. Due to its small strength, young o.l. (oceanic
lithosphere) requires stability in order to survive the initial period
of its existence. On the other hand, subduction of old lithosphere in
a Benioff zone, which is dominated by negative buoyancy, requires a
lithosphere, which, before subduction, is unstably layered with
respect to the upper mantle. It was inferred that young o.l. might
not age sufficiently to reach the transition between stable and
unstable stratification, and that consequently negative buoyancy upon
subduction would be reduced to the extent that "normal" Benioff subduction should not take place. Instead, lacking the necessary negative
buoyancy, o.l. younger than a certain critical age was assumed to
remain afloat at a shallow depth in the upper mantle or even might be
subducted subhorizontally, if very young. Vlaar and Wortel (1976), by
studying global seismicity, found a unique relationship between the
depth of the deepest earthquakes in a specific subduction zone and the
age of the subducted lithosphere concerned. The relationship proved to

65

be valid in general. The few exceptions could be explained "ad hoc".

They took this as sufficient evidence for the age dependence of subduction behaviour as described above. It should be envisaged, however, that the occurrence of earthquakes in a descending slab is dependent on the strength of the material involved, and hence on the resorption time of the subducted lithosphere (Deffeyes, 1972). This resorption time depends on the thermal contrast between the subducted lithosphere and the surrounding mantle. Resorption therefore also would cause vanishing of seismicity in a subduction zone below the depth of resorption, and this depth also depends on age. Therefore, the hypothesis of the age-dependent behaviour as described above might become difficult to substantiate. However, the subduction behaviour of the younger part of the o.l. in the Peruvian subduction zone can be explained only by assuming age varying buoyancy of the subducted slab. The slab has a length of 800 km and sinks to a depth of 250 km, as is delineated by its seismicity. Though the lithosphere concerned is not very old, it is not very young either. The o.l. concerned is about 40 Ma when it reaches the convergence zone. This explains its large resorption time, which is defined as the quotient of down-dip length and convergence rate. If subduction behaviour would not be age-dependent and when variable buoyancy would not exist, the slab should show deep seismicity. Only age-dependent buoyancy can explain these observations.

In adjacent parts of the S.American subduction zone, in cases where older lithosphere is involved, deep seismicity occurs (Wortel and Vlaar, 1978). According to the findings of Vlaar and Wortel (1976) the 40 Ma old Peruvian slab should be weakly negatively buoyant. If these findings are extrapolated to the subduction of still younger o.l., it is to be expected that lithosphere younger than 30 Ma is residing horizontally in the upper mantle just below the overriding strong upper plate.

The initial stable stratification of young o.l. is due to the petrological differentiation of uprising mantle material underneath an oceanic spreading center. Fertile mantle peridotite, upon diapiric ascent is differentiated and segregated in a less dense depleted peridotite or harzburgite layer, topped by a still lighter basalt layer. Initially, this layering is inherently stable, density increasing with depth. Cooling of the spreading and ageing lithosphere, however, decreases the stability of the layered system in the sense that the average density of a segregated column of harzburgite plus basalt decreases and even may reach the density of undepleted mantle material. At the age that this occurs the stable stratification changes into unstable stratification. In the latter case, by cooling, the average density of the segregated layers, has surpassed the density of the underlying undifferentiated mantle.

Oxburgh and Parmentier (1976) determined the transition age to be about 30 Ma. This agrees well with observational evidence concerning age-dependent subduction behaviour (Vlaar and Wortel, 1976). These findings imply however, that most of the o.l. which is older than 30 Ma should founder in the underlying lower density asthenosphere, at least if unstable stratification is the only requirement for foundering. In a study of the initiation of subduction of o.l., (Cloetingh et al., 1982), it could be demonstrated that the increasing strength of the ageing lithosphere prevents it from sinking by unstable stratification. It was found that the age of o.l. alone is not a sufficient condition for the initiation of its subduction. Instead, young lithosphere, by its small strength is a more likely candidate, particularly in a compressive regime. Cool and strong adjacent lithospheric plates can override young o.l. and even oceanic ridge systems. The rise of mantle material at a spreading center is the passive result of the diverging of the adjacent o.l. on which ridge push forces act (Vlaar, 1975). Once a ridge has been overthrusted by a strong lithospheric plate, ridge push must vanish and hence large scale spreading should come to a halt. It should be remarked that spontaneous foundering of a dense layer in a weaker one is only possible if the dense layer is sufficiently weak, either by fraction or by ductility, to allow the material of the lower layer to flow on top of the upper one. It can be inferred that strength and mechanical coherency is the important property preventing a gravitationally instable horizontal layering from collapsing. This applies in general to the lithosphere, whether oceanic or continental.

If thermal convection occurs in the upper mantle, it requires fluid properties and vanishing strength. The connection between the lithosphere and possible convective flow underneath can only be loosely coupled. Rather should convection be induced by lithospheric evolution than the opposite, the driving of lithospheric plates by convection currents. Lithospheric strength prohibits the plates to act as a viscous layer. Plate tectonics can be understood in terms of the forces which are inherent: ridge push and slab pull, the latter being strongly age-dependent.

2. Subduction

"Normal" or Benioff subduction, i.e. subduction of deeply penetrating old oceanic lithosphere, requires large negative buoyancy forces acting on the descending slab. These forces arise from the density excess due to the thermal contrast of the slab and the surrounding mantle. In addition, the supposed olivine-spinel transition, normally being situated at some 400 km depth, and which now has been shifted to a shallower level in the descending cool slab, may add another pulling force of about equal magnitude. The slab pull forces act on a mechanically coherent lithospheric plate. Due to its old age when subducted at the trench, it has acquired sufficient strength up to a thickness of some 100 km, or more, such that it largely remains coherent before resorption annihilates coherency. In this case, the average density surplus of the plate determines the average downward slab pull which is obtained by integration of the negative buoyancy of the parts. Oxburgh and Parmentier (1976) deduced that once subduction has taken place, former o.l. of any age should be negatively buoyant as a consequence of the phase change of basalt into the denser eclogite taking place shortly after subduction at shallow depth. They had to assume, of course, that this transition is not delayed and also, that melting does not counteract the densification. Moreover, deeply penetrating

Benioff subduction under the action of negative buoyancy requires that the slab remains sufficiently strong to allow earthquakes to occur. Foregoing conditions may not be fulfilled when young o.l. is subject to subduction. A necessary condition for slab pull is mechanical coherency of the slab, such that negative buoyancy of the separate layers can be averaged. Another condition is that the basalt (or eclogite) does not reach its melting point when descending. Ahrens and Schubert (1975) presented evidence that the basalt-eclogite phase change is strongly rate limited and depends on the degree of hydration. They showed that dry basalt could not transform to eclogite within less than 10 - 100 Ma at shallow mantle temperatures, and hence within the residence time of the coherent slab when subducted. Hydrated basalt, however, was found to accomplish the transition shortly after the descent started, at normal plate velocities, within a depth range up to 50 km. However, hydration also results in a lowered melting point of basalt or eclogite. At relatively low temperatures, in the reheating slab, hydrated basalt or eclogite may become melted. The hydrated part of the basaltic crust is here assumed to be the top 1 - 2 km and also that the underlying crust is dry. Consequently, if the basalt-eclogite phase change is to contribute to the negative buoyancy, it is required to take place within the residence time in the upper mantle of the subducted coherent lithosphere, i.e. for old o.l. less than 10 Ma. This requirement may prove to be difficult to meet for dry basalt, but certainly applies to hydrated crust. In the latter case, however, the melting point is considerably lowered relative to the dry solidus, and melting of either basalt or eclogite may set in shortly after the beginning of descent.

The subduction behaviour of young and very young o.l. differs from that of old o.l. as can be observed by its shallower seismicity. The latter is due to higher temperatures which affect the rheology. As young o.l. is gravitationally stably stratified before subduction and not subject to slab pull forces at the beginning of descent, mechanical resistence against subduction is larger than for old o.l. (England and Wortel, 1980; Wortel and Cloetingh, 1986).

As the strength of young o.l. is small, and in the absence of strong negative buoyancy, active subduction by ridge push alone, in a resistive environment, may become difficult to enforce. It appears preferable, therefore, to envisage convergence to take place by overriding of the upper strong plate of the subducted young o.l., which then is supposed to be not moving with respect to the upper mantle. The upper, strong plate, in this case, therefore, overthrusts a horizontally layered lithosphere-upper mantle layering. If the overridden o.l. remains sufficiently buoyant after subduction, the lower system will maintain its horizontally layering and will reside just underneath the upper strong lithospheric plate. This mechanism, subhorizontal subduction, has been termed "lithospheric doubling" (Vlaar, 1982, 1983).

For subducted o.l. younger than about 30 Ma, seismicity vanishes below the depth of 180 km, and with an age less than 20 Ma below 100 km depth. Mechanical coherency has vanished below this depth and thus buoyancy pertains to the separate parts in the subducted system. We, therefore, have to envisage buoyancy of the separate layers with respect to each other.

Flexure of the o.l. indicates that the depth limit of its strong part should be placed at the $800^\circ C$ isotherm (McAdoo et al., 1985). Also the deepest seismicity in young o.l. gives a temperature of $750^\circ C$ (Stein et al., 1986). A study by Wortel (1982) of subducted o.l. places this critical temperature at $700^\circ C$ at a depth of 100 km.

The wet basalt solidus at this depth is at about $750^\circ C$. Hence, only a small rise of temperature caused by conduction of heat from the surrounding mantle, or by mechanical friction, should suffice to start melting of hydrated basalt or eclogite. The fate of the dry basalt which is supposed to constitute the lower part of the oceanic crust, is dependent on the reaction rate of the basalt-eclogite transition, which increases with increasing temperature. Ahrens and Schubert (1975) state that the dry basalt-eclogite reaction can not take place within 10 Ma at a depth of 100 km, below $600^\circ C$, and probably needs a much longer reaction time. As the initially cool basalt layer remains considerably below the temperature of surrounding layers for a prolonged period after subduction, it may only reach the phase transition after lithospheric doubling has been effective and thus remains buoyant with respect to the overlying strong lithosphere and the lower harzburgite layer. The harzburgite layer with a thickness of about 25 km has acquired ductile behaviour, and as such is not mechanically coherent with a possible strong basalt layer. Its temperature, in any case above $750^\circ C$, causes the harzburgite to remain buoyant with respect to the lower lherzolite.

The only candidate for negative buoyancy then is the former ocean crust, when hydrated, has not yet melted and if the basalt-eclogite phase change has taken place. As the latter may be delayed by several tens of million years, the former o.l. may remain gravitationally stably stratified, internally and with respect to the underlying layers during this period, which would prohibit the sinking of lithospheric material to larger depths and should lead to lithospheric doubling as the only mode of subduction. Should the transition take place earlier, it remains to be seen whether the large density of the eclogite layer is a sufficient condition for its foundering into the underlying harzburgite and lherzolite layers. It should also have acquired sufficient ductility and fluid behaviour, which, because of the lower temperatures of the original basaltic crust may not be reached.

The crucial question here is the remixing time of the unstable layering and can not be immediately answered. It could be envisaged that remixing proceeds so slowly that even on a longer time scale the eclogite remains afloat near its position after doubling. Seismological evidence for a high density layer at a depth of 80 - 150 km under Europe is emerging (Nolet et al., 1986), which appears to be only interpretable in terms of a partly eclogitic composition. In view of the foregoing this should be the result of freezing in of eclogite in the cooling lithosphere after a period of lithospheric doubling in an earlier geological period.

Lithospheric doubling requires the overriding lithosphere to be strong enough to resist downbuckling into the lower, less dense, and weaker subducted oceanic structure. This unstable state, as has been

argumented before, is not uncommon for the lithosphere. Gravitational
instability can only lead to downward motion of the upper plate if the
lighter material of the underlying layers is allowed to flow on top of
the upper plate. This is possible if lithosphere transecting conduites
like faults are present. A candidate for the outflow is basaltic
magma, originating from the hydrated basalt of the former oceanic
crust. Also weakening of the upper lithosphere can be caused by
diapiric uprising of the peridotite of the subducted structure, its
temperature being higher than the one of the upper plate, and hence
its density smaller.

3. Thermal consequences of lithospheric doubling.

Oceanic lithosphere is thermally and rheologically varying as a function of its age. In particular, its thickness increases as an effect
of cooling. The thermal state is characterized by high temperatures at
shallow depth for young o.l., compared to the geothermal situation in
old o.l. or continental shield and platform areas. Also, the geothermal gradient is very high near a spreading center and decreases with
age, as suming "normal" values for old o.l. After subduction of young
o.l. in the mode of lithosperic doubling, the geothermal state of the
doubled system is thoroughly perturbed relative to the "normal" geotherm, characteristic for the upper lithospheric plate in the absence of
doubling.
    Firstly, the cooled top layer of the former o.l., and in particu-
lar its crust, introduces a strong temperature inversion below the
upper plate. By heat conduction or local convective flow this will
eventually loose its extreme character.
    Secondly, the large age dependent geothermal gradient at the sur-
face of young o.l. and the high temperatures prevailing at shallow
depths in it, together with its cool top layer, causes a kinky geotherm in the doubled system. A typical geotherm, during a prolonged
period of lithospheric doubling could be exemplified as follows: If
the level of detachment of the upper plate is fixed at $1000°C$ at 100 km
depth in the unperturbed situation, after doubling has taken place,
the geotherm rises to a maximum of about $900°C$ at a depth of some 90
km, then flattens or is subject to a temperature inversion in a depth
interval of some 20 - 30 km, and with increasing depth rises strongly
to reach a temperature of $1300-1350°C$ at a depth of some 140 km. The
kinky structure favours local diapirism and convection. It may take
100 Ma before this anomalous state has been annihilated by vertical
conductive or convective heat exchange.
    We note that in the coolest part, which comprises the former oce-
anic crust, temperatures are well above the wet basalt solidus and
magma may be generated at depth. An apparent thermal anomaly may persist in the upper mantle over a geological period, long after manifest
subduction at the earth's surface has ceased. If the upper plate is
moving at plate tectonic rates over the underlying mantle which once
constituted an oceanic lithosphere-mantle system, a thermal anomaly,
to be associated with the former spreading center, may manifest itself
as a "hot spot" which is moving relative to the lithosphere.

4. Some consequences of lithospheric doubling.

Though lithospheric doubling at this stage should be considered as a tentative scenario, mainly because of the absence of observational evidence in the form of seismicity, the circumstantial evidence of the correctness of the hypothesis is growing. In fact, lithospheric doubling can explain many geological, geochemical, and geophysical phenomena, which are not yet understood in the framework of conventional plate tectonic theory. In the following, a concise overview is presented.

(1) If isostatic equilibrium of the Pratt type to a depth of some 200 km is to be maintained during lithospheric doubling, this must cause uplift of the earth's surface relative to the situation in which doubling has not taken place.

The material of the mantle below the upper plate has been substituted by lower density material of the subducted o.l. This density deficiency at depth must be compensated by uplift of the upper lithospheric plate. The uplift may amount to over 1500 m over large areas. Accompanying the uplift, a negative Bouguer anomaly may be expected and also a shift of the geoid. This applies, of course, to plateau uplift as exists in the western part of North and South America, Tibet, East Africa, the German "Mittelgebirge" and many other tectonic provinces where epeirogenetic uplift is involved.

(2) The geochemical signature of basaltic magmas originating from former oceanic crust after lithospheric doubling, are to be characterized by contamination of former ocean bottom sediments and deviating isotope ratios as compared to magma derived from fertile upper mantle rocks. In particular, the scenario is suited for explaining lavas such as carbonatites and ultrapotassic lavas. As a source of magma, the melted hydrated basalt layer of the former ocean crust, could be considered, which, by rising temperature after doubling, has been melted "in situ" at depth. Another source may be constituted by diapirs of basalt or lherzolite of the lower stack, which have risen through the upper lithosphere in the solid state and intersect the solidus at a shallower depth, thus, giving rise to suites of magma. Kimberlites, in particular, are well suited to be explained in the context of lithospheric doubling (Vlaar, 1983). They must originate from a low temperature source ($\sim 650^\circ C$) , which at the prevailing depths can only be explained by introduction of material having much lower temperature than normal mantle ($> 1000^\circ C$) . Their matrix contains a complete suite making up a column of oceanic crust and upper mantle, i.e.: (1) original ocean bottom sediments rich in incompatible elements and fluids and volatiles such as H_2O, CO_2, CH_4, Cl and F and also organic remnants, (2) basalt (eclogite) of the former ocean crust, (3) harzburgite of the next deeper layer, (4) undepleted peridotite of the former oceanic mantle, and (5) for sufficiently deep sources, diamonds, which have to be explained as being derived from remnants of ocean bottom organic matter.

Xenoliths in kimberlites consist typically of (Boullier and Nicolas, 1975): (1) lower temperature ($< 1000^\circ C$), undepleted peridotite which is sampled along the conduite wall in the upper plate, (2) lower temperature course grained depleted harzburgite (~1000°C) to be ascribed to originate from the former oceanic harzburgite layer, (3) undepleted garnet peridotite which has been subject to large deformation in the solid state and having high temperatures (~1400°C) , originating in the present model, from the high temperature lower undepleted mantle beneath young o.l.

The low temperature at the source of kimberlites can be explained if eruption took place shortly enough after subduction to prevent the material to have acquired higher temperatures. The large temperature differences can be explained by the pronounced irregular shape of the geotherm shortly after doubling, in particular the large thermal gradients prevailing. A large gradient or density stratification should have been required anyhow for explaining the strong deformation of solid undepleted peridotite nodules. The kinky shape of the present model geotherm agrees with Boyd's pyroxene geotherm sampled from kimberlite xenoliths (Boyd, 1973).

(3) Continental rift structures usually can be associated with older weakness zones in the lithosphere, which are reactivated by episodic causes. In the present context, reactivation is supposed to be caused by shifting of the upper, locally weakened lithosphere over an originally young subducted oceanic system. By assuming that reactivation occurs by diapiric uprise of hot undepleted or depleted peridotite of the former oceanic lithosphere-upper mantle system, a number of characteristic features of graben formation can be explained by lithospheric doubling. Though diapiric uprise is due to the density deficit of the rising material relative to the material in the upper plate of essentially identical petrological constitution, once the diapir reaches the Moho, its uprise is halted because of the low density of the crust. The peridotite diapir forms a cushion below the Moho, and its excess heat can be transported into the lower crust by conduction or magma transport. The magma involved and associated vulcanism can be ascribed to differentiation of the peridotite blob residing at the Moho. Increasing temperature in the lower crust causes anatexis, metamorphism and possibly granitisation. Increased ductility and flow of the lower crust together with fracture of the upper crust causes crustal stretching and thinning. Lithospheric thinning had already been caused by diapirism of peridotite and associated elevation of temperature, thus resulting in shallowing of the asthenosphere.

(4) Orogeny of fold belts can be associated with the closing of small ocean basins. These basins, in general, are generated by pull apart in a transform system which accommodates shear motion of adjacent more rigid plates. The basins, in general, concern the generation of young oceanic lithosphere, and thus create an anomalous thermal situation, which, when subducted upon closing of the basin must cause a number of features as described above (Vlaar and Cloetingh, 1984).

a. In order to account for anatexis and metamorphism necessary to cause the tectonic mobility of the lower crust of the overriding platform, a heat source must be available. Standard concepts in plate tectonics as to be associated with the Wilson cycle involve old and thus cool o.l. and thus can not explain the raised temperatures, common in orogenic old belts.

b. The transport of nappes over large distances is to be ascribed to gravity gliding. This requires elevation of the source area and f.e. in the Alpine belt, an apparent migration of this source area. This can be well explained by shifting of the platform lithosphere over an anomalously hot upper mantle to be associated with the over-ridden oceanic structure. Ophiolites, whether emplaced as nappes, or squeezed between colliding rigid plates, after closing of the ocean basin, are remnants of oceanic crust and upper mantle. The thickness of the ophiolitic pile of less than 10 km, indicates detachment of this layer from underlying ductile upper mantle. As has been observed by Nicolas and Le Pichon (1980), young oceanic crust is involved. This is also indicated by the shallow level of detachment which gives an age of 10 - 20 Ma (Cloetingh et al., 1983).

c. Post-orogenic uplift of the mountain belt and its foreland can be understood as a consequence of lithospheric doubling.

(5) In order to gain greater insight in the role of plate tectonics in the geological past, the present author applied some of the concepts of this abstract to the Precambrian (Vlaar, 1986a, 1986b). The main factor influencing the evolution of global tectonics in the geological history is the increased mantle temperature towards the past. Plate tectonic style changes as a consequence of the greater thickness of the segregated oceanic layering. Proterozoic ocean crust was thicker and, therefore, the oceanic lithosphere had greater gravitational stability. This may have influenced plate tectonic style considerably and lithospheric doubling may have occurred on a larger scale than at present. The higher mantle temperature involved becomes very critical when considered relative to the melting point of the rocks involved. A small rise of temperature relative to the present may cause much larger scale melting and alternative petrologies. Stable ocean basins, together with lithospheric doubling can account for growth of continental area by sediment platforms, and growth of thickness by diapirism from the subducted o.l. Mantle temperature in the Archaean is considered too high to allow an organized style of plate tectonics as we know it today.

(6) Seismic tomography (Spakman, 1986) reveals an image of the upper mantle which favours horizontal layering over Benioff type subduction. Tomography, in fact, may prove to constitute an observational basis for verifying the concepts of age dependent subduction and its geological consequences as developed at present.

References

Ahrens, Th.J. and G.Schubert - Gabbro-eclogite transition and its geophysical significance, *Rev.Geophys.Space Phys.*, **13** , pp. 383-400, 1975.
Boullier, A.M. and A. Nicolas - Classification of textures and fabrics of peridotite xenoliths from South African kimberlites, *Phys.Chem.Earth*, **9** , p. 467, 1975.
Boyd, F.R. - A pyroxene geotherm, *Geochim.Cosmochim.Acta*, **37** , p. 2533, 1973.
Cloetingh, S. - Evolution of passive continental margins and

74

initiation of subduction zones", *Geologica Ultraiectina*, **29** , pp. 1 - 111, 1982.

Cloetingh, S.A.P.L., M.J.R.Wortel and N.J.Vlaar - State of stress at passive margins and initiation of subduction zones, *Am.Assoc.Pet.Geol.Mem.* **34** , pp. 717 - 723, 1983.

Deffeyes, K.S. - Plume convection with an upper-mantle temperature inversion, *Nature*, **240** , pp. 539 - 544, 1972.

England, Ph. and R. Wortel - Some consequences of the subduction of young slabs, *Earth Planet.Sci.Lett.*, **47** , pp. 403 - 415 , 1980.

McAdoo, D.C., C.F. Martin and S.Poulouse - Seasat observations of flexure: evidence for a strong lithosphere, *Tectonophysics*, **116** , pp. 209 - 222, 1985.

Nicolas, A. and X. Le Pichon - Thrusting of young lithosphere in sub-duction zones with special reference to structure in ophiolite peridotites, *Earth Planet.Sci.Lett.*, **46** , pp. 397 - 406 , 1980.

Nolet, G., B.Dost and H.Paulssen - The upper mantle under Europe: an interpretation of some preliminary results from the NARS project, *Geologie en Mijnbouw*, **65** , pp. 155 - 165 , 1986.

Oxburgh, E.R. and E.M. Parmentier - Compositional and density stratif-ication in oceanic lithosphere: causes and consequences, *J.Geol.Soc.Lond*, **133** , pp. 343 - 355 , 1977.

Stein, S., S.Cloetingh, D.A. Wiens and R.Wortel - Why does near ridge extensional seismicity occur primarily in the Indian Ocean? , *Earth and Plan.Sci.Lett.*, **82** , pp. 107 - 113 , 1987.

Spakman, W. - Subduction beneath Eurasia in connection with the Meso-zoic Tethys, *Geologie en Mijnbouw*, **65** , pp. 145 - 153 , 1986.

Vlaar, N.J. - The driving mechanism of plate tectonics, a qualitative approach, in: *Progress in Geodynamics*, North Holland Publishing Coy., Amsterdam, New York, 1975.

Vlaar, N.J. and M.J.R.Wortel - Lithospheric aging, instability and subduction, *Tectonophysics*, **32** , pp.331, 1976.

Vlaar, N.J. - Lithospheric doubling as a cause of intra-continental tectonics, *Proceedings Series B*, **85** (4) , pp. 469 - 483 , 1982.

Vlaar, N.J. - Thermal anomalies and magmatism due to lithospheric dou-bling and shifting, *Earth and Planet.Sci.Lett.*, **65** , pp. 322 - 330 , 1983.

Vlaar, N.J. and S.A.P.L.Cloetingh - Orogeny and ophiolites: plate tec-tonics revisited with reference to the Alps, *Geologie en Mijnbouw*, **63** , pp. 159 - 164 , 1984.

Vlaar, N.J. - Geodynamic evolution since the Archaean, *Proceedings Series B*, **89** , pp. 159 - 164 , 1984.

Vlaar, N.J. - Archaean global dynamics, *Geologie en Mijnbouw*, **65** , pp. 91 - 101, 1986b.

Wortel, M.J.R. and N.J. Vlaar - Age-dependent subduction of oceanic lithosphere beneath western South America, *Phys.Earth.Planet.Inter.*, **17** , pp. 201 - 208 , 1978.

Wortel, R. - Seismicity and rheology of subducted slabs, *Nature*, **296** , no. 5857, pp. 553 - 556 , 1982.

Wortel, M.J.R. and S.A.P.L.Cloetingh - Accretion and lateral varia-tions in tectonic structure along the Peru-Chile trench, *Tectonophysics*, **112** , pp. 443 - 462 , 1985.

A REVIEW OF THE THERMAL-MECHANICAL STRUCTURES AT CONVERGENT PLATE BOUNDARIES AND THEIR IMPLICATIONS FOR CRUST/MANTLE RECYCLING.

M. Nafi Toksöz and Albert T. Hsui*
Earth Resources Laboratory
Dept. Earth, Atmospheric and Planetary Sciences
Massachusetts Institute of Technology
Cambridge, MA 02142 USA

*On leave from the Department of Geology, University of Illinois, Urbana, Illinois 61801, USA

Oceanic plates are derived from the mantle at mid-ocean ridges. They migrate away until they reach convergent zones where they subduct into the mantle and in time they will be assimilated through heating. This represents the large scale crust/mantle recycling process. There are other processes that contribute to recycling at smaller scales. Island arc volcanism and back-arc spreading at convergent plate boundaries are some of the examples. Dynamic motions responsible for recycling are driven largely through thermal mechanisms. In this paper we review the thermal mechanical processes at zones of plate convergence and their implications to geological recycling processes. From a thermal point of view, plate subduction represents the cooling of the mantle since "cold" lithospheric plates are being returned to the "hot" interior. However, surface geological observations do not support the notion that extensive cooling has occurred within the mantle surrounding a descending slab, although the slab itself remains relatively cool. Active island arc volcanism and the high heat flows observed in back arc regions suggest the existence of a hot mantle above the subducting slab, produced by induced motions or secondary processes, that play an important role in influencing local geological evolution.

It has been pointed out by many investigators (see Hsui and Toksöz, 1979 for a summary) that a natural consequence of plate subduction is to induce a convective current immediately adjacent to each side of a slab. The configuration of this convective current is strongly dependent on the viscosity structure of the mantle. For a uniform viscosity upper mantle, the horizontal scale of this induced cell will be approximately the same as its vertical

S. R. Hart and L. Gülen (eds.), Crust/Mantle Recycling at Convergence Zones, 75–80.
© 1989 by Kluwer Academic Publishers.

scale. If we consider an asthenosphere with a low-viscosity channel, mechanical energy of the convective motion will preferentially concentrate within this low viscosity region. As a result, the horizontal scale of the induced cell will be dictated by the thickness of this low viscosity layer. Models calculated with realistic viscosity profiles (Figure 1) indicate that locations of back arc spreading centers can be correlated to the upwelling of the secondary induced flows. The vertical scale of the cell in this case can be many times larger than the horizontal scale. Due to the larger viscosity at the lower part of the cell, convective velocities at depth are orders of magnitude smaller than those within the low viscosity region.

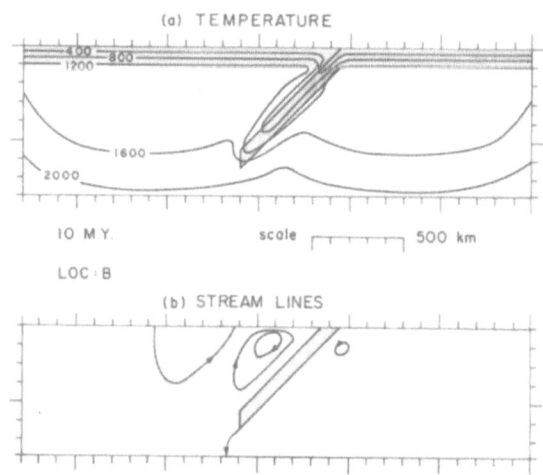

Figure 1: Thermal-mechanical structure of a subduction zone after 10 m.y. of evolution. Notice the cold isotherms are mostly confined to the vicinity of the slab in the top diagram and the strong induced convective cell in the bottom diagram.

Plate subduction does provide cooling to the mantle materials immediately adjacent to the slab. However, because of the existence of the induced flow, the chilled materials are dragged downward by the slab. Warmer mantle materials move into their places to satisfy mass conservation. Consequently, the upper part of the wedge corner is kept warm continuously as long as plate subduction is in progress. Mantle cooling occurs only at the bottom of the convective layer. It is of interest to study how close the high temperature regime of the mantle wedge can approach the subducting slab and whether this can melt the surface of a slab to produce island arc magma (March 1979). A local model based on a more regional study has been developed (Hsui et al., 1983) to address these problems. As indicated in Figure 2, it is found that the induced flow is sufficiently strong so that the surface of a subducting slab can reach dry melting temperature at a distance about 10 km for the wedge corner. Melting of a subducting slab can possibly occur sooner if wet solidus is assumed.

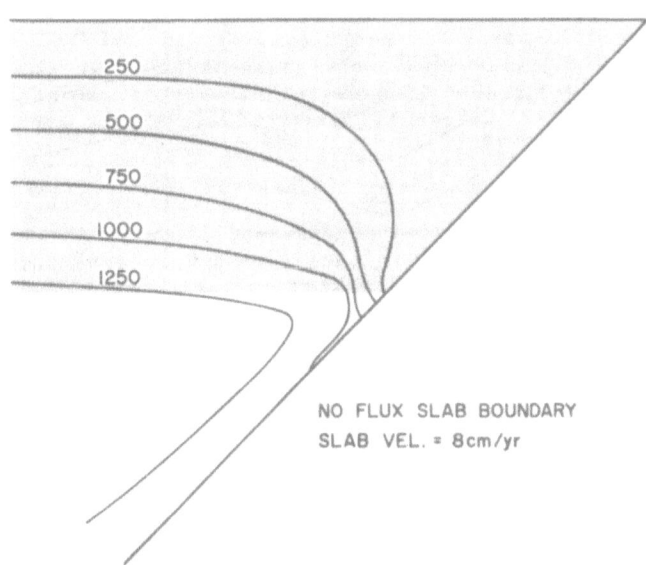

Figure 2: Steady-state, regional thermal structure of a wedge region above a descending plate. The inclined surface represents the upper surface of a subducting slab. The shaded area represents the lithosphere which is about 70 km thick. Solid lines are isotherms and temperatures in C are given by the number associated with each isotherm. Notice that the 1000°C isotherm intercepts the slab surface at a distance about 10 km from the wedge corner.

In the case of continental convergence, crustal thickening generally occurs. Very often, it is also found that the lower crust can reach partial melting during thickening. Neither the viscous heating nor the upward concentration of radiogenic isotopes can produce the necessary thermal event to generate crustal melting (Toksöz et al., 1981). It appears that anomalously high mantle heat fluxes are essential. Figure 3 illustrates the thermal history of a thickening crust under different conditions. As evident from these diagrams, the only case in which the lower crust can reach melting at about 800°C is when there exists a mantle heat flux that is about 1/3 higher than normal. One possible mechanism to have high mantle heat flux is through small scale convection within the mantle (Buck and Toksöz, 1983). It has been demonstrated to be an effective way to remove the cold thermal boundary layer beneath the lithosphere and maintain a relatively warm mantle underneath.

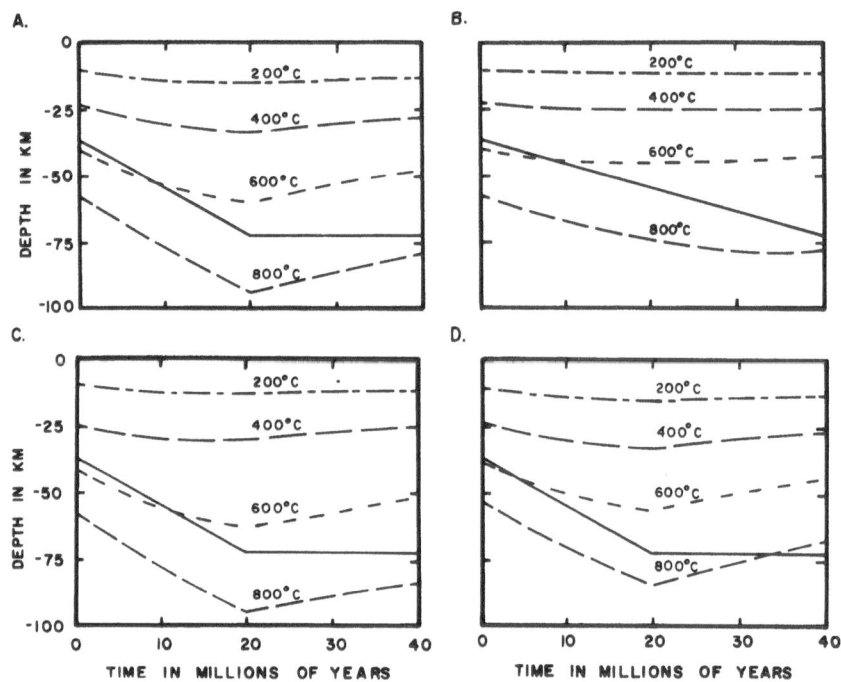

Figure 3:   Results for the thermal evolution of thickening crust. Dashed lines are isothersm expressed as a function of time since the beginning of crustal thickening.   The solid line is the crust-mantle boundary position. All cases have the doubling of the crust taking 20 m.y. except case B which is for twice that time.   Case A is for a uniform radioactive distribution in the crust and 1 kbar shear stress during deformation. Case B differs only in that the shear stress is 2 kbar. Case C shows that when the concentration of crustal radioactive sources is twice as large in the upper crust than the lower that temperatures remain lower in the crust. Case D has one third more mantle heat flux than the others.

The induced flow, the melting of a subducting slab and the melting of a thickened lower crust are mechanisms that contribute to the dynamic recycling of crustal and mantle materials at convergent zones. Recycling time scales for the processes are different. Material recycling through the induced cell varies from 1 m.y. to over hundreds of millions of years depending on which flow path the materials follow, the maximum depth that the convective regime can extend to and the magnitude of subduction velocity. Generally speaking, recycling within the low viscosity region immediately below the over-riding plate will take a shorter time interval because the flow velocity is higher and the total length of a flow circuit is shorter. On the other hand, if materials follow an outer flow path of the convective cell, it will take a much longer time to complete a circuit. Recycling time for the island arc volcanism is of the order of a few to tens of millions of years. Much of that time is taken up by the travel of the materials from the surface to the mantle wedge corner. Once they reach partial melting, the melts will migrate through the lithosphere within a relatively short time. Recycling of granite melts has a time scale that is equivalent to the time scale of crustal thickening which is of the order of a few tens of millions of years. Heating and assimilation of a subducted slab takes as much as 100 m.y. (Toksöz et al., 1973). Recycling time for the large scale mantle convection is generally greater than 100 m.y. throughout the mantle (Arkani-Hamed and Toksöz, 1984). Since the total length of the circuit is much longer in this case, it is not surprising that recycling time is orders of magnitude higher than that for the local recycling processes.

In summary, the major dynamic transport processes for crust/mantle recycling are the induced wedge flow, melting of slabs to produce island arc volcanism and the melting of thickening crusts to generate granites. Typical time scales for all these local recycling processes are of the order of a few to tens of millions of years which is orders of magnitude smaller than that due to the large scale mantle convection.

REFERENCES

Arkani-Hamed, J. and M. N. Toksöz. 'Thermal evolution of Venus,' Phys. Earth and Planet. Int., 34, 232-250, 1984.

Buck, W.R. and M.N. Toksöz, 'Thermal effects of continental collisions: Thickening a variable viscosity lithosphere,' Tectonophysics, 100, 53-69, 1983.

Hsui, A.T., B.D. March and M.N. Toksöz, 'On melting of the subducted oceanic crust: Effects of subduction induced mantle flow,' Tectonophysics, 99, 207-220, 1983.

Hsui, A.T. and M.N. Toksöz, 'The evolution of thermal structure beneath a subduction zone,' Tectonophysics, 60, 43-60, 1979.

Marsh, B.D., 'Island arc development: some observations, experiments, and speculations,' J. Geol., 87, 687-713, 1979.

Toksöz, M.N., W.R. Buck and A.T. Hsui, 'Crustal evolution and thermal state of Tibet', in D.S. Liu (ed.), Geological and Ecological Studies of Qinghai-Xizang Plateau, Gordon and Breach, New York, NY 847-857, 1981.

Toksöz, M.N., N.H. Sleep and A.T. Smith, 'Evolution of the downgoing lithosphere and the mechanisms of deep focus earthquakes.' Geophys. J.R. Astr. Soc., 35, 285-310, 1973.

SEDIMENT RECYCLING AT CONVERGENT MARGINS:
CONSTRAINTS FROM THE COSMOGENIC ISOTOPE ^{10}BE

JULIE MORRIS, FOUAD TERA, I. SELWYN SACKS, LOUIS BROWN
Carnegie Inst. of Washington, Dept. Terrestrial Magnetism,
5241 Broad Branch Rd., N.W.,
Washington, DC   20015

JEFF KLEIN and ROY MIDDLETON
Tandem Accelerator Lab,
University of Pennsylvania,
Philadelphia, PA   19104

ABSTRACT.   The question of whether sediments are subducted at convergent margins and subsequently incorporated in arc magmas has long been controversial.   The cosmogenic isotope ^{10}Be ($T_{1/2}$ = 1.5 million years) suggests that sediments are indeed recycled at some volcanic arcs.   Further, the ^{10}Be data suggest that sediment subduction may be enhanced where topographic steps in the oceanic plate (fracture zones, seamount chains, grabens) are subducted.   If all sediments contained within grabens are presumed to be subducted, then the global sediment subduction rate is 2-5 × 10^5 km^3/Ma; model calculations suggest that 11-14% of the subducted sediment is recycled in volcanic arcs and that the remainder is available for subduction deep into the earth.

Over the last 25 years, an accumulating body of geochemical data relevant to the issue of whether sediments subduct has been interpreted by different workers as weak, permissive or somewhat supportive but rarely compelling evidence of a sediment contribution to arc volcanism.   This lengthy debate reflects, in part, real differences in the chemistry and dynamics of different arcs.   The problem is further compounded by the difficulties in distinguishing uniquely between chemical effects due to sediment subduction and those that may result from pre-existing heterogeneity in the mantle source region.

Although interpretation of ^{10}Be data for arc lavas is subject to its own limitations, this cosmogenic isotope provides compelling evidence for sediment recycling at some arcs.   Further, as a result of its unique characteristics, ^{10}Be data may be used to understand better the balance between sediment subduction and accretion at convergent margins.

S. R. Hart and L. Gülen (eds.), Crust/Mantle Recycling at Convergence Zones, 81–88.
© *1989 by Kluwer Academic Publishers.*

Figure 1 schematically illustrates the [10]Be cycle from the atmosphere through subduction zones. As a result of its atmospheric source and highly adsorptive behavior during sedimentation (*Valette-Silver et al.*, 1985; *Measures and Edmond*, 1983), [10]Be concentrations in pelagic sediments are typically more than 10,000 times greater than in MORB or OIB. Thus, [10]Be concentrations in arc lavas can be very sensitive to incorporation of sediment, while being totally insensitive to the type of mantle source (MORB, OIB, lumpy, etc.) assumed in model calculations.

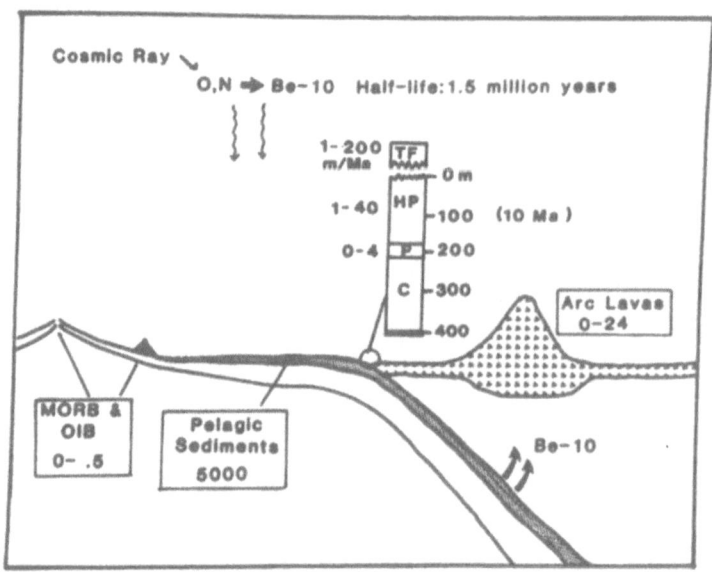

Figure 1. Schematic illustration of the [10]Be cycle from the atmosphere through to volcanic arc lavas. Numbers in boxes are [10]Be concentrations in units of 10^6 atoms/gram. Half-life from *Yiou and Raisbeck*, 1972. Globally averaged [10]Be deposition rate is $\sim 1.2 \times 10^6$ atoms/cm^2-yr (*Monaghan et al*, 1986), leading to predicted [10]Be inventory ([10]Be integrated over depth of sediment column) of 2×10^{12} atoms/cm^2. Measured inventories in the N. Pacific show as much as 3×10^{13} atoms/cm^2 (*Ku et al*, 1985), suggesting that major departures from uniform deposition are to be expected. A typical near-trench sediment column (DSDP 495 near Guatemala) is sketched to show relative proportion and lithology of sediments (TF, trench fill; HP, Hemi-pelagic; P, pelagic; C, Carbonate). Significant amounts of [10]Be are present only in 10 Ma and younger hemi-pelagic sediments. Sub-bottom depths and sedimentation rates are indicated.

The short half-life of ^{10}Be has several implications for interpretation of the data. The first is that sediment contamination must be a recent event (< 8-10 Ma) involving Pliocene or younger sediments; any ^{10}Be introduced in an older event would long since have decayed. This observation requires that some fraction of the uppermost sediments (and presumably the underlying column) be subducted. In terms of ^{10}Be assimilation during magma ascent, only a very thin veneer of young marine or continental sediment on the upper plate (missing in some arcs) is capable of contaminating arc lavas. This is in contrast to the Sr, Nd and Pb isotope systems, where contamination along the entire path through the upper plate is possible.

^{10}Be data for ~ 200 volcanic rocks are shown in Figure 2. The most striking feature is that lavas from many volcanic arcs contain ^{10}Be, while lavas from mid-ocean ridges, oceanic islands and continental rifts never do. Multiple analyses of several standards indicate that these results are not an experimental artifact. Nor do the results reflect sample alteration, as a number of arc lavas and ashes collected upon eruption contain ^{10}Be, while older (but < 500,000 yr), somewhat altered OIB lavas do not.

The data in Figure 2 indicate that the distribution of ^{10}Be in arc lavas is not a simple issue. Some arcs which have geochemical characteristics thought indicative of sediment involvement (e.g. Lesser Antilles, Sunda) do not contain ^{10}Be. Although the arcs summarized in Figure 2 cover much of the Pacific rim, there are no obvious geographic distinctions between regions that contain ^{10}Be and those that do not. Further, contiguous arc systems (e.g. Aleutians-E. Alaska, Mexico-Central America; Kuriles-Kamchatka) can be separated into ^{10}Be-rich and -poor areas.

In some locations, the absence of ^{10}Be reflects the unique geochemistry of the cosmogenic isotope, and cannot be used as evidence for or against sediment recycling. For example, descriptions of DSDP cores outboard of the Mariana Trench suggest that much of the Pliocene (i.e. ^{10}Be-bearing) sediments may have been eroded away. In the E. Alaska, Cascade and Lesser Antilles arcs, the absence of ^{10}Be in the lavas is probably due to the long subduction times (≥ 8 Ma) during which the ^{10}Be has decayed.

In all other arcs summarized in Figure 2, subduction times are short enough that arc lavas should contain ^{10}Be if young pelagic sediments are subducted and incorporated into the source of arc magmas. Precise quantitative calculations of the volumes of sediment subducted or accreted will have to wait until our measurements of ^{10}Be in near-trench sediment cores become available. In the meantime, however, it is possible to use the data from selected arcs to draw some general conclusions about sediment recycling.

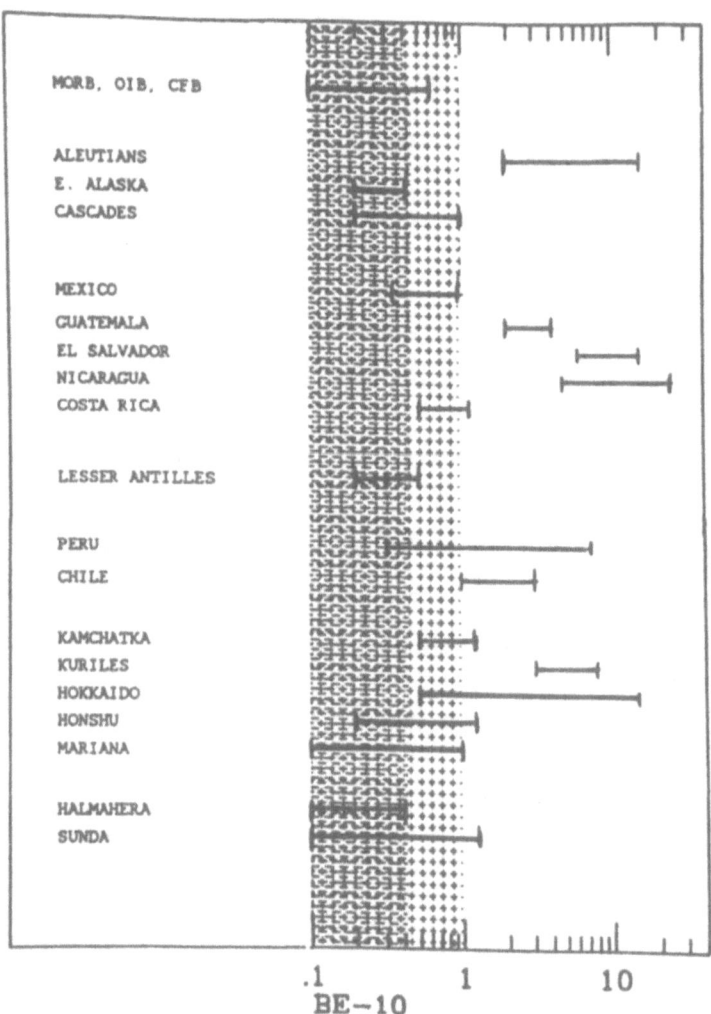

Figure 2. ^{10}Be plotted on a log scale in units of 10^6 atoms/gram for arc and non-arc lavas. Ranges for arc segments are based on 4-15 samples; MORB, OIB and CFB group reflects 30 measurements. Reproducibility on individual analyses is 12% (1 σ). Cascade samples studied in collaboration with S. Church. Stippled region indicates measurements that cannot be reliably distinguished from analytical baseline; region is lightly stippled to indicate range of ^{10}Be values amenable to study following accelerator upgrade.

The data for southern Chile, plotted in Figure 3a, indicate that sediment may be subducted even where the slab is young (10-40 Ma), buoyant and tightly coupled to the overriding plate (*Kanamori*, 1986). Note also that there is a small but real spike in ^{10}Be values at Villarrica volcano, which lies above the subducted extension of the Valdivia Fracture Zone. The fracture zone forms a topographic step in the subducting sea floor and may represent an effective sediment trap. A similar effect is seen in S. Hokkaido, where uniquely high ^{10}Be values overlie a subducting seamount chain.

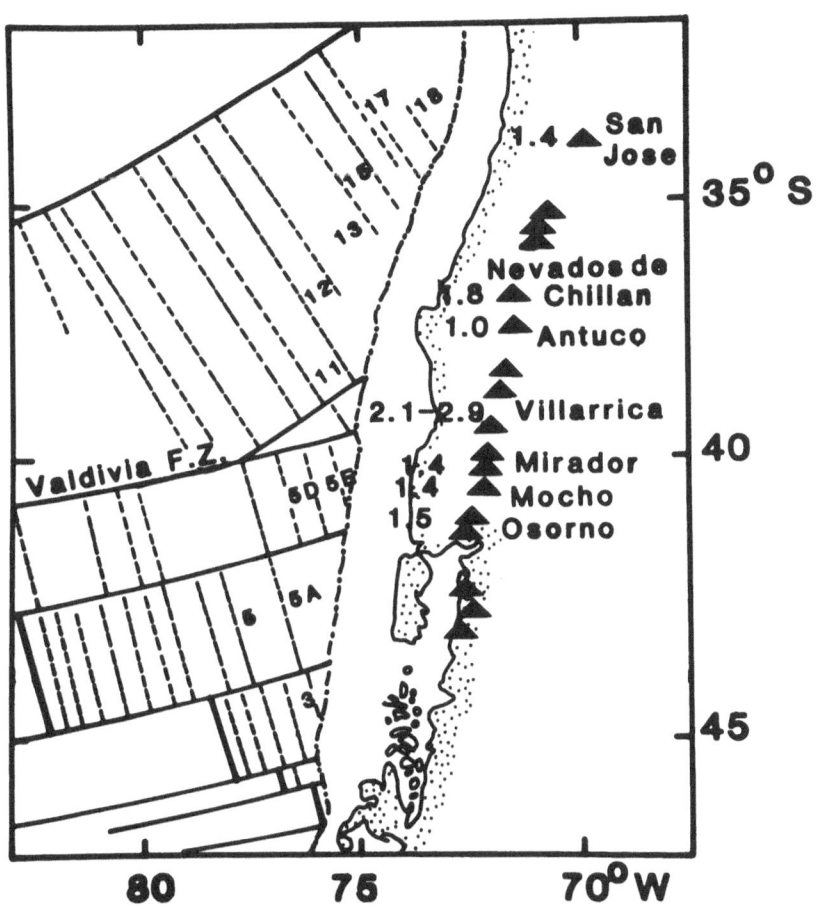

Figure 3a. Map of the Southern Volcanic Zone of southern Chile, showing distribution of ^{10}Be values along the length of the arc. Marine magnetic anomalies show that plate age varies from ∼ 40 Ma in north to ∼ 10 Ma in south (*Herron et al*, 1981). Note range of ^{10}Be concentrations in 5 basalts from Villarrica Volcano.

Although the data for Chile indicate that sediment recycling can occur even at tightly coupled margins, the results from Mexico and Central America indicate that this isn't always the case. Seismic surveys (*Shipley and Moore*, 1986; *Moore et al*, 1979) suggest that half of the Pliocene section is accreted in the Mexican fore-arc, while all of the Pliocene and younger sediments are apparently accreted off Costa Rica. In contrast, all sediments appear to be subducted beneath Guatemala. These differences in subduction style are reflected in the ^{10}Be data for the region, shown in Figure 3b. Calculations show that >10% sediment incorporation would be necessary to impart a ^{10}Be signature to the Mexican lavas while the Central American lavas can be modelled by 2-8% sediment incorporation (*Tera et al*, 1986). When the profile of ^{10}Be distribution in near trench sediment columns can be combined with seismic imaging of the fore-arc region, it will be possible to study the means by which sediment is incorporated in arc magmatism and to estimate more precisely the volumes involved.

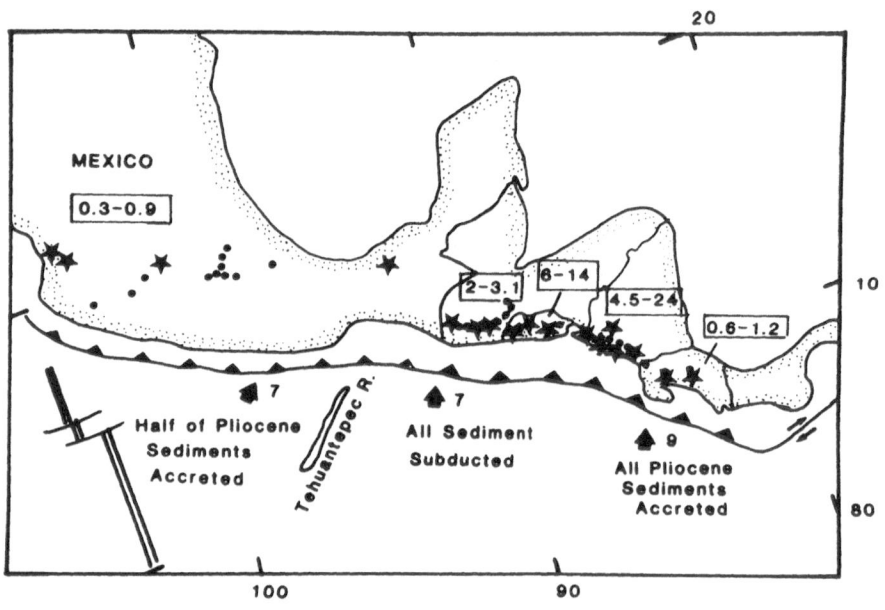

Figure 3b. Map of Middle America showing volcanoes analyzed for ^{10}Be (stars) and range of ^{10}Be values in units of 10^6 atoms/gram for Mexico, Guatemala, El Salvador, Nicaragua and Costa Rica. Note segmentation of arc into regions of low and high ^{10}Be concentrations. Drawing also shows regions of sediment subduction and accretion based on seismic studies across the fore-arc in Mexico (*Watkins et al*, 1981), Guatemala (*Moore et al*, 1979), and Costa Rica (*Shipley and Moore*, 1986).

The Central American seismic studies are also interesting in that they highlight the role that grabens in the subducting plate may play in permitting sediment subduction. A number of studies indicate that down-dropped sediment basins form in sea-floor as old as ~ 135 Ma (Marianas, Japan) and as young as 10 Ma (Mexico) (*Hilde*, 1983; *Maurauchi and Ludwig*, 1980). These studies indicate that grabens are 150-750m deep, typically 1-2 km wide, 2-10 km apart, and may extend for 15-100 km, with their long dimension sub-parallel to the trench. The role of these subducted basins in determining the ^{10}Be budget will depend on sediment thickness relative to graben displacement, and on the detailed behavior of the décollement. Data from the Aleutians and central Middle America suggest that it is certainly possible to subduct to depth sediments that lie outside of grabens. Nonetheless, if grabens have been as important in the past as they appear to be today, it is possible to use graben volume to make a conservative estimate of sediment recycling. Presently, the earth is girdled with 37,000 km of subduction zones, where plates are subducting at an average rate of 80 km/Ma (*Reymer and Schubert*, 1984). Assuming that grabens average 0.3 km deep, 1 km wide, and cover 20% of the subducted sea floor, then approximately 2×10^5 km^3 of sediment may be subducted per million years. This translates to subduction of 5×10^8 km^3 of sediment since the Archaean (2.6 by), ~ 7% of the volume of the continental crust. An upper limit for this number, using maximum realistic graben volume, would be ~ 20%.

The volumes of sediment recycled through volcanic arcs may be compared against the amount subducted. Assuming a globally averaged volcanic production rate of 30 km^3/km arc length per million years (averaged from 16 arcs summarized in *Reymer and Schubert*, 1984) and using 2-8% sediment incorporation in arc lavas leads to the suggestion that 11-44% of the subducted sediment returns to the surface in arc volcanism, with the remainder available for subduction deep into the earth. If grabens are the primary mechanism for sediment subduction, then continentally-derived materials are continuously injected into the mantle and are recycled as long ribbons embedded within a matrix of oceanic crust. These ''continental'' ribbons comprise ~ 1.5% of the volume of the basaltic fraction of the subducting crust.

References

Herron, E.M., S.C. Cande, and B.R. Hill, 1981, 'An active spreading center collides with a subduction zone: a geophysical study of the Chile Margin Triple Junction', *Geol. Soc. Amer. Mem.*, **154**, 683-702.

Hilde, T.W.C., 1983, 'Sediment subduction versus accretion around the Pacific', *Tectonophys.*, **99**, 381-397.

Kanamori, H., 1986, 'Rupture process of subduction-zone earthquakes', *Ann. Rev. Earth Planet. Sci.*, **14**, 293-322.

Ku, T.L., J.S. Southon, Z.C. Vogel, M. Leang, M. Kusakabe, and D.E. Nelson, 1985, '^{10}Be distributions in DSDP site 576 and 578 sediments studied by accelerator mass spectrometry', *Initial Rep. Deep Sea Drilling Proj.*, **58**, 539-546.

Maurauchi, S., and W.J. Ludwig, 1980, 'Crustal structure of the Japan Trench: The effect of subduction of oceanic crust', *Initial Rep. Deep Sea Drilling Proj.*, **56-57**, 463-472.

Measures, C.I., and J.M. Edmond, 1983, 'The geochemical cycle of ^{9}Be: A reconnaissance', *Earth Planet. Sci. Lett.*, **66**, 101-110.

Monaghan, M.C., S. Krishnaswami, and K.K. Turekian, 1986, 'The global average production rate of ^{10}Be', *Earth Planet. Sci. Lett.*, **76**, 279-287.

Moore, J.C., J.S. Watkins, T.H. Shipley, and the shipboard party, 1979, 'Progressive accretion in the Middle America Trench, Southern Mexico', *Nature*, **281**, 638-642.

Reymer, A., and G. Schubert, 1984, 'Phanerozoic addition rates to the continental crust and crustal growth', *Tectonics*, **3**, 63-77.

Shipley, T.H., and G.F. Moore, 1986, 'Sediment accretion, subduction and dewatering at the base of the trench slope off Costa Rica: A seismic reflection view of the décollement', *Jour. Geophys. Res.*, **91**, 2019-2028.

Tera, F., L. Brown, J. Morris, I.S. Sacks, J. Klein, and R. Middleton, 1986, 'Sediment incorporation in island arc magmas: Inferences from ^{10}Be', *Geochim. Cosmochim. Acta*, **50**, 535-550.

Valette-Silver, J.N., F. Tera, M.J. Pavich, L. Brown, J. Klein, and R. Middleton, 1985, '^{10}Be contents of natural waters', *EOS*, **66**, 423.

Yiou, F., and G.M. Raisbeck, 1972, 'Half-life of ^{10}Be', *Phys. Rev. Lett.*, **29**, 372-375.

GEOPHYSICAL CONSTRAINTS ON RECYCLING OF OCEANIC LITHOSPHERE

I. Selwyn Sacks and John F. Schneider
Department of Terrestrial Magnetism
Carnegie Institution of Washington
Washington, DC 200145

The path of an oceanic slab as it subducts beneath a continent or ocean is generally illuminated by a plane of earthquakes. In most subduction zones the earthquake activity is probably confined to the upper tens of kilometers of the plate and also to depths less than about 680 km. However, despite the apparent rigidity suggested by the presence of earthquakes, recent studies have shown that the subducted lithosphere can contort without tearing and/or without substantial release of seismic energy (e.g., Schneider and Sacks, 1987, Hasegawa and Sacks, 1981). In addition, slabs can have significant aseismic extent. In many regions, therefore, the path of the slab or even its existence is not obvious.

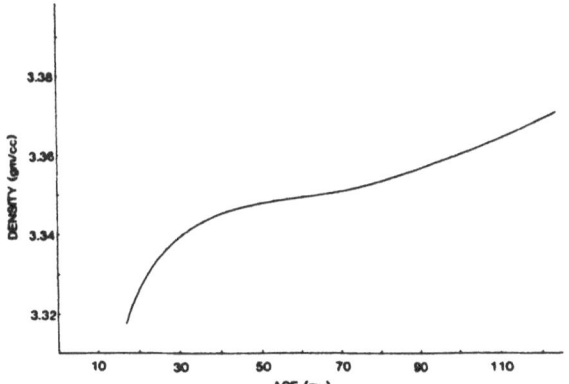

Figure 1:   Density of the oceanic plate as a function of age.   The structure, density data, and lithosphere thickness are the same as those of Leeds et al. (1975). The specifics of the model are not important, only the general pattern of density increasing with age is. This trend is not sensitive to the details of the model, though the actual values are.

S. R. Hart and L. Gülen (eds.), Crust/Mantle Recycling at Convergence Zones, 89–96.
© 1989 by Kluwer Academic Publishers.

The density of an oceanic slab is a function of its age (Figure 1). Prior to subduction, slabs lose heat through the sea floor; old slabs are cooler and thicker than young slabs (through underplating of asthenospheric material), and are denser than the surrounding asthenosphere. Slabs younger than about 50 m.y. can be buoyant if the (crustal) basalt-to-eclogite phase change is retarded by low temperature (Sacks, 1983). In order for the transformation of basalt to eclogite to occur, there must be diffusive transport, both into and out of appropriate species to grow new minerals. At temperatures less than about 700 degrees, transport is by solid diffusion less and the phase transformation can take many millions of years (Sacks, 1983). If the basalt (density 2.9) transforms to eclogite (density 3.5), slabs of any age will sink.

Buoyant subduction is observed where a subducted slab remains in contact with the overlying (continental) lithosphere, with the subducted slab travelling horizontally for a significant distance from the trench (up to 400 km beneath central Peru). Buoyant slabs are characterized by:

1) absence of recent volcanism
2) low heat flow, and
3) sparse seismicity in the near-horizontal section

Buoyant subduction is recognized in at least three regions; central Peru, central Chile, and south-west Honshu, Japan (Sacks, 1983). Figure 2 shows seismicity sections and the hypocentral trend for central Peru, indicating the apparent buoyancy of the slab from the horizontal trend of hypocenters. Figure 3 shows seismic velocity data (seismicity is too sparse) beneath south-west Honshu indicating a horizontally travelling slab. In these areas, sediment which might be partially returned in volcanic magmas in other subduction zones, remains in the mantle.

The slab is most obvious when seismically active. However, no slabs have earthquakes below about 680 km (Stark and Frohlich, 1985) and many slabs become aseismic below some hundred or so kilometers. In neither case is it clear that the slab terminates or is discontinuous. Beneath western South America, the slab is predominantly aseismic between depths of 200 to 540 km. However, there is a high-Q (low attenuation) path in this region (Sacks and Okada, 1974) indicating continuity of the slab. There are at least two reasons for aseismic slabs.

1) Absence of stress. The worldwide decrease in seismicity in the 300 to 500 km range is probably due to a balancing of the sinking force due to gravity (which dominates at intermediate depth) with the resistance to penetration into higher viscosity material below 500 km or so. Figure 4 shows the absorption of shear waves as a function of depth, indicating a rapid increase of Q and by inference, viscosity at about this depth.

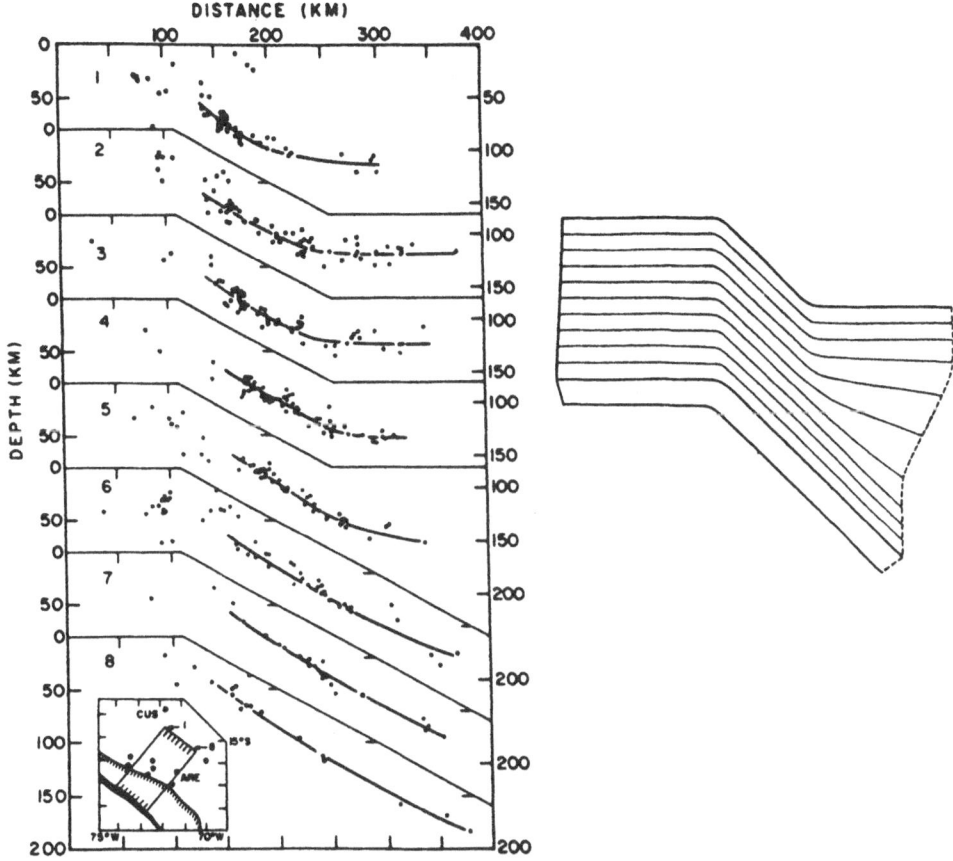

Figure 2: Divided vertical cross sections of relocated earthquakes located inside the rectangles from region 2 to region 8 on the inserted map. Solid curves show the shape of the seismic zone (below 50-km depth) inferred from the seismicity in each vertical section (after Hasegawa and Sacks, 1981). Lower schematic model of the descending Nazca plate near the boundary between the nearly horizontal subduction zone beneath central Peru and the more steeply dipping subduction zone beneath souther Peru. In both regions the initial subduction angle is 30° and persists to at least 100 km.

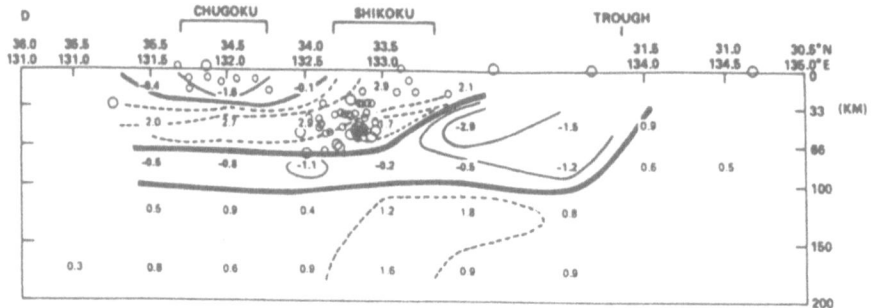

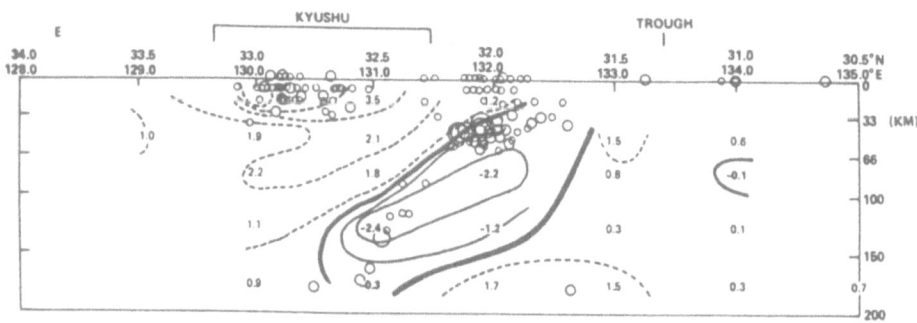

Figure 3: Three-dimensional seismic structure beneath southwest Honshu (upper) and Kyushu (lower). The slowness perturbations and hypocenters are projected onto vertical cross sections approximately normal to the Nankai trough. Thicker lines and negative signs indicate higher velocities, which are assumed to show the subducted plates. The younger east Philippine Sea plate dips down to about 60-km-depth, then flattens to horizontal (top figure). The older, denser, west Philippine Sea plate has a dip which is maintained to at least 200-km depth (bottom figure) (after Hirahara, 1981).

2) Temperature of the slab above a critical brittle-ductile failure level. Observations of maximum earthquake depth in oceanic slabs suggests that seismic activity does not occur if the slab temperature is above 600-700 degrees (e.g., Chen and Molnar, 1983). It is not presently well understood how this critical temperature is affected by pressure; but, the disappearance of the upper seismic zone beneath central Honshu, and the limited maximum seismic depth of young slabs, are examples consistent with this general concept.

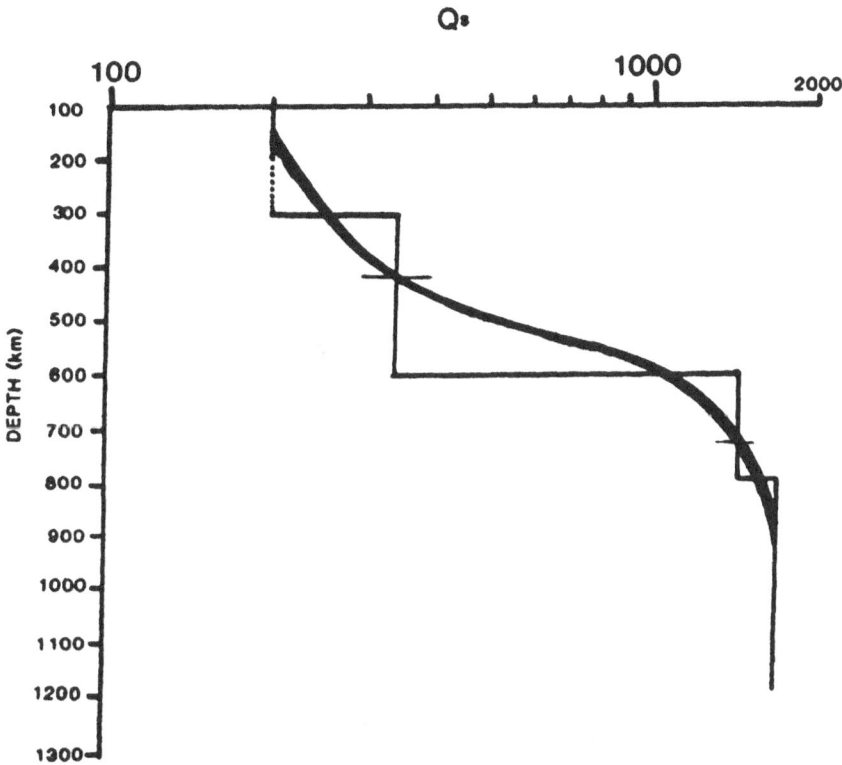

Figure 4: Q_s as a function of depth in the sub-oceanic mantle (after Chan and Sacks, 1987).

The Rayleigh number of the upper mantle is certainly above critical, so that convection is indeed possible in the asthenosphere adjacent to the subducting slab. Intrusion of a cold slab produces a large thermal perturbation sure to dominate the local asthenosphere flow, even if the asthenosphere were not otherwise convecting (Hsui and Toksoz, 1979). The asthenosphere overlying a slab will tend to follow the slab downward for two reasons:

a) the motion of the slab induces physical drag; and

b) the cold slab cools the adjacent mantle, which makes the mantle more dense, causing it to sink. Therefore the upper surface of the slab heats more slowly than it would if the asthenosphere remained fixed (Figure 5). It therefore seems unlikely that there is significant melting of even the uppermost part of the slab until well beyond volcano-source depths. Thus, magma generation results from localized melting and may be triggered by fluxing (of water or other release) from subducted sediments. This would be consistent with the observation of [10]Be from a third rank volcano, Oshima Oshima, in the Japan Sea (Tera et al., 1986). If the

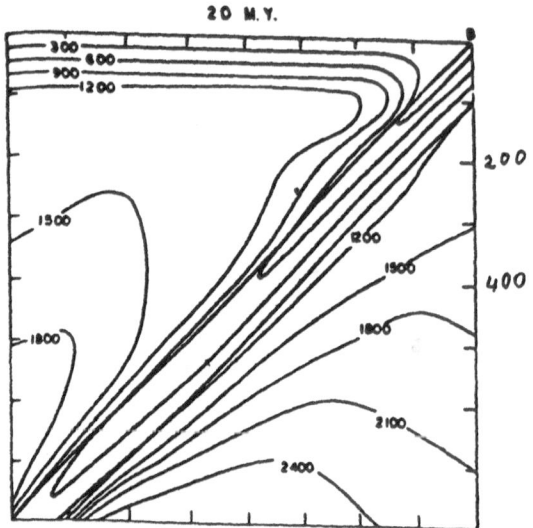

Figure 5: Section through subducted slab showing temperature in the slab and surrounding mantle (after Hsui and Toksoz, 1979). Tick marks are at 100 km spacing. Upper surface of slab is diagonal of figure, from B (upper right corner).

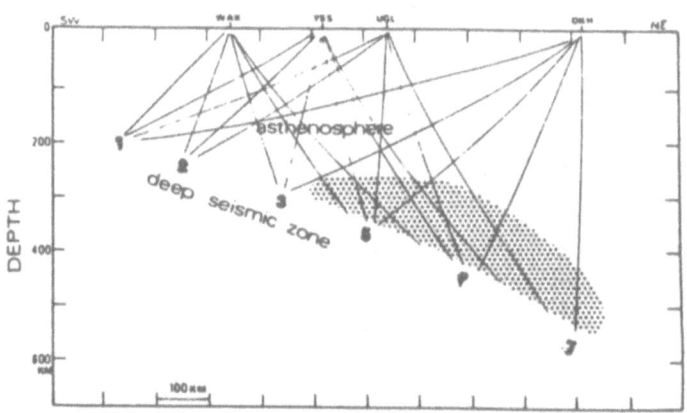

Figure 6: Approximate seismic ray geometry for southern Okhotsk Sea region. The cross section is taken in NE-SW direction. The stippled zone indicates the lower velocity region needed to explain the data. Some of the longer distance paths travel partially in the higher velocity subducted lithosphere.

uppermost part of the slab were significantly melted by the time it reached the position of the first rank volcano, Esan on the Pacific coast of Hokkaido, it is unlikely that any ^{10}Be-bearing sediments would survive further mantle penetration of more than 100 km.

The cooled asthenosphere, which follows the subducted slab down to considerable depths (up to 600 km), cannot return to shallower levels until it heats sufficiently to regain its buoyancy. However, high heat flow and even spreading in back arc regions indicates upwelling of hot material. For example, there is generally high heat flow in the Sea of Japan 300-500 km above the slab. One cause may be the return flow driven by subduction itself. This process may be aided by water release from the slab at depth, possibly from phase B of the hydrous-magnesium-silicate system $M_gO-SiO_2-H_2O$. Laboratory experiments show that this system is stable from about 100 kb to at least 160 kb at temperatures of 750-1200 degrees (Akaogi and Akimoto, 1979). Suyehiro and Sacks (1983), from a study of velocity structure in the Japan subduction zone, found decreased velocity in the asthenosphere above the slab in the 300-500 km depth range (Figure 6). Consistent with this result, they also find an increased absorption of shear-wave energy in this region. The dehydration of phase B, if it exists in the subducting plate, could explain qualitatively the zone of anomalously low velocity and low Q inferred from the above study. The release of water would at least decrease viscosity and might produce flux melting in the overlying asthenosphere, thus aiding upwelling beneath the back arc. (Is there a relevant geochemical signature in back arc volcanism?)

Two conclusions are:

1) Sediment is not substantially returned to the surface by volcanos, but is carried into the mantle, particularly in regions of buoyant subduction.

2) Even at depths of 600 km, a substantial region of older subducted slabs is far cooler than the surrounding mantle.

REFERENCES

Akaogi M. and S. Akimoto, 'High-pressure stability of a dense hydrous magnesium silicate $Mg_{23}Si_8O_{42}H_6$ and some geophysical implications,' Tech. Rep. ISSP, Univ. of Tokyo, Ser. A, 997, 1979.

Chan, W. Winston and I.S. Sacks, 'Surface waves attenuation beneath the Iceland Plateau', in preparation, to be submitted to J. Geophys. Res., 1987.

Chen, W.-P. and P. Molnar, 'Focal depths of intracontinental and intraplate earthquakes and their implications for the thermal and mechanical properties of the lithosphere,' J. Geophys. Res., 88, 4183-4214, 1983.

Hasegawa, A., and I.S. Sacks, 'Subduction of the Nazca plate beneath Peru as determined from seismic observations,' J. Geophys. Res. 86, 4971-4980, 1981.

Hirahara, K., 'Three-dimensional seismic structure beneath southwest Japan: The subducting Philippine Sea plate,' Tectonophysics, 79, 1-44, 1981.

Hsui, A.T., and M.N. Toksoz, 'The evolution of thermal structures beneath a subduction zone,' Tectonophysics, 60 43-60, 1979.

Leeds, A.R., L. Knopoff, and E.G. Kausel, 'Variations of upper mantle structure under the Pacific Ocean,' Science, 186, 141-143, 1975.

Sacks, I.S., 'The subduction of young lithosphere,' J. Geophys. Res., 88, 3355-3366, 1983.

Sacks, I.S., and H. Okada, 'A comparison of the anelasticity structure beneath western South America and its tectonic significance,' Phys. Earth Planet. Interiors, 9, 211-219, 1974.

Schneider, John F., and I. Selwyn Sacks, 'Stress in the contorted Nazca plate beneath southern Peru from local earthquakes,' in press, J. Geophys. Res., 1987.

Stark, Philip B., and Cliff Frohlich, 'The depths of the deepest deep earthquakes,' J. Geophys. Res. 90, 1859-1869, 1985.

Suyehiro, K., and I.S. Sacks, 'An anomalous low velocity region above the deep earthquakes in the Japan subduction zone,' J. Geophys. Res., 88, 10429-10438, 1983.

Tera, F., L. Brown, J. Morris, I.S. Sacks, J. Klein, and R. Middleton, 'Sediment incorporation in island-arc magmas; inferences from ^{10}Be,' Geochim. Cosmochim. Acta, 50, 535-550, 1986.

INFLUENCE OF H_2O AND CO_2 ON MELT AND FLUID CHEMISTRY IN SUBDUCTION ZONES

David H. Eggler
Geosciences Department
The Pennsylvania State University
University Park, PA 16802

ABSTRACT. Volatiles can be transferred from slabs to wedge asthenosphere either by silicate melts of by hydrous fluids. Experiments defining the nature of these agents, and expected ratios of transported elements, are not well-constrained, especially for fluids. Phase equilibria of basaltic slab material and probable geotherms suggest that slabs devolatilize or melt, or both, at depths shallower than the average 136 km beneath arc fronts. Volatiles probably are transported into immediately overlying asthenosphere and then carried by flow beneath arcs. Simple addition of probable fluid to MORB or OIB-type asthenosphere cannot explain observed H_2O and K_2O to asthenosphere and more complicated histories of melt generation; the melts generated at 100-140 km beneath arcs bear little chemical relation to arc basalts. Volatiles thus must pass numerous "filters" before they emerge in arc volcanics.

TRANSFER OF VOLATILES FROM SLABS TO ASTHENOSPHERE

Thermal models of descending plates vary considerably. If te peratures at the mean depth to subducting plate, 136 km (Gill 19), exceed about 800°C, then H_2O-bearing eclogitic plates will be partially molten. The H_2O cannot be derived from amphibole, however, inasmuch as the maximum pressure of stability of amphibole, at the solidus, is about 26 kbar (86km). Such hydrous, low-volume melts could rise into overlying asthenosphere, reacting with and metasomatizing peridotite (Ringwood, 1974).

It has been argued, nevertheless, that a descending slab could release H_2O to depths of perhaps 125 km without causing melting of the slab itself (e.g., Delaney and Helgeson 1978). A series of minerals dehydrate as slabs descend, leading to continuous dehydration (H_2O accompanied by CO_2 and methane) as the slab penetrates the mantle. Much of the volatiles may actually travel back up the slab some

f

97

S. R. Hart and L. Gülen (eds.), Crust/Mantle Recycling at Convergence Zones, 97–104.

98

distance, inasmuch as asthenosphere above the slab is
hotter, in most models, than the slab (Fig. 1). Thus
volatiles probably enter the asthenosphere above slabs at
depths considerably shallower than 136 km (Tatsumi and
Hamilton 1986).

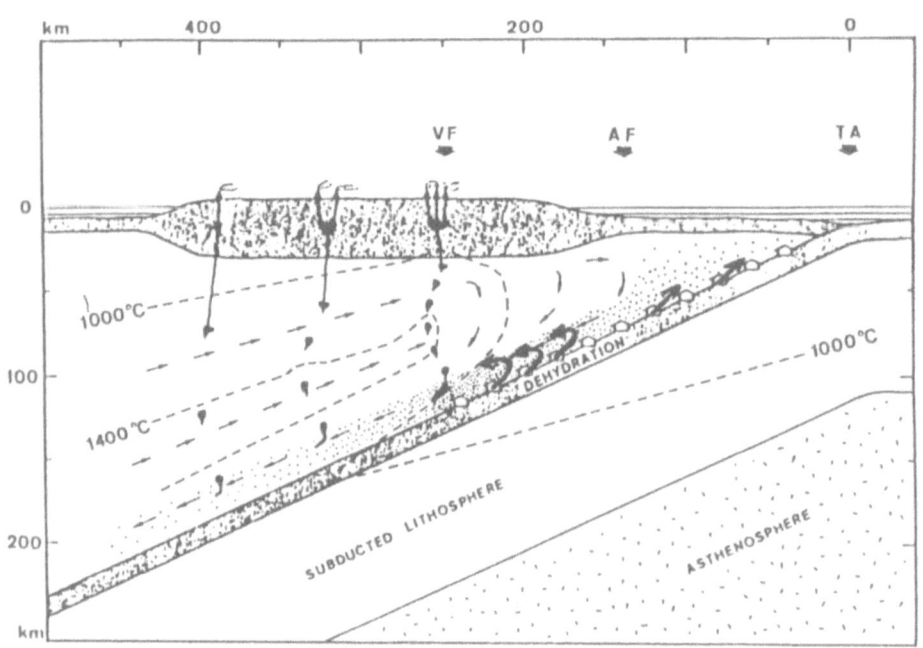

Fig. 1: Cartoon of convergent plate margin, after
Tatsumi and Hamilton (1986). Thick arrows show
possible direction of movement of fluids off slabs.

Volatiles could amphibolitize, carbonatize, or melt the
asthenospheric wedge, depending upon pressure and
temperature (Fig. 2). Distance of penetration into
asthenosphere would depend on the a amount and migration
characteristics of a fluid (H_2O-CO_2) or melt phase.
Penetration into partially-molten asthenosphere would be
quite limited, because of slow diffusion in melt, unless
flux-induced melt fractions were large enough to cause
escape. Penetration into subsolidus asthenosphere would
also be limited, because of high wetting angles of fluids in
peridotite (Watson and Brenan, 1987), unless accumulated

volume of fluid were enough to cause interconnectiveness or
hydraulic fracturing.

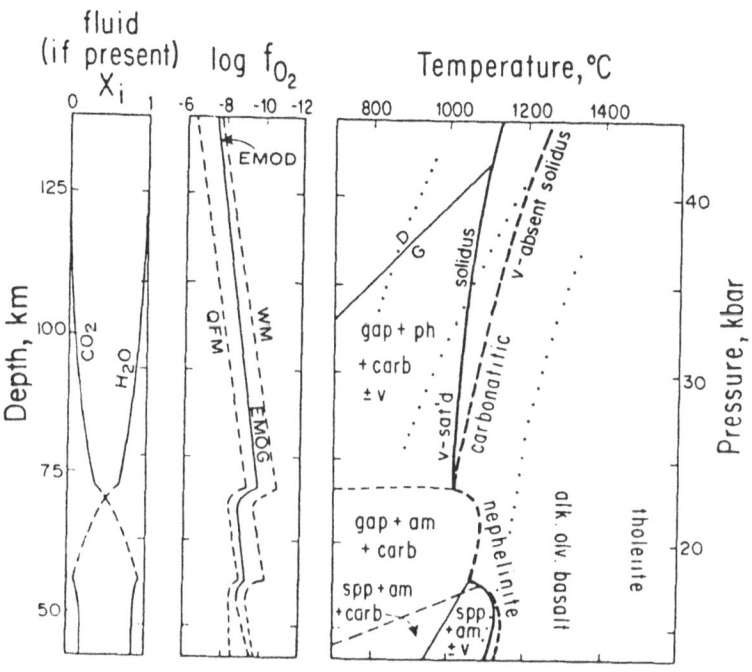

Fig. 2: Phase relations of a peridotite composition in
the presence of small amounts of H_2O and CO_2, after
Olafsson and Eggler (1983). Fluid-buffered solidus is
shown as heavy solid lines and fluid-absent solidus as
heavy dashed lines. Melts produced from buffered
peridotite are labelled with approximate chemistry.
Dotted lines are geotherms for oceanic lithospheric
ages of 30 m.y. (right) and 180 m.y. (left). Offset
boxes show compositions of fluids at the fluid-buffered
solidus and the fO_2 at the solidus. For details see
Eggler (1987).

ARE ARC BASALTS INDICATIVE OF SUBARC PROCESSES?

Over the last ten years it has been realized that parental
magmas of arc volcanism, at crustal levels, are basaltic
rather than andesitic (e.g., Arculus and Johnson, 1978;
McBirney, 1978; Marsh, 1982). Andesites can be related to
basalts by processes including crystallization fractionation
(Perfit et al., 1980; Grove and Baker, 1984; Baker and

Eggler, 1987), crustal contamination, magma mixing, and
liquid fractionation (McBirney et al., 1985).

Opinion is divided on the nature of primitive basaltic
magmas. In several oceanic arcs, such as the Aleutians,
high-alumina (calcalkaline) arc tholeiites dominate. Such
tholeiites are at least plausibly derived by extensive
partial melting of subducted eclogitic slabs (e.g., Marsh,
1982; Myers, et al. 1985; Brophy and Marsh, 1986). Other
workers recognize multiple primitive basaltic series--
olivine tholeiite, high-alumina, and alkali olivine basalts
(e.g., Kuno 1966; Jakes and Gill, 1970; Tatsumi et al.,
1983). These series characterize arcs at various distances
from the trench and also characterize evolution of arcs with
time. Still others relate primitive high-alumina basalts to
primary MORB-like olivine tholeiites by crystal
fractionation at the base of island arc crusts (e.g., Perfit
et al. 1980; Gust and Perfit 1987). All of these primitive
tholeiites or alkali basalts are presumably derived by
partial melting, at 50 to 100 km depth, of asthenosphere
above slabs.

If arc basalts represent high-degree melts of quartz
eclogite slabs, then their chemistry directly reflects slab
chemistry at the point of separation, save for crystal-melt
partitioning in the course of diapir/melt ascent through
asthenosphere. If they represent partial melting of
asthenosphere above slabs, then the chemistry of slabs is
reflected in basalt chemistry through indirect and
complicated processes.

WHAT ARE THE CHEMISTRIES OF ASTHENOSPHERIC PARTIAL MELTS?

The asthenospheric mantle wedge above the slab is at
temperatures hot enough, over most of its volume, to melt
(compare Figs. 1 and 2) if H_2O or $H_2O + CO_2$ are present.
The solidus temperature is not dependent on $H_2O/(H_2O+CO_2)$
over a considerable range of $H_2O/(H_2O +CO_2)$; that is
particularly true at pressures exceeding 23 kbar, where
carbonate minerals buffer fluid compositions to very H_2O-
rich values (Fig. 2). The asthenosphere presumably contains
sufficient volatiles, apart from any volatiles added from
the slab, for such melting to occur. Addition of volatiles
would serve chiefly to increase the degree of melting.

Near-solidus asthenospheric melts at depths immediately
above the slab (75 to 125 km) are not well-characterized in
experimental studies on natural peridotite compositions.
For a general review see Olafsson and Eggler (1983).
Existing experiments and analogous synthetic systems do
indicate, however, that melts are highly alkalic and silica-
poor--e.g., carbonatitic, alkali carbonatitic, or
melilititic compositions. With higher degrees of melting,

melts will become more olivine-rich, trending toward alkali picrites.

It is apparent, then, that melts generated immediately above the slab, in the region where slab-derived fluids or melts, or both, enter asthenosphere, have little in common with melts parental to arc volcanism. The nature of such melts was summarized above. In the model of Tatsumi et al. (1983), such melts initiate diapirism (Fig. 1). Diapiric rise and melt separation, at depths between 100 and 35 km (Tatsumi et al. 1983; Nye and Reid 1986), provide yet another filter to transfer of elements from slab to arc volcanics.

H_2O-K_2O RELATIONS

Figure 3 shows H_2O and K_2O contents of MORB basalts and of basalts and andesites from arcs. Some of the arc values represent careful measurements on fresh glasses (Garcia et al. 1979), and some represent estimates from phase equilibria studies. The data can be interpreted in a straightforward manner if it is assumed that within magmatic series, at least up to silica contents representative of andesites, both H_2O and K_2O behave as incompatible elements; by extension, H_2O/K_2O of primitive arc basalts should reflect source regions.

Various estimates of primitive tholeiitic arc volcanic magmas contain 0.3-0.9 wt percent K_2O. Assuming that arc, MORB, and OIB tholeiites are formed by equivalent degrees of melting, it is apparent that H_2O and K_2O contents of arc volcanic source regions considerably exceed MORB sources (and somewhat exceed OIB sources). The H_2O/K_2O ratios, however, do not appear to exceed or greatly exceed MORB ratios. This observation can be interpreted in several ways:

(1) Arc volcanic sources represent substantial addition of H_2O and K_2O, from subducted basalt and sediment, to a MORB-type asthenosphere. Similarity of H_2O/K_2O ratio to MORB is coincidental.

(2) Arc volcanic sources represent less-substantial addition of H_2O and K_2O to OIB-type asthenosphere, with higher H_2O and K_2O contents (about 0.5% H_2O and 0.4% K_2O for intermediate-K Hawaiian tholeiites: Moore 1970).

(3) Arc volcanic sources represent modest addition of H_2O and K_2O to MORB-type asthenosphere, but substantial differences in processes of melting-extraction-differentiation produce much higher K_2O and K_2O contents in arc basalts.

A final question regards the nature of addition of H_2O and K_2O to MORB-type or OIB-type asthenosphere. The nature of low-volume melts that could ascend from the slab, either at about 136 km depth or shallower, is not well-constrained,

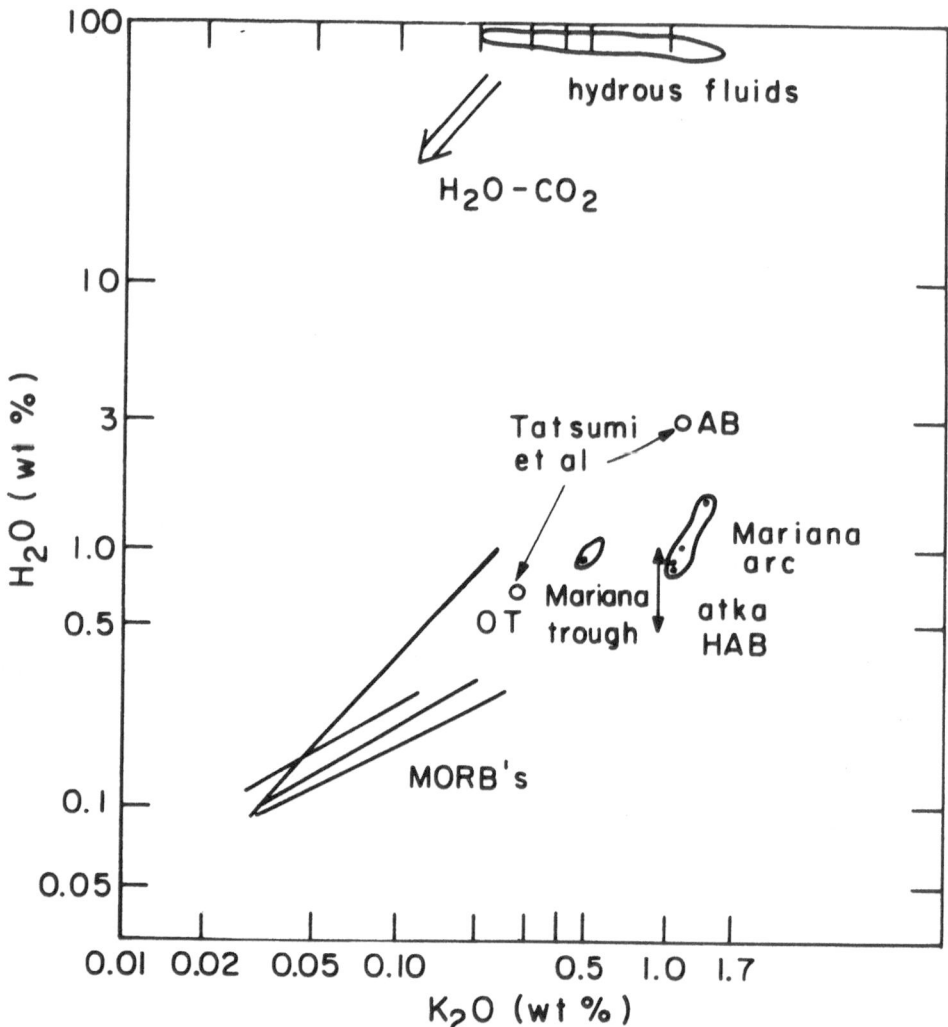

Fig. 3: Contents of H$_2$O and K$_2$O in fresh MORB glasses (Byers et al. 1986), in fresh glasses of Mariana arc and trough (Garcia et al. 1979), and estimate for primitive Atka (Aleutians) high-Alumina basalt (Baker and Eggler 1983), and estimates for olivine tholeiite and alkali basalt of the Japan arc (Tatsumi et al. 1983). For fluids see text.

but they might contain 1-2 percent K_2O and 1-3 percent H_2O. These could produce trends shown. The more likely fluid leaving the slab, at depths shallower than 136 km, would be H_2O-rich. Limited experimental data exist on the nature of hydrous fluids (summarized by Eggler 1987), but application of even those limited data to slabs is problematic because of dissimilar mineralogies of experiments and slabs. The K_2O content of any fluid probably will not exceed that of fluid in equilibrium with andesitic siliactate melts, however. Such data (summarized by Eggler 1987) for fluids, with or without chloride, indicate that only modest K_2O contents would be expected because fluids contain limited solute (about 20 wt percent at 20kbar) and because hydrous solutes are dominated by other cations including silica, alumina, and soda. Addition of CO_2 will severely depress solute contents (Figure 3). Thus simple addition of fluid from slab to Morb-type asthenosphere cannot produce the arc signatures (Figure 3). Precipitation of fluids or reaction of fluids with wallrocks is not a simple process, however (see, for example, Schneider and Eggler, 1986); it is possible that in asthenosphere above slabs H_2O and K_2O could be decoupled sufficiently to produce regions with relatively low H_2O/K_2O.

REFERENCES

Arculus R.J. and Johnson R.W. (1978) Earth Planet. Sci. Lett. 39, 118-126.

Baker D.R. and Eggler D.H. (1987) Amer. Mineral.

Brophy J.G. and Marsh B.D. (1986) J. Petrol. 217, 763-789.

Byers C.D., Garcia M.O. and Muenow D.W. (1979) Earth Planet. Sci. Lett. 79, 9-20.

Delaney J.M. and Helgeson H.C. (1978) Amer. J. Sci. 278, 638-686.

Eggler D.H. (1987) in: Menzies M.A. and Hawkesworth C.J. (eds.) Mantle Metasomatism, pp. 21-41.

Garcia, M.O., Lui N.W.K. and Muenow D.W. (1979) Geochim. Cosmochim. Acta 43, 305-31.

Grove T.L. and Baker M.B. (1984) J. Geophys. Res. 89, 3253-3274.

Gust D.A. and Perfit M.R. (1987) Contrib. Mineral. Petrol.

Jakes P. and Gill J.B. (1970) Earth Planet. Sci. Lett. 9, 17-28.

Kuno, H. (1966) Bull. Volc. 29, 195-222.

McBirney A.R. (1978) Ann. Rev. Earth Planet. Sci. 6, 437-456.

McBirney A.R., Baker B.H. and Wilson R.H. (1985) J. Volc. Geotherm. Res. 14, 1-24.

Marsh B.D. (1982) in: Thorpe, R.S. (ed.) Andesites, pp. 9-114.

Moore J.G. (1970) Contrib. Mineral. Petrol. 28, 272-279.

Myers, J.D., March B.D. and Sinha A.K. (1985) Contrib. Mineral. Petrol. 91, 221-234.

Nye C.J. and Reid M.R. (1986) J. Geophys. Res. 91, 10271-10287.

Olafsson M. and Eggler D.H. (1983) Earth Planet. Sci. Lett. 64, 305-315.

Perfit M.R., Gust D.A., Bence A.E., Arculus R.J. & Taylor S.R. (1980) Chem. Geol. 30, 227-256.

Ringwood A.E. (1974) J. Geol. Soc. Lond. 130, 183-204.

Schneider M.E. and Eggler D.H. (1986) Geochim. Cosmochim. Acta 50, 711-724.

Tatsumi Y., Sakuyama M., Fukuyama H. and Kushiro I. (1983) J. Geophys. Res. 88, 5815-5825.

Tatsumi Y. and Hamilton D.H. (1986) J. Volc. Geotherm. Res. 29, 293-310.

ANHYDROUS HFSE DEPLETED PERIDOTITE AS A UBIQUITOUS MANTLE COMPONENT.

Vincent. J.M. Salters
Center for Geoalchemy
Department of Earth, Atmospheric and Planetary
Sciences
Massachusetts Institute of Technology
Cambridge, Massachusetts, 02139, USA.

ABSTRACT.   We report here the widespread occurrence of
mantle peridotites in both suboceanic and subcontinental
lithosphere that show HFSE depletions similar to those of
island arc volcanics. The generation of these depletions is
not related to the subduction process. These anhydrous
peridotites, which reside at high levels in the mantle,
cannot generate, nor be in equilibrium with, mid-ocean ridge
or ocean island basalts, and have a complex geochemical
history involving partial melting, melt extraction and
metasomatism.

1. INTRODUCTION

Island   arc   magmas   characteristically   exhibit   low
concentrations of High Field Strength Elements (HFSE): Ti,
Ta, Nb, Zr and Hf, and high concentrations of Sr, compared
to the REE (Gill, 1981; Hickey et al, 1986). The depletion
of the HFSE relative to the REE is considered as one of the
most diagnostic geochemical features of island arc magmas
(Pearce and Cann, 1973; Pearce and Norry, 1979). The origin
of calc-alkaline volcanics, the HFSE depletion and Sr-
enrichment is debated; one school argues that island arc
volcanics are formed by melting of an ocean island basalt
(OIB)-type source mantle (Morris and Hart, 1983; Stern,
1981, DePaolo and Wasserburg, 1977) with a Ti-rich residual
phase causing the HFSE depletions(Morris and Hart, 1983;
Green, 1981). Other workers argued that island arc volcanics
are formed by melting of a mid-ocean ridge basalt (MORB)-
type mantle enriched in incompatible elements by a component
from the subducted slab (Gill, 1981; Nicholls and Ringwood,
1973; Kay, 1980). The HFSE depletion is thus interpreted as

S. R. Hart and L. Gülen (eds.), Crust/Mantle Recycling at Convergence Zones, 105–119.
© 1989 by Kluwer Academic Publishers.

caused by the addition of slab component. The Sr-enrichment over the REE is explained by preferential incorporation of Sr in a fluid from the subducted slab (Hawkesworth et al, 1977; DePaolo and Johnson, 1979, Hickey et al., 1986).

We found that clinopyroxenes from some peridotites and harzburgites are depleted in HFSE relative to REE, and show different degrees of Sr-enrichment (Shimizu, 1987; Salters, 1987). Subsequent analyses of samples as well as a literature survey has revealed that HFSE-depleted lherzolites are indeed very common. Here we present our data bearing on REE and HFSE abundances in clinopyroxenes from four phase anhydrous lherzolites from subarc, subcontinental and suboceanic lithosphere and review the literature. These data allow us to address the origin and extent of HFSE depletions in both volcanics and mantle materials and investigate the relationship between HFSE depletion and Sr-enrichment.

2. RESULTS

The HFSE depletion is best illustrated on an incompatibility (or spider) diagram where the elements are arranged according to increasing compatibility in the system peridotite-melt (Wood, 1979; Thompson et al.,1983; Sun 1980), Fig. 1. The HFSE depletion is measured relative to the neighboring REE. HFSE/HFSE* notation as with Eu-anomalies is most useful in quantifying the depletion. Zr^* and Hf^* are defined as the average of the chondrite normalized values of its neighboring REE: Sm and Nd $(Hf^*=Zr^*=(Sm_N+ Nd_N)/2$. Ti^* is defined in the same manner with Eu^* and Tb as its neighboring REE.

REE and HFSE of clinopyroxenes were analyzed in situ using a Cameca IMS 3f ion microprobe. Energy filtering was used to exclude molecular ion interferences (Shimizu and Hart, 1982). Concentrations were determined through the use of working curves based on a set of well determined standards. Precision of the analyses is always better than 10% and in most cases better than 5%.

We use the REE and HFSE abundance pattern of clinopyroxenes from anhydrous four-phase spinel lherzolite as an accurate representation of those of bulk rock anhydrous four-phase spinel lherzolites. This has been argued, for REE, by many investigators (Downes and Dupuy, 1987; Stosch and Seck, 1980; M.F. Roden et al., 1984) The data of Jagoutz et al. (1979a,b) and our data (Salters and Shimizu, 1988) clearly demonstrates the REE/HFSE of clinopyroxenes and their peridotites are similar, and that the REE and HFSE budget of a whole rock spinel lherzolite is largely contained in the clinopyroxene.

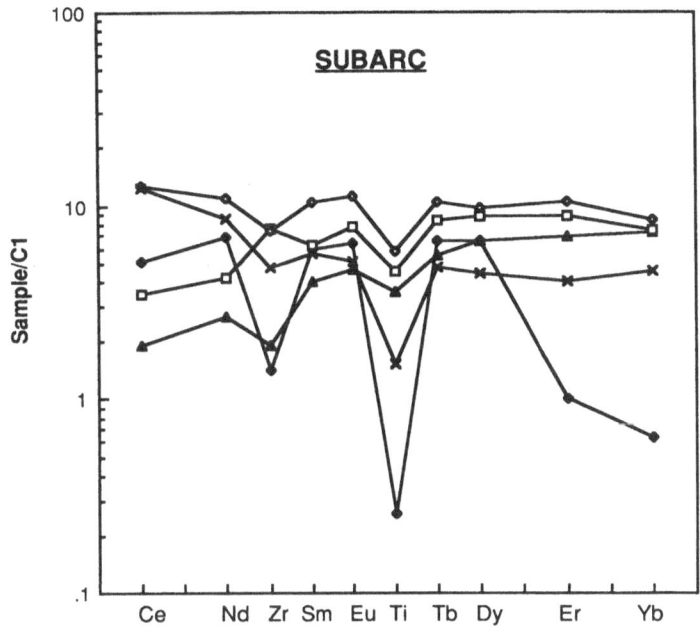

Figure 1: Incompatibility diagram of clinopyroxenes from four xenoliths in alkali basalts from the Carpathians. Crosses represent sample LS 71 from Pali Aike, Patagonia, Chile. Tb values are interpolated.

2.1. SUBARC MANTLE

Anhydrous lherzolite nodules from subarc mantle are found in Quaternary alkali basalts in the Pannonian basin, Hungary. The alkali basalts are part of the Carpathian volcanic province, and postdate Eocene and Miocene calc-alkaline volcanism (Burchfiel and Royden, 1982; Royden et al. 1983). The clinopyroxenes from all nodules have Ti/Ti*<1, and all except for one have Zr/Zr*<1, Fig. 1. The nodules are depleted character as is evidenced by the high Cr and Mg/Fe ratio of the clinopyroxenes, and the low modal clinopyroxene content of the nodules (2-8%).

The Carpathian clinopyroxenes are extremely variable in Nd-isotopes (ϵ_{Nd}= -5.1 to +19.5), and REE content, but restricted in Sr-isotopic composition. The degree of HFSE-depletion is not correlated with the Nd-isotopic composition. Both the isotopes and the trace elements exclude equilibrium between host alkali basalts and the lherzolite nodules (Salters et al., 1988 and Salters unpublished data).

Both garnet lherzolites as well as spinel lherzolites occur as xenoliths in alkali basalts from the Patagonian Plateau, Chile. The proximity of the alkali basalts to the subduction zone plus the depleted isotopic character of these lherzolites makes them very likely candidates for subarc mantle (Stern et al., 1986; Stern, pers.comm.). Except for one-garnet lherzolites, all lherzolites (one-garnet-bearing and five-spinel-bearing) show HFSE depletions.

2.2 SUBCONTINENTAL MANTLE

There is significantly more data available on peridotites from continental regions, especially spinel lherzolites. Localities of lherzolites with HFSE-depletions and references for the data are: Dreiser Weiher, Germany; San Carlos, Arizona (Jagoutz et al., 1979a); San Quintin, Mexico (Basaltic Volcanism Study Project, 1981); Silezia, Poland (Blusztajn and Shimizu 1988); Massif Central (Downes and Dupuy, 1987); and Lherz (Bodinier et al., 1987), France; Spitsbergen (Amundsen, 1988); Assab, Ethiopia (Ottonello, 1980); and Nunivak, Alaska, Kilbourne Hole and Potrillo Maar, New Mexico (this paper). Garnet-bearing lherzolites with HFSE-depletions occur in S. Africa (Nixon et al., 1981) and on the Colorado Plateau (Ehrenberg, 1982). Some examples of HFSE depleted nodules are given in Fig. 2. With the exception of the S. The S. African lherzolites show the HFSE depletions only in modally metasomatized peridotites depleted in major elements, but enriched in trace elements (Nixon et al., 1981). The garnet lherzolites which are fertile based on their major element chemistry are also undepleted in Ti and Zr; some show enrichments in these two HFSE. This agrees with the general trend that modally metasomatized hydrous lherzolites can be enriched in (Menzies et al., 1987). At all continental lherzolite occurrences where more than two nodules are analyzed; i.e. Silezia, Nunivak, Kilbourne Hole, Potrillo Maar, Massif Central, Colorado Plateau and S. Africa, both HFSE depleted and HFSE undepleted nodules occur. Single depletions in either Ti or Zr also occur. The HFSE-depleted samples typically show a wider range in Ti/Zr than samples classified as undepleted and not metasomatised (Shimizu and Allegre, 1978).

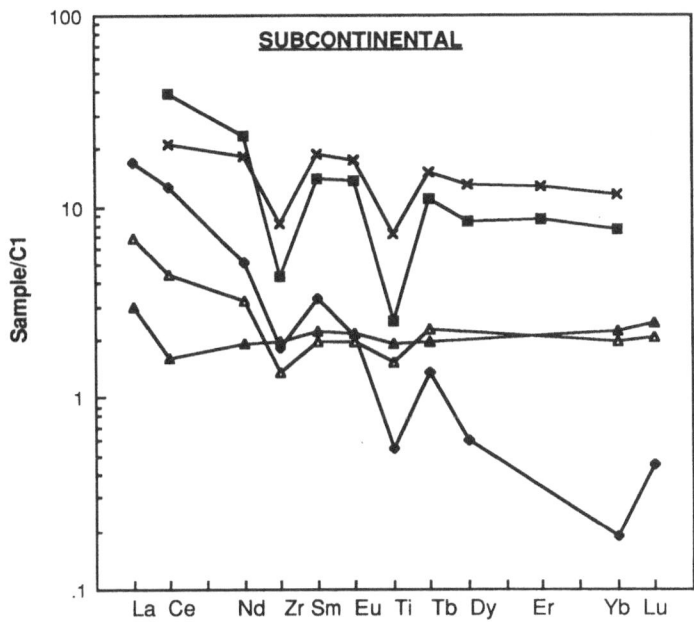

Figure 2: Incompatibility diagram of selected localities from continental regions illustrating the wide range in trace element chemistry of the depleted nodules. Diamond is sample 2814 from Jagersfontein, S. Africa, Nixon et al. (1981). Triangles are samples Ms17 and RP70 from Massif Central, Hf is plotted instead of Zr, (Downes and Dupuy, 1987) Nunivak (squares, sample 1004 in M.F. Roden et al., 1984) and Kilbourne Hole (crosses) is clinopyroxene data rest is whole rock data. Normalization as in Fig. 1.

2.3 SUBOCEANIC

At present only twelve localities of mantle samples from an oceanic setting have been investigated. Of these twelve the Ronda peridotite massif (Frey et al., 1985); abyssal peridotites from the South Indian Ridge (Johnson et al., 1987), from Mid Indian Ocean Ridge (MIOR) and from the Romanche fracture zone, Atlantic ocean; peridotite nodules from Salt Lake Crater, Oahu (Shimizu, 1987); Charles Island, Galapagos; Anjouan, Comores; and Ascension island; St. Paul's Rock peridotite, Atlantic Ocean (Frey, 1970; Melson et al., 1972; M.K. Roden et al., 1984), and the peridotite of Zabargad Island, Red Sea (Bonatti et al., 1986) all show

the HFSE depletion, Fig. 3. The Azores, and Kerguelen do not
show the depletion. It must be stressed however, that we
only analyzed clinopyroxenes from one nodule for Galapagos,

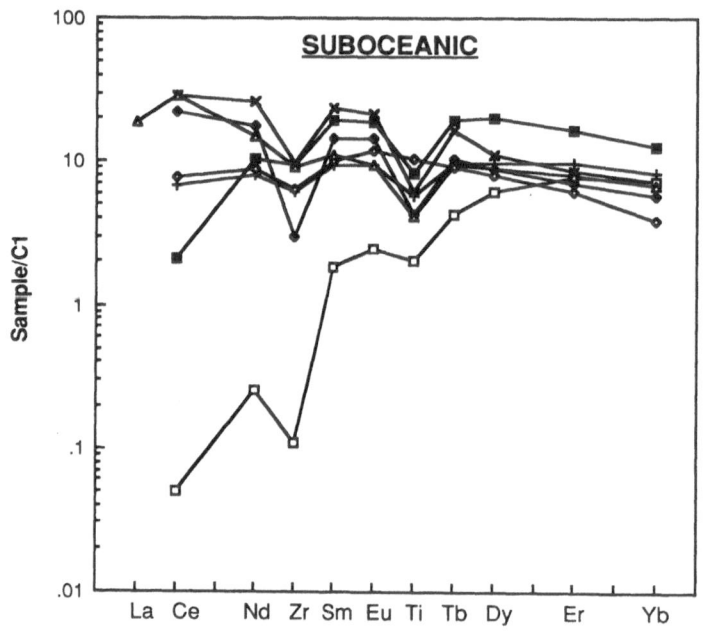

Figure 3: Incompatibility diagram of selected oceanic
occurrences of HFSE depleted peridotites. Symbols are:
open triangles=St. Paul's, whole rock (Data of Frey,
1979, and M.K. Roden, 1984). Rest of analyses is
clinopyroxene data: crosses = Galapagos, open squares
= Mid Indian Ocean Ridge (Shimizu and Hart, 1974),
diamonds = Hawaii, closed square=Romanche FZ, plusses =
Ascension. Normalization as in Fig. 1.

Ascension, Azores, Kerguelen, MIOR and Romanche. From the
available data it seems that the HFSE-depleted mantle is
ubiquitous in the suboceanic mantle: all Salt Lake Crater
lherzolites analyzed (12) and all St. Paul's Rock peridotite
analyses (6) show HFSE depletions. 70% of the South West
Indian Ridge (10) (K.T.M. Johnson, pers. comm.), 80% of
Ronda (15) and 85% of Zabargad Island peridotites (19)
exhibit HFSE/HFSE*<1. Numbers in parentheses are number of
samples investigated. These data strongly suggest that HFSE
depleted mantle is of widespread occurrence in the
suboceanic mantle.

3. DISCUSSION

The order of the elements on the incompatibility diagram is based on the gradual increase in compatibility from left to right in the system peridotite-melt. However, very few data is available where HFSE and REE distribution coefficients, D, are determined on the same samples. There is a definite need for simultaneous determination of the D's for clinopyroxene/melt for the HFSE and REE. Only one study (Fujimaki et al., 1984) determined D's for Hf, Zr and REE for clinopyroxene. These data indicate that the Ds for Hf and Zr are indeed intermediate between Sm and Nd. One study indicates that Hf and Ti are more incompatible than the REE, although the absolute D's are close to other results (Watson et al., 1987). This would cause melts to have $HFSE/HFSE^*>1$ and the residual solid $HFSE/HFSE^*<1$. A consistent set of D's can be derived from the literature with D's from .1 to .3 for the light to heavy REE respectively (Grutzeck et al., 1974). A distribution coefficient of .20-.25 for Zr is indicated by several studies (Fujimaki et al., 1984; McCallum et al, 1978), while a D of .30 for Ti is most consistent with experimental data (Huebner et al, 1976; Grove and Bryan, 1983; Tormey et al, 1987). These Ds indicate that in a single-stage melting process $(HFSE/HFSE^*)_{melt}= (HFSE/HFSE^*)_{clinopyroxene}$, i.e. the HFSE trait of the source is inherited by the melt. However, the observed range in D's is quite significant, and accurate determination of the D's for REE and HFSE on the same samples is necessary.

Fig. 4 shows Ti/Ti^* versus Zr/Zr^* for MORB, OIB and basic calc-alkaline volcanics. Although the scatter on this diagram is large, MORB and OIB are clustered around one for both Ti/Ti^* and Zr/Zr^* while the maximum Ti/Ti^* for calc alkaline volcanics is less than one. Very few OIB or MORB have both Ti and Zr depletions while the majority of the anhydrous spinel lherzolites show both Zr and Ti depletions. With the above Ds it is impossible for MORB and OIB to be in equilibrium with HFSE-depleted mantle, $HFSE/HFSE^*$ as low as .02. Contrary to what is suggested by several workers (Dick and Fisher, 1984; Shibata and Thompson, 1986; Prinzhofer and Allegre, 1985) the occurrence of HFSE depletions in abyssal peridotites indicates that these peridotites and MORB are not related by a simple melting process. With the present mineralogy, the widespread occurrence of HFSE-depleted peridotites and the predominance of this type over HFSE-undepleted peridotites at all known oceanic occurrences suggests that HFSE-depleted peridotite is a volumetrically significant constituent of the suboceanic lithosphere.

The occurrence of HFSE-depleted mantle obviates the need to call upon a residual titanate phase in order to generate

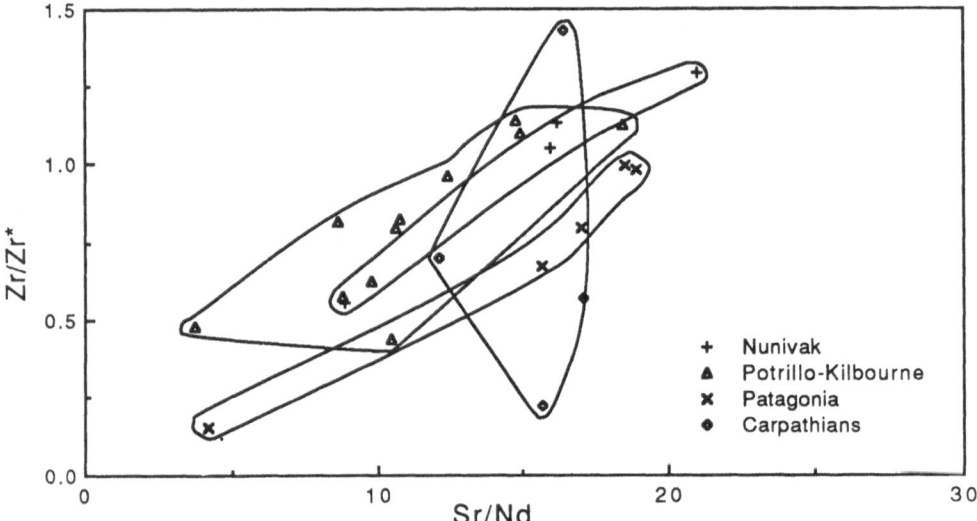

Figure 4: Zr/Zr* versus Ti/Ti* for volcanics from three different tectonic settings. MORB data used is of le Roex et al. (1985); Price et al. (1986); Wood et al. (1979). OIB field compiled of the data of Basaltic Volcanism Study Project (1981); D. Weis, unpublished data; Liotard et al. (1986); Wood (1978); Leeman et al. (1980). Plusses are calc-alkaline volcanics, data of Morris (1984); Fujimaki (1986); Dupuy et al. (1982), Basaltic Volcanism Study Project (1981) and Gerlach (1986). Solid triangles are data for clinopyroxenes from anhydrous spinel lherzolites. Data of Frey et al. (1985); Blusztajn and Shimizu (1988); Frey (1970), Melson et al. (1972); Roden, M.K. et al. (1984); Jagoutz et al. (1979 a+b) and unpublished data.

island arc type volcanics. Also recent experimental data (Ryerson and Watson, 1988) indicates a melt in equilibrium with a residual Ti-rich phase will have significantly higher TiO_2-contents than island arc basalts. Magmas with $HFSE/HFSE^*<1$ can be generated by direct melting of HFSE-depleted mantle.

The possible existence of a slab component in some of the more volatile elements of island arc volcanics is well documented (Morris and Hart, 1983; Kay, 1980; Tera et al., 1986). However, if the slab component creates the HFSE depletions and Sr-enrichments in both island arc basalts and in peridotites one would expect Sr/Nd and Zr/Zr^* ratios to be anticorrelated. Fig. 5 shows the Sr/Nd and Zr/Zr* values of clinopyroxenes from nodules which ascended close by a

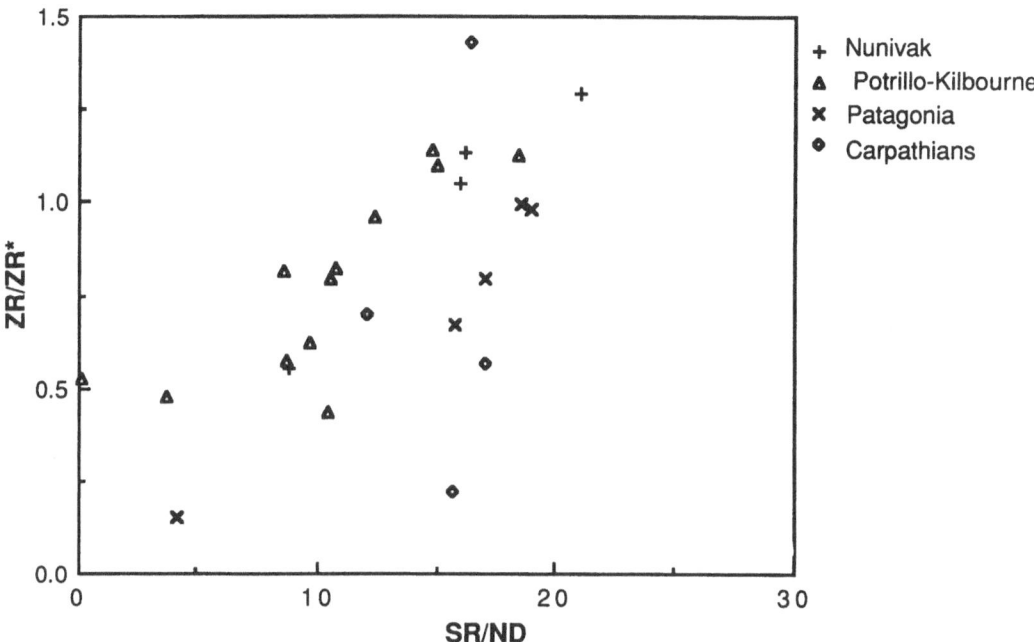

Figure 5: Zr/Zr* versus Sr/Nd for clinopyroxenes from lherzolites nearby a subduction zone. Plusses are data from Nunivak, triangles are from Potrillo Maar and Kilbourne Hole, crosses are from Patagonia and diamonds are from Carpathians.

subduction zone: Kilbourne Hole and Potrillo Maar, USA; Nunivak, Alaska; Patagonia, Chile and Carpathians, Hungary. The positive correlation between Sr/Nd and Zr/Zr* indicates that the Sr enrichment and Zr depletion are not generated by the same process.

Recent data suggest that high pressure phases as perovskite, majorite and garnet can fractionate HFSE from the REE in melting processes (Irving and Frey, 1984; Fujimaki et al, 1986; Kato et al, 1987; Ringwood, 1988; Salters and Shimizu, 1988). It is thus suggested that these high pressure phases play an important role in the creation of HFSE depletions. How these phases can effect spinel lherzolites at shallow levels in mantle is unclear to us.

The range in Ti/Ti* and Zr/Zr* in MORBs and OIBs suggests their sources might also be affected to some extent by HFSE depleted material. Clague and Frey (1982) suggested inhomogeneity in the source for the HFSE compared to the REE to explain the trace element variations in the Honolulu volcanics. However the HFSE depleted mantle is more pronounced in IAB than in MORB or OIB.

HFSE-depleted material is until now only observed in materials that reside at shallow levels in the mantle just before transportation to the surface. The lack of HFSE depletions in other types of volcanics than island arc type indicates the HFSE depleted component could be restricted to shallow levels in the mantle; below its solidus. In an arc-type setting where the mantle wedge can be fluxed with volatile elements as Be, Cs, K, Ba, B, this addition can lower the solidus and thus cause melting of HFSE-depleted peridotite.

4. CONCLUSIONS

- HFSE depleted, anhydrous, four phase lherzolites are ubiquitous in both subarc, suboceanic and subcontinental mantle.
- HFSE depleted mantle is not related to MORB or OIB by simple melting.
- HFSE depletions in island arc magmas can be generated by simple melting of HFSE depleted mantle.
- HFSE depletions in anhydrous four phase lherzolites are not created by addition of a slab component, but is most likely the result of a more complex process involving high pressure phases.

ACKNOWLEDGEMENTS.

Ken Burrhus is thanked for keeping the ion microprobe in excellent condition. M.F. Roden, S.R. Hart, C.R. Stern, J. Wadsworth and F.A. Frey are thanked for providing samples. Comments by and fruitful discussions with N. Shimizu and S.R Hart greatly improved this manuscript and is greatly appreciated. This work was supported by grants EAR 8419832 to N. Shimizu and S.R. Hart, and EAR 8219853 to S.R. Hart from NSF.

REFERENCES

Amundsen H.E.F. (1987) 'The lithosphere beneath northwestern Spitsbergen: Evidence from upper mantle and lower crustal xenoliths.' M.Sc. Thesis, Univ. Oslo.
Anders E. and Ebihara M. (1982) 'Solar system abundances of the elements.' Geochim. Cosmochim. Acta. 46, 2363-2380.
Basaltic Volcanism Study Project (1981) Pergamon Press, 1286pp.
Blusztajn J. and Shimizu N. (1988) 'The trace element variations in clinopyroxenes from spinel peridotite xenoliths from SW Poland (abstr.)'. V.M. Goldschmidt Conf. Abstr., 30.

Bodinier J.L., Dupuy C. and Dostal J. (1987) 'Geochemistry and petrogenesis of Eastern Pyrenean peridotites.' Geochim. Cosmochim. Acta, submitted.

Bonatti E., Ottonello G. and Hamlyn P.R. (1986), 'Peridotites from the island of Zabargad (St. John), Red Sea: petrology and geochemistry.' J. Geophys. Res. 91-B1, 599- 631.

Burchfiel B.C. and Royden L. (1982) 'Carpathian foreland fold and thrust belt and its relation to the Pannonian and other basins.' AAPG Bull. 66, 1179-1195.

Clague D. A. and Frey F. A. (1982) 'Petrology and trace element geochemistry of the Honolulu volcanics, Oahu: Implications for the oceanic mantle below Hawaii.' J. Petrol. 23, 447- 504.

DePaolo D. J. and Johnson R. W. (1979) 'Magma genesis in the New Britain island arc: constraints from Nd and Sr isotopes and trace-element patterns.' Contrib. Mineral. Petrol. 70, 367- 380.

DePaolo D.J. and Wasserburg G.J. (1977) 'The sources of island arcs as indicated by Nd and Sr isotopic studies.' Geophys. Res. Lett. 4, 465- 468.

Dick H.J.B. and Fisher R.L. (1984) In: Kimberlites II: The mantle and crust- mantle relationships, Ed. J. Kornprobst, Elsevier, Amsterdam, 295- 308.

Downes H. and Dupuy C.(1987) 'Textural, isotopic and REE variations in spinel peridotite xenoliths, Massif Central, France.' Earth Plan. Sci. Lett 82, 121- 135.

Dupuy C., Dostal J., Marcelot G., Bougault H., Joron J. L. and Treuil M. (1982) 'Geochemistry of basalts from central and southern New Hebrides arc: implication for their source rock composition.' Earth Planet. Sci. Lett. 60, 207- 225.

Eggler D.H. (1987) 'Solubility of major and trace elements in mantle metasomatic fluids: Experimental constraints.' In: Mantle Metasomatism, ed. C.J. Hawkesworth and M.A. Menzies, Academic Press, 21- 41.

Ehrenberg S.N. (1982) 'Rare earth element geochemistry of garnet lherzolite and megacrystalline nodules from minette of the Colorado Plateau province.' Earth Planet. Sci. Lett. 57, 191- 210.

Frey F.A. (1970) 'Rare earth and potassium abundances in St. Paul's rocks.' Earth Planet. Sci. Lett. 7, 351-360.

Frey F.A., Suen J. and Stockman H.W. (1985) 'The Ronda high temperature peridotite: Geochemistry and petrogenesis.' Geochim. Cosmochim. Acta 49, 2469- 2491.

Fujimaki H. (1986) 'Fractional crystallization of the basaltic suite of Usu volcano, southwest Hokkaido, Japan, and its relationships with the associated felsic suite.' Lithos 19, 129- 140.

Fujimaki H., Tatsumoto M. and Aoki K. (1984) 'Partition coefficients of Hf, Zr and REE between phenocrysts and groundmasses.' J. Geophys. Res. 89-B, 662- 672.

Gerlach D.C. (1986) 'Geochemistry and petrology of recent volcanics of the Puyehue-Cordon Caulle area, Chile (40.5°).' Ph.D. dissertation, MIT.

Gill J.B. (1981) Orogenic andesites and plate tectonics, Springer, Berlin, 390 pp.

Green T.H. (1981) 'Experimental evidence for the role of accessory phases in magma genesis.' J. Volc. Geotherm. Res. 10, 405- 422.

Grove T.L. and Bryan W.B. (1983) 'Fractionation of pyroxene-phyric MORB at low pressure: An experimental study.' Contrib. Mineral. Petrol. 84, 293- 309.

Grutzeck M., Kridelbaugh S. and Weill D. (1974) 'The distribution of Sr and REE between diopside and silicate liquid.' Geophys. Res. Lett. 1, 273- 275.

Hawkesworth C. J., O'Nions R. K., Pankhurst R. J., Hamilton P. J. and Evensen N. M. (1977) 'A geochemical study of island arc and back arc tholeiites from the Scotia Sea.' Earth Planet. Sci. Lett. 36, 253- 262.

Hickey R. L., Frey F. A. and Gerlach D. C. (1986) 'Multiple sources for basaltic arc rocks from the southern volcanic zone of the Andes (34°- 41°S): trace element and isotopic evidence for contributions from subducted oceanic crust, mantle and continental crust.' J. Geophys. Res. 91-B6, 5963- 5983.

Huebner J.S., Lipin B.R. and Wiggins L.B. (1976) 'Partitioning of chromium between silicate crystals and melts.' Proc. Lunar Sci. Conf. 7th, 1195-1220.

Irving A.J. (1978) 'A review of experimental studies of crystal/liquid trace element partitioning.' Geochim. Cosmochim. Acta 42, 743- 770.

Irving A.J. and Frey F.A. (1984) 'Trace element abundances in megacrysts and their host basalts: Constraints on partition coefficients and megacryst genesis.' Geochim. Cosmochim. Acta. 48, 1201- 1221.

Jagoutz E., Lorenz V. and Wanke H. (1979) 'Major and trace elements of Al- augites and Cr- diopsides from ultramafic nodules in European alkali basalts,' In: The mantle sample, Inclusions in kimberlites and other volcanics, Ed. F.R. Boyd & H.A.O. Meyer, Am. Geophys. Union, Washington D.C., 382- 390.

Jagoutz E., Palme H., Baddenhausen H., Blum K., Cendales M., Dreibus G., Spettel B., Lorenz V. and Wanke H. (1979) 'The abundances of major, minor and trace elements in the earth's mantle as derived from primitive ultramafic nodules.' Proc. Lunar Planet. Sci. Conf. 10th, 2031- 2050.

Jeanloz R. (1987) 'Mineral physics and mantle evolution.' NATO Adv. Res. Workshop Abstr., 137- 138.

Johnson K.T.M., Shimizu N. and Dick H.J.B. (1987) 'Rare earth and other trace element concentrations in clinopyroxenes from abyssal peridotites.' EOS 68, 448.

Kato T., Irifune T. and Ringwood A.E. (1987) 'Majorite partition behavior and petrogenesis of the earth's upper mantle.' Geophys. Res. Lett. 14, 546- 549.

Kay R.W. (1980) 'Volcanic arc magmas: implications of a melting- mixing model for element recycling in the crust- upper mantle system.' J. Geol. 88, 497- 522.

le Roex A. P., Dick H. J. B., Reid A. M., Frey F. A., Erlank A. J. and Hart S. R. (1985) 'Petrology and geochemistry of basalts from the American-Antarctic Ridge, Southern Ocean: implications for the westward influence of the Bouvet mantle plume.' Contrib. Mineral. Petrol. 90, 367- 380.

Leeman, W.P., Budahn J.R., Gerlach D.C., Smith D.R. and Powell B.N. (1980) 'Origin of Hawaiian tholeiites: trace element constraints.' Am.J. Sci. 280-A, 794-819.

Liotard J.M., Barsczus H.G., Dupuy C. and Dostal J. (1986) 'Geochemistry and origin of basaltic lavas from Marquesas Archipelago, French Polynesia.' Contrib. Mineral. Petrol. 92, 260-268.

McCallum I.S. and Charette M.P. (1978) 'Zr and Nb partition coefficients: implications for the genesis of mare basalts, KREEP, and sea floor basalts.' Geochim. Cosmochim. Acta 42, 859- 869.

Melson W.G., Hart S.R. and Thompson G. (1972) 'St Paul's rocks, equatorial Atlantic: petrogenesis, radiometric ages, and implications on sea-floor spreading.' Geol. Soc. Am. Mem. 132, 241- 272.

Menzies M.A., Rogers N., Tindle A. and Hawkesworth C.J. (1987) 'Metasomatic and enrichment processes in the lithospheric peridotites, an effect of asthenosphere-lithosphere interaction.' In: Mantle metasomatism, Ed. C.J. Hawkesworth and M.A. Menzies, Academic Press, London, 313- 361.

Morris J.D. (1984) 'Enriched geochemical signatures in Aleutian and Indonesian arc lavas: an isotopic and trace element investigation.' Ph.D. dissertation, MIT.

Morris J.D. and Hart S.R. (1983) 'Isotopic and incompatible element constraints on the genesis of island arc volcanics from Cold Bay and Amak Island, Aleutians and implications for mantle structures.' Geochim. Cosmochim. Acta 47, 2015- 2030.

Nicholls I.A. and Ringwood A.E. (1973) 'Effect of water on olivine stability in tholeiites and the production of silica- saturated magmas in the island arc environment.' J. Geol. 81, 285- 300.

Nixon P.H., Rogers N.W., Gibson I.L. and Grey A. (1981) 'Depleted and fertile mantle xenoliths from South African kimberlites.' Ann. Rev. Earth Planet. Sci. 9, 285- 309.

Ottonello G. (1980) 'Rare earth abundances and distribution in some spinel peridotite xenoliths from Assab (Ethiopia).' Geochim. Cosmochim. Acta 44, 1885- 1901.

Pearce J.A. and Cann J.R. (1973) 'Tectonic setting of basic volcanic rocks determined using trace element analyses.' Earth Planet. Sci. Lett. 19, 290- 300.

Pearce J.A. and Norry M.J. (1979) 'Petrogenetic implications of Ti, Zr, Y and Nb variations in volcanic rocks.' Contrib. Mineral. Petrol. 69, 33-47.

Price R. C., Kennedy A. K., Riggs-Sneeringer M. and Frey F. A. (1986) 'Geochemistry of basalts from the Indian Ocean triple junction: implications for the generation and evolution of Indian Ocean ridge basalts.' Earth Planet. Sci. Lett. 78, 379- 396.

Prinzhofer A. and Allegre C.J. (1985) 'Residual peridotites and the mechanism of partial melting.' Earth Planet. Sci. Lett. 74, 251- 265.

Ringwood A.E. (1988) 'Constitution of and evolution of the Earth's mantle.' Ingerson distinguished lecture of the International Association of Geochemistry and Cosmochemistry.

Roden M.F., Frey F.A. and Francis D.M. (1984) 'An example of consequent mantle metasomatism in peridotite inclusion from Nunivak, Alaska.' J. Petrol. 25, 546- 577.

Roden M.K., Hart S.R., Frey F.A. and Melson W.G. (1984) 'Sr, Nd and Pb isotopic and REE geochemistry of St. Paul's rocks: the metamorphic and metasomatic development of an alkali basalt mantle source.' Contrib. Mineral. Petrol. 85, 376- 390.

Royden L., Horvath F. and Rumpler J. (1983) 'Evolution of the Pannonian basin system, 1.' Tectonics 2, 63-90.

Ryerson F.J. and Watson E.B. (1988) 'Rutile saturation in magmas: implications for Ti-Nb-Ta depletions in island-arc magmas.' Earth Planet. Sci Lett. 86, 225, 239.

Salters V. J. M. and Shimizu N. (1988) 'World-wide occurrence of HFSE depleted mantle.' Geochim. Cosmochim. Acta., in press.

Salters V.J.M. (1987) 'Characteristics of the mantle beneath the Carpathians, Hungary, as inferred from calc-alkaline volcanics,' NATO Adv. Res. Workshop Abstr., 120- 121.

Salters V.J.M. and Shimizu N. (1988) 'HFSE depletions in peridotites, local variations and possible origin (abstr.).' V.M. Goldschmidt Conference Abstr. 71.

Salters V.J.M., Hart S.R. and Panto G. (1988) 'Origin of late Cenozoic volcanic rocks from the Carpathian Arc, Hungary.' In L. Royden and F. Horvath, Pannonian Basin: a study in basin evolution, Chapter 18, AAPG and Hungarian Geol. Soc. (in press).

Shibata T. and Thompson G. (1986) 'Peridotites from the Mid-Atlantic Ridge at 43° N and their petrogenetic relation to abyssal tholeiites.' Contrib. Mineral. Petrol. 93, 144-159.

Shimizu N. & Allegre C.J. (1978) 'Geochemistry of transition elements in garnet lherzolite nodules in kimberlites.' Contrib. Mineral. Petrol. 67, 41- 50.

Shimizu N. (1987) 'Trace element abundance patterns of pyroxenes in spinel lherzolite nodules from Salt Lake Crater, Oahu.' EOS 68, 447.

Shimizu N. and Hart S.R. (1974) 'Rare earth element concentrations in clinopyroxenes from an ocean ridge lherzolite.' Carnegie Inst. Wash. Yearb., 73, 964 -967.

Shimizu N. and Hart S.R. (1982) 'Applications of the ion microprobe to geochemistry and cosmochemistry.' Ann. Rev. Earth Planet. Sci. 10, 483- 526.

Stern A.J. (1981) 'A common mantle source for Western Pacific arc and "hot-spot" magmas- Implications for mantle structure.' Carnegie Inst. Yearb. 81, 455- 462.

Stern C.R., Saul S., Futa K. and Skewes M.A. (1986) 'Nature and evolution of the subcontinental mantle lithosphere below southern South America and implications for Andean genesis.' Revista Geologica de Chile 27, 41- 53.

Stosch H.-G. and Seck S.A. (1980) 'Geochemistry and mineralogy of two spinel peridotite suites from Dreiser Weiher,' West Germany. Geochim. Cosmochim. Acta 44, 457- 470.

Sun S.-S. (1982) 'Chemical composition and origin of the Earth's primitive mantle.' Geochim. Cosmochim. Acta 46, 179-192.

Tera F., Brown L., Morris J., Sacks I.S., Klein J. and Middleton R. (1986) 'Sediment incorporation in island-arc magmas: Inferences from [10]Be.' Geochim. Cosmochim. Acta 50, 535-550.

Thompson R.N., Morrison M.A., Dickin A.P. and Hendry G.L. (1983) 'Continental flood basalts...Arachnids rule OK?' In: Continental basalts and mantle xenoliths, Shiva, Ed. C.J. Hawkesworth and M.J. Norry, 158-185.

Tormey D.R., Grove T.L. and Bryan W.B. (1987) 'Experimental petrology of normal MORB near the Kane Fracture Zone: 22°-25°N, mid-Atlantic ridge.' Contrib. Mineral. Petrol. 96, 121- 139.

Watson E.B., Othman D.B., Luck J.M. and Hofmann A.W. (1987) 'Partitioning of U, Pb, Cs, Yb, Hf, Re and Os between chromian diopsidic pyroxene and haplobasaltic liquid.' Chem. Geol. 62, 191- 208.

Wood D. A., Tarney J., Varet J., Saunders A. D., Bougault H., Joron J. L., Treuil M. and Cann J. R. (1979) 'Geochemistry of basalts drilled in the North Atlantic by IPOD Leg 49: implications for mantle heterogeneity.' Earth Planet. Sci. Lett. 42, 77- 97.

Wood D.A. (1978) 'Major and trace element variations in Tertiary lavas of Eastern Iceland and their significance with respect to the Iceland geochemical anomaly.' J. Petrol. 19, 393- 436.

Wood D.A. (1979) 'A variable veined suboceanic upper mantle-Genetic significance for mid-ocean ridge basalts from geochemical evidence.' Geology 7, 499- 503.

VOLATILES AND STABLE ISOTOPES IN RECYCLING

M. JAVOY, F. PINEAU and P. AGRINIER
Université Paris VII and Institut de Physique du Globe de Paris,
Laboratoire de Géochimie des Isotopes Stables
2, place Jussieu - 75251 Paris Cedex 05
France

The use of stable isotopes is pertinent to the study of recycling in several ways, among which the fact that oxygen is the main constituent of terrestrial materials, and the fact also that volatiles, which can be both outgassed from the mantle and recycled, are traced by three more stable isotope ratios, namely those of carbon, nitrogen and hydrogen. In addition to stable isotope ratios, some chemical ratios in the volatiles can be used to link their fluxes if they are not too much affected by the processes of dissolution and outgassing. We shall review in the next pages, first some of the isotope aspects, then some chemical characteristics of the fluids involved in outgassing of the mantle, and of those possibly recycled.

OXYGEN ISOTOPES

The isotopic composition of fresh mantle magmas, either from mid-ocean ridges or from hot spots is remarquably little variable (1, 2, 3) with $\partial^{18}O$ of 5.5 + 0.2 ‰. The same is true for the bulk of peridotites, from massive orogenic massifs as well as from nodules in basalts, with the same range of variations (4, 5). A very small proportion of peridotites in both types display different values, up to 8.5 ‰ (4, 5). These values had been tentatively explained by Kyser et al (5) to represent the result of complex fractionation curves of mantle minerals at high pressures but these curves do not obey the basic thermodynamic rules outlined for example by Bottinga & Javoy (6), and the deviations from the restricted mantle range can be explained by alteration models such as that of Gregory & Taylor (7). It is then remarquable that the major constituent of the mantle shows so little variability whereas virtually all the other isotopic tracers display conspicuous differences among mantle components.

Probably the main reason for this resides in the way in which the original mantle composition is first altered at ridges. As shown first by Javoy (1970), (8) in the Pindos ophiolite, the hydrothermal interaction of sea water enriches the first 2 or 3 km of the crust in ^{18}O, then, because of the increase in temperature with depth, depletes the underlying rocks on a similar thickness. Gregory & Taylor (9), have postulated that this enrichment and depletion exactly balance each other, so that the mid ocean ridge interaction acts as a buffer for sea water composition, with a mantle-ocean

S. R. Hart and L. Gülen (eds.), Crust/Mantle Recycling at Convergence Zones, 121–138.
© 1989 by Kluwer Academic Publishers.

fractionation of about 6 ‰. This fact is supported by the study of various ophiolite massifs (8, 9, 10, fig. 1), at least qualitatively. This change in sign of the isotopic effect, impossible for radiogenic isotopes, can lead to a steady state of the ^{18}O flux from mantle to ocean and is probably the major reason for explaining the surprising constancy of mantle $\partial^{18}O$. However deviations from this general scheme may appear, due to the spreading rate which imposes the thermal regime at ridge axes. For slow spreading ridges such as that represented by the ophiolites from the Tsang Po suture zone (11) there is a net ^{18}O enrichment of the crust, which means in turn a net depletion of the oceans. If fast spreading ridges are exactly "^{18}O balanced", as postulated by Gregory & Taylor, then the possible changes in the $\partial^{18}O$ of the oceans are governed by the amount of slow spreading ridges. Hence, at present, ^{18}O changes could be due to the activity of the mid Atlantic ridge.

If we take into account the ^{18}O uptake by sediments, we can write the evolution of the ocean's isotope composition as :

$$\partial_w = \{(\partial_{wo}-(\Delta - C))\} \ \exp(-t/8 \ 10^7) - (\Delta - C)^*,$$

where δ_{wo} is the starting composition fo the water and Δ is the mean isotope fractionation oceanic crust-ocean water. $\Delta = 7x + 5.7(1-x)$ where x is the contribution of the slow spreading ridge to the total spreading. The half life of this variation is similar to that found by Gregory & Taylor and is governed in particular by the ratio of the mantle oxygen output to the ridge to the flux of hydrothermal water. It is about 10^8 years, that is, comparable to the life time of a spreading cycle. The maximum range of the isotope variation of the ocean depends also of the sediment "flux" and is about 3‰. This corresponds to a maximum variation of the $\partial^{18}O$ of the upper mantle of only 0.0026.

The very characteristic imprint displayed in fig. 1 is a striking marker of origin, as had been stressed by JAVOY (1970, 1971) (8, 12), who had shown the identity of distribution of $\partial^{18}O$ for eclogites and ophiolites. This path has now been followed down to more than 30 kbars by AGRINIER et al (13) in the coesite-magnesite eclogites of NORWAY. The same origin is now postulated for the variations shown in eclogite nodules from kimberlites. Although not so evident and charactristic, the scarce occurence of excentric $\partial^{18}O$ in massive HT peridotites (5), unexplained by mantle mechanisms, could represent the last souvenir of crustal life in these materials before complete homogeneisation by the combination of mechanical deformations and chemical diffusion.

---

* $C = (\Phi+1)/\Phi \ ^* \Delta + (r'/r)$ where Φ is the water/rock ratio during mid ocean ridge hydrothermal activity ($\Phi \approx 1.5 \pm 0.5$). Δ' is the average sediment water fractionation ($\Delta \approx 20$). r and r' arc the yearly oxygen fluxes, rom mantle when forming the new oceanic crust and of sediment into the mantle. If the present state of the ocean represents the steady state situation ($\partial_w \approx - 0.5$) then $r'/r \approx 6 \ 10^{-3}$ and $r' \approx 2.5 \ 10^{-4}$ g/year. This means that approximately 1/10 of the sediments are subducted.

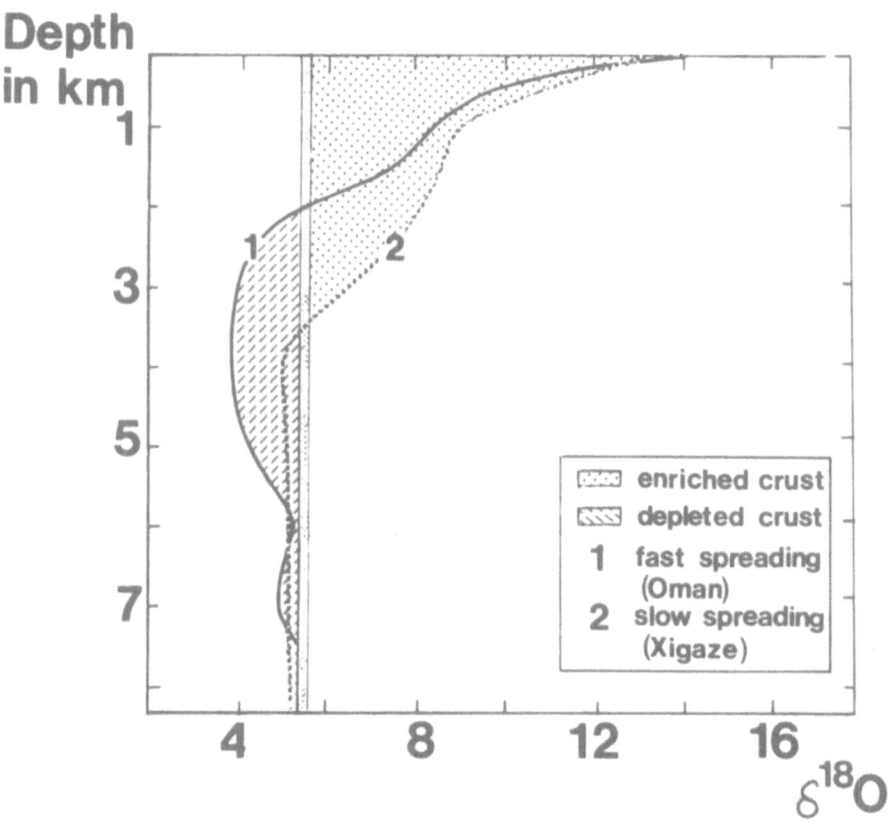

FIGURE 1 : ^{18}O profile in oceanic crust.
A : enrichment of crust
B : depletion of crust
1 : fast spreading (Oman)
2 : slow spreading (Xigazé)

CARBON ISOTOPES

Since the discovery of "carbonatitic carbon" in MORB tholeiites by PINEAU et al (1976) (14), several studies have been conducted on these materials, especially in the last four years. All have shown the presence of heavy carbon (-9 < $\partial^{13}C$ < -3), extracted by crushing or by heating above 700°C, contained in CO_2 vesicles, and of light carbon (-30 < $\partial^{13}C$ < -20), extracted, as first shown by PINEAU et al, at low temperature (below 700°C) (14, 15, 16, 17). All the authors agree on the fact that the heavy carbon has the isotopic composition of the bulk of the carbon brought out of the mantle by basaltic magmas and outgassed to various degreees from these magmas. There is also an agreement on the fact that the CO_2 in equilibrium with these magmas is richer in ^{13}C than the dissolved carbon.

There are however serious discrepancies on the question of how far this fractionation can drive down the $\partial^{13}C$ of the residual dissolved carbon and on the significance of the low T-low $\partial^{13}C$ fraction, wether it represents indigenous carbon or, because of this low T-low $\partial^{13}C$ character, a mere organic contamination. Here we just want to point out that : i) carbon stable in the melt is not stable in the corresponding glass, hence there is no reason why it should be extracted at high temperature. ii) cooling of a silicate system generally produces a strong decrease in oxygen fugacity which in turn favours the formation of various reduced carbon species, among which organic molecules. The presence of complex carbon compounds in cooled silicate systems has been evidenced by MATHEZ (18). The isotopic relationships between coexisting carbon species may permit to give indications on the initial carbon content of a melt (14, 15). However, if there are doubts about the exact nature of some carbon species, chemical ratios can help to relate different volatile fluxes as we shall see below.

There is presently no conclusive evidence that hot spot magmas have isotopic characteristics different from those of MORB (19, 20). For subduction zones however the case is eventually different since some authors claim that they have recorded in back arc basins basalts (BAB), a distinc signature (17), around -12 %, which could be due to contamination by subducted organic material. This "typical" signature however is obtained by discarding about half of the results, corresponding to both low concentrations of carbon and low $\partial^{13}C$ but extracted at high temperatures, hence no attributable to the so called low T organic contamination. when all the results are taken into account, they fit well a model of ^{13}C depletion due to outgassing.

Contrary to BAB, the CO_2 from aerial volcanoes in both island arcs and continental margins (19, 21, 22), record a majority of high $\partial^{13}C$, up to -2,5 ‰. This can be explained by contamination with crustal or sedimentary carbonates. However there is absolutely no correlation with other isotopic tracers which record indeed such a contamination like strontium isotopes (20). At the opposite there seems to be a smooth correlation with the thickness of continental crust which does not accord with the idea of crustal contamination, which would be more dominated by the variability of the encountered crustal compositions. We propose an alternative model of variable outgassing due to the variable depth of the deep magma chambers.

In this model, we estimate the saturation depth from the highest $\partial^{13}C$ to be $\approx$ 50 km. Using as a conservative estimate the solubility data of Fine and Stolper (23) we

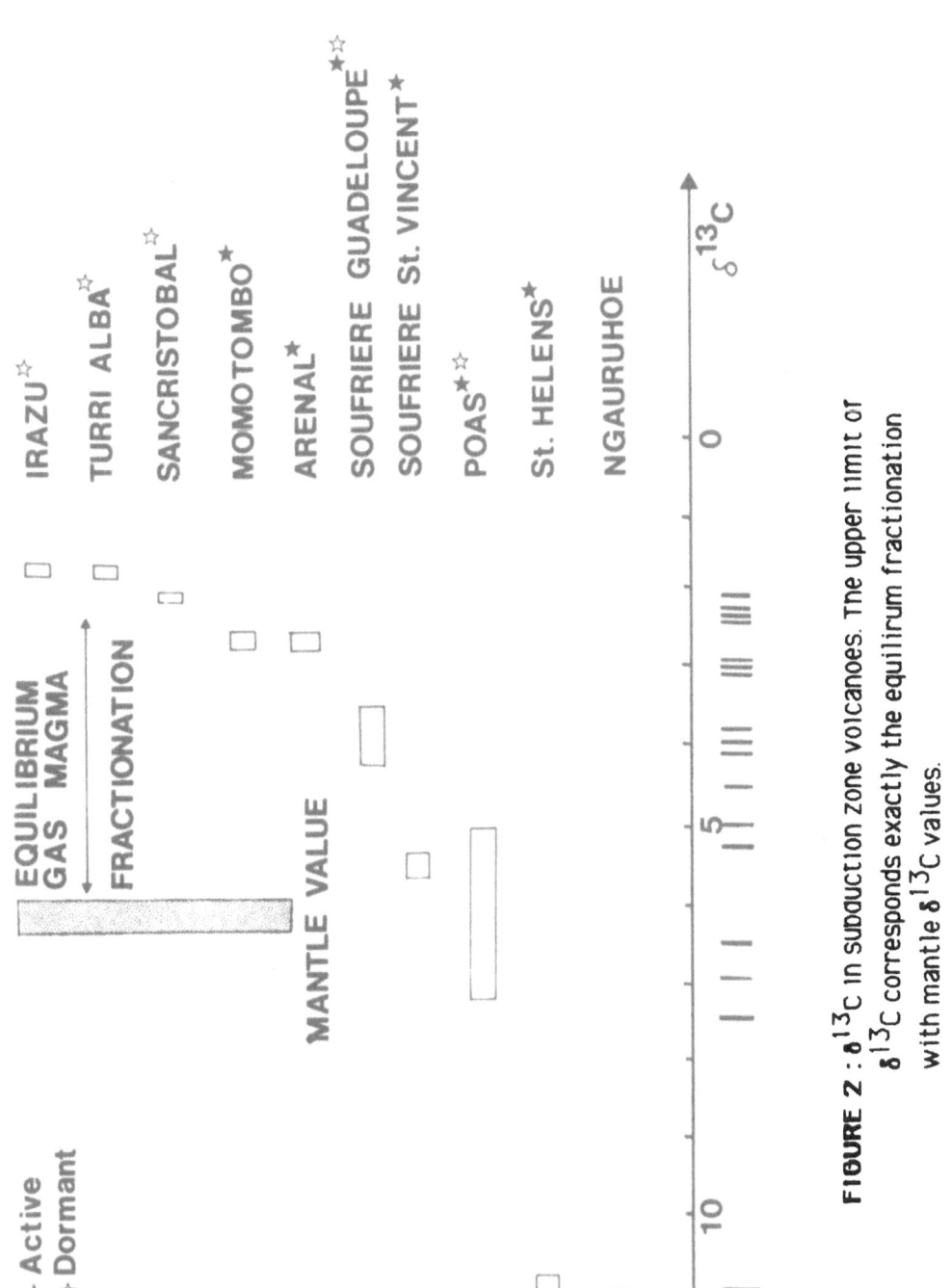

FIGURE 2 : $\delta^{13}C$ in subduction zone volcanoes. The upper limit of $\delta^{13}C$ corresponds exactly the equilirum fractionation with mantle $\delta^{13}C$ values.

can write :

$$C_{carbon\ ppm} = 265\ p_{kbar}^{1.04}$$

(at a temperature of around 1000°C, justified by the fractionation factor found below). then the $\partial^{13}C$-depth (=thickness of crust) relationship can be deduced from :

$\partial - \partial o = \Delta\ Lnf$ (Rayleigh distillation of the magma chamber)

$f = C/Co = (p/po)^{1.04}$

where Co is the initial concentration and po the saturation depth.

Then : $\partial - \partial o = 1.04\ \Delta\ Lnp/po$

$$= -1.04\ \Delta\ Lnpo + 1.04\ \Delta Lnp$$

From the linear regression in fig.

$1.04\ \Delta = 6.95$ or $\Delta = 6.7$

(which corresponds (19) to T $\approx$ 1000°C) and $\partial o = -0.94$. this value is the $\partial^{13}C$ of CO_2 in equilibrium with a non outgassed magma chamber. The ∂o of the chamber uself

is $-0.94 - \Delta = -7.6$, very similar to that observed in MORB and hot spots.

This is also coherent with the fact that in granulite facies rocks (Pineau et al, 1981), we find carbonate witnesses of this outgassing with $\partial^{13}C$'s corresponding to the range implied by the above equation. The mafic rocks in the granulite facies contain up to 1600 ppm carbon extracted only upon fusion, coherent with a 50 % outgassing at 20 km depth of a magma intially contaming 3300 ppm, the initial concentration derived from the $\partial^{13}C$ depth correlation.

Hence, for carbon, it does not seem that contamination by continental crust plays a significant role. However, the initial $\partial^{13}C$ obtained corresponds both to normal MORB and to the balance of sedimentary carbon. The concentration of 3 300 ppm significantly higher than the initial concentration of MORB (see below) could be due to a sizeable enrichment of the mantle wedge by fluids of sedimentary origin. This does not preclude the existence of contaminations conspicuously demonstrated by other tracers but means that these contaminations are selective. This is also in accordance with the fact that it is difficult to decompose carbonates in a cold subducting plate submitted to higher and higher pressures.

This last feature is well demonstrated by the existence of very significant amounts of carbonates in the coesite-magnesite eclogites of Norway, with isotopic characteristics similar to those of ophiolites (13) and aged DSDP samples of oceanic crust (24), that is a mixture of mantle carbon and marine carbonate (fig. 3).

Less has been done up to now on peridotites, in particular because of their high melting temperatures. Recent results point out to a very wide range of $\partial^{13}C$ even at high extraction temepratures ($-32 < \partial^{13}C < -2$). These variations occur in virtually any environment, from subduction related material to nodules in hot spot magmas (25, 26, 27).

The study of carbonatites, kimberlites and diamonds is a subject in itself and these materials have often serve as references for other mantle carbon samples because their generally accepted mantle origin and their large carbon concentrations made them "uncontaminable" mantle specimens. We just want to point out here that most of their values are in the range $-9 < \partial^{13}C < -2$ with a mean value very close to -7, with again the lowest and most extreme values shown by the reduced species, that is the deamonds, but these extreme values correspond to a very small (< 1%) proportion. Wether these extreme values correspond to reinjected heterogeneities or to extreme

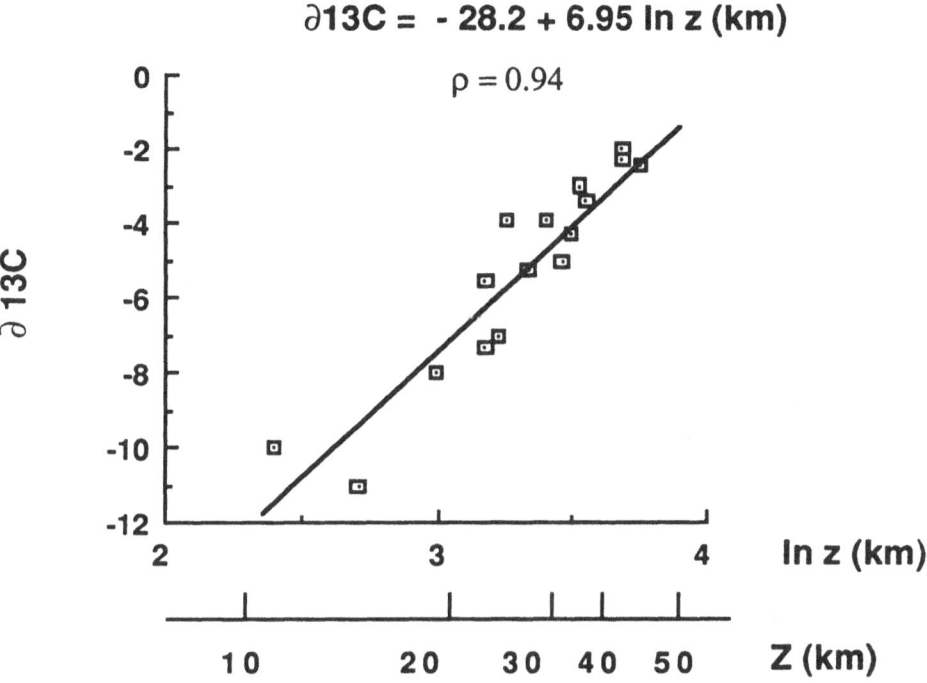

FIGURE 3 : Correlation of $\delta^{13}C$ with continental crust thickness in subduction zone volcanoes (After Allard).

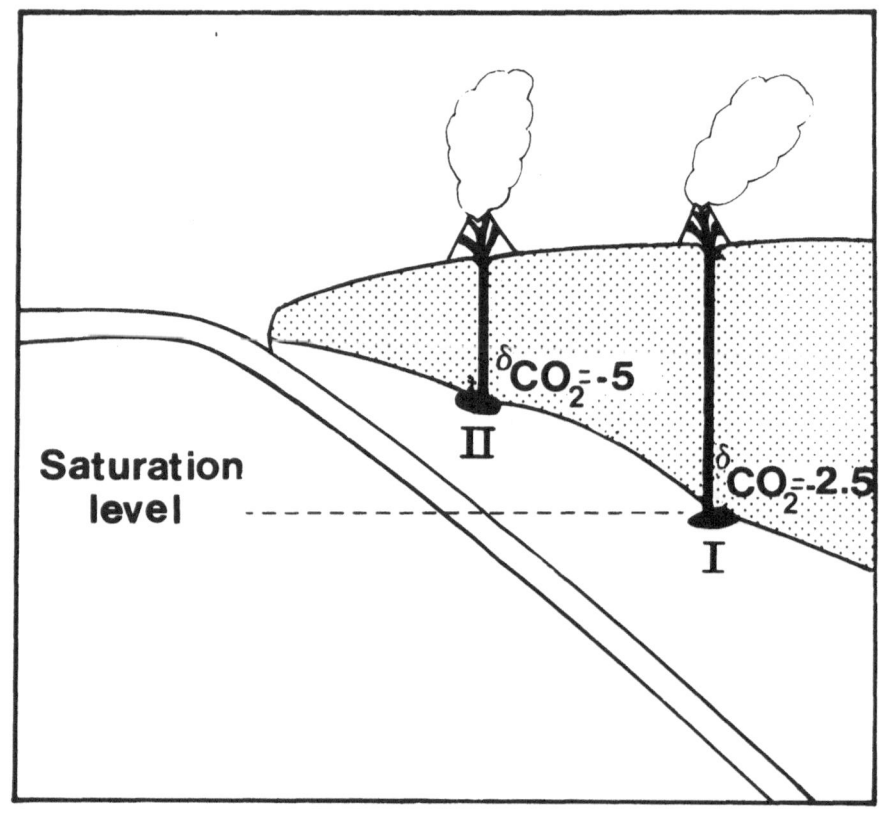

FIGURE 4 : Sketch of an outgassing model explaining the correlation $\delta^{13}C$-thickness of crust in subduction zones.

differenciation processes is still open to question (20).

Finally the sediments show the whole range of values between the lowest $\partial^{13}C$ of the organic products (-30 ‰) up to the highest values of marine carbonates (+3%).

The mean isotopic composition lies in the range -12 to -6 % as recognized long ago by CRAIG (28), very close to -7 ‰ in most cases, which, in the new light of recycling models, means that the recycled carbon has the same isotopic composition as the carbon outgassed by the mid ocean ridges, a situation very similar, isotopically, to the one described above for oxygen, albeit for somewhat different reasons.

In conclusion to the isotopic aspect of carbon recycling, we can say that the odds are in favour of a steady state for the isotopic composition of mantle carbon. A point central to that question is the amount and isotopic composition of carbon outgassed above subdudction zones. We have now, as said above, a large amount of isotopic data but we lack the precise evaluation of the fluxes which would allow us to make an acceptable balance. We shall come back to that question when speaking of volatile fluxes.

NITROGEN ISOTOPES

Until recently very little was known about nitrogen isotopes in the mantle. The situation has improved with a few studies on diamonds, nitrogen in basalts, and nitrogen in volcanic gases, mainly from subduction zones (29, 30, 31, 32).

JAVOY et al (29, 30) have found essentially only negative to very negative values (relative to the atmosphere ($\partial^{15}N$ down to -12‰) in a suite of very large diamonds from the MBUJI MAYI district in ZAIRE, which produces a very sizable proportion of the world's diamonds (12 millions caracts). From this they concluded that mantle nitrogen was similar to that of enstatite chondrites and, for that reason, predicted the possible existence of mantle $\partial^{15}N$ down to -40 ‰. The Open University group shortly after reported such values (33) but were apparently unable to reproduce them.

This group has found until then values similar to those of JAVOY et al, together with some positive $\partial^{15}N$ in coated diamonds whose coating has positive $\partial^{15}N$ whereas the core has negative values. They have also found $\partial^{15}N$ from -6 to +14 in various basalt glasses (MORB & OIB), with the highest values in hot spot samples. JAVOY et al. (19) have found negative (down to -8‰) in volcanic gases from Central America and Iceland, as well as a value of -2,5 ‰ in vesicles from a MORB popping rock at 15°N in the Atlantic.

Hence, unlike the case of oxygen or carbon, part (or the majority ?) of mantle nitrogen has $\partial^{15}N$ outside the range of surficial nitrogen, whose $\partial^{15}N$'s are essentially all positive. Again, due to the fact that nitrogen fractionates in a way opposite to CO_2 relative to possible dissolved species, the positive values can be explained as residual from outgassing or as organic contamination, either experimental or genuine. Among other features the interest of the enstatite chondrite hypothesis is that nitrogen is very stable in these materials even when heating under vacuum (34) and shall be even more stabilized under mantle pressures raising the possibility that most of terrestrial nitrogen is still in the mantle.

Finally the negative $\partial^{15}N$ found in several of the subduction zone volcanoes studied to

130

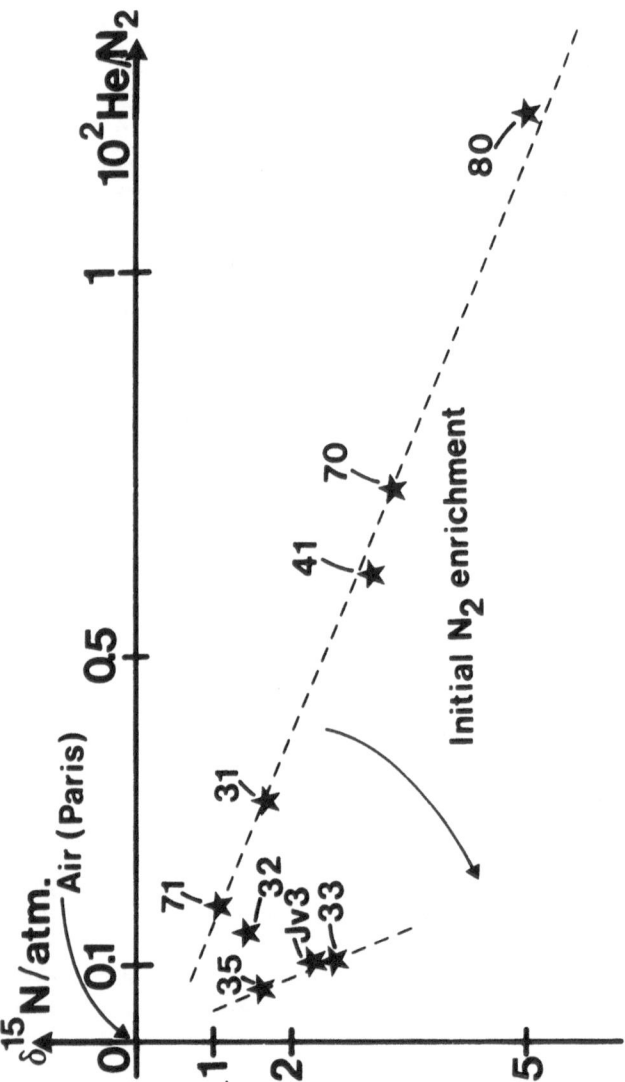

FIGURE 5 : Correlation $\delta^{15}N$-He/N_2 in subduction zone volcanoes.
The distinctly negative $\delta^{15}N$ preclude significant
contribution of sediments ($\delta^{15}N > 0$) except from
particular locations (e.g. Momotombo).

date is a strong argument against a sedimentary origin for most of the volatiles.

HYDROGEN ISOTOPES

Due to the very large isotope fractionations of this element event at high temperature, the range of variation observed even in mantle products is large. Tentatively it can be assumed that the most characteristic mantle range would be -80 -10 (35,36) whereas hydrothermal circulations would impose ∂D in the range -50 to -40. However low temperature circulations in the aging oceanic crust can further extend this range down to -100 -120 ‰. Hence, although a recent study by CRAIG et al. (37) in the vicinity of Iceland can suggest a mixing pattern between two different sources, the general situation is that there is more or less complete overlapping between fresh basaltic glasses from MORB or hot spots and the various kinds of possible subducted materials in the range -40 to -80 ‰. Water from sub aerial subduction zone volcanoes has a very distinct local meteoric water signature.

CHEMICAL RATIOS, SOLUBILITIES AND FLUXES

The chemical and isotopic analysis of vesicles from submarine basalts is one of the best ways to arrive at the characteristics of the original mantle fluids, since the glasses have likely suffered a large and variable degree of outgassing. On the other hand it is useful (but difficult) to analyze the concentration of the various elements in glasses free of vesicles of any size to determine the solubilities of the various gases, which are the necessary parameters to reconstruct the ougassing patterns. Very seldom care has been taken to discuss the significance of the actual gas content of a basaltic glass and very frequently raw measurements are taken at face value to represent the original content. It is obviously not so. Vesicle free glasses likely represent a much too low estimate. Taking vesicles into account will give a better estimate if they have not traveled a long way through the glass since their formation, which may be achieved if they have formed rapidly (that is, with a large degree of supersaturation and close to the surface). Even if it is so this will be a sensible estimate only if one takes care of the fact that a large proportion of them may now be empty. Then the original concentration is obtained through the chemical analysis of well preserved vesicles, a measurement of the vesicularity and of the hydrostatic pressure.

We have analysed vesicles both in MORB and hot spot glasses by crushing under vacuum and analysing the resulting gases in a twenty four collector CAMECA mass spectrometer (15, 19, 39, 40). Typical analyses are as follow :

MORB Gases : 80 to 94 % CO_2, 20 TO 6 % H_2O, 0.1 % of non condensible gases of which 95 % N_2, 1 to 2.5 % He, Ar in proportions relative to He from 1/1.5 to 1/100 and subordinate amounts of H_2, CO and CH_4.

HOT SPOT GASES : 30 to 50 % CO_2, 70 to 50 % H_2O, 0.1 to 0.2 % incondensible gases of which 50 to 85 % H_2, 50 to 15 % N_2, 0.1 to 1 % He and He/Ar ratios close to 1.5.

The most striking differences between the two gases are the H_2O and H_2 contents, which are both linked to the water content of the glasses, typically o.2 to 0.5 % for the MORB, 1.3 to 1.5 % for the hot spot gases.

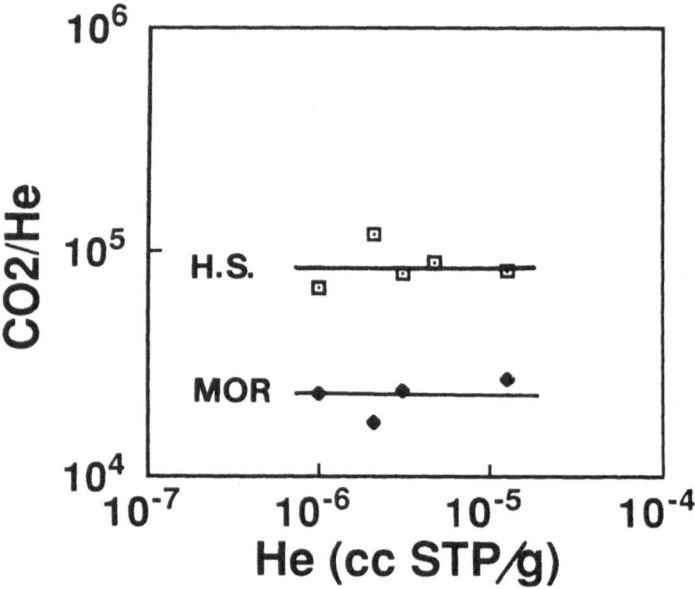

FIGURE 6 : CO_2/He and He/Ar ratios in submarine basalts as a function of gas content.

The water vapour pressure is essentially in equilibrium with the dissolved water according to a BURNHAM (41) type equation : $\%H_2O = 0.1 \sqrt{PH_2O}$

From the vesicles' free carbon content the solubility of carbon is about 200 ppm/KBar.

The CO_2/N_2 ratio in MORB gas is around 750 for the glasses richest in gas and has not been measured accurately for the others. For hot spot glasses this ratio varies from 650 to 2550.

In MORB vesicles the CO_2/He ratio is essentially independent of the gas content (mean 25000), which we attribute to roughly equal solubilities, whereas He/Ar ratios vary by almost a factor 100 (fig. 3).

In hot spot gases the CO_2/He ratio is very different (mean 82000) but rather constant for the same solubility reasons. We don't have the $^3He/^4He$ ratio on these gases but if they have typical hot spot values then one may infer that the $CO_2/^3He$ ratios are the same to better than 20 % in both MORB and hot spot gases.

From the above values one can calculate the original carbon content in a glass of a given vesicularity by the formula : $Cppm = 52.4 * P_{meters} * \% \text{ vesicles} * x_{CO2}$ in gas $/T°K$. For the same vesicularity, the carbon contents of the hot spot glasses shall be about half that of the MORB content at the same depth. For example a popping rock of 20 % vesicularity at 3 500 m contained originally about 2 900 ppm C whereas a glass from Tahiti with the same vesicularity at 2600 meters contained only 1200 ppm. The first popping rocks which we analyzed in 1976 with 2 % vesicles contained 185 ppm in the vesicles, close to the amount really recovered.

From the characteristics of the fluids outlined above, it is striking that the main diffeence between hot spots and MORB from the C-He point of view is the dilution of ^3He in ^4He in MORB. There is a significant difference, however, in the water contents, which may point out to the recycling of large amounts of water down to the boundary layer. Loihi glasses give the same type of concentrations but the interest of Tahiti samples is to demonstrate that these concentrations are in equilibrium with the fluids outgassed during the formation of the vesicles, hence are truly magmatic.

The chemical ratios above allow the calculation of mantles fluxes from the knowledge of one of them. since ^3He flux is actually considered to be the best known, we can derive from it the carbon flux : $1100 \times 25000 \times 86000 = 2.36 \ 10^{12}$ moles/year. Then the nitrogen flux is about $3.15 \ 10^9$ moles/year. However if we consider the last estimates of ALLEGRE et al (42) for the ^3He content of the upper mantle ($1.8 \ 10^{-10}$ cc/g or $1.2 \ 10^{-13}$ ples) we have reasons to think that this flux is underestimated. If $4.5 \ 10^{16}g$ of new basaltic crust are created/year with a degree of partial melting of 10 % they will carry out $1.8 \ 10^{-9} \times 4.5 \ 10^{16}/22400 = 3620$ moles.

Then the carbon flux shall be $3620 \times 25000 \times 86000 = 7.8 \ 10^{12}$ moles and the nitrogen flux about 10^{10} moles. The nitrogen flux is very small relative to the mass of the present atmosphere. However this is not the case for the carbon flux, which is of the same order as that calculated by JAVOY et al (1982) (43), and can build the present supergene reservoir of $6.7 \ 10^{21}$ moles in $8.6 \ 10^8$ years. Then recycling of carbon has to occur to a large scale.

This problem of carbon recycling has been brought out recently in a different and

134

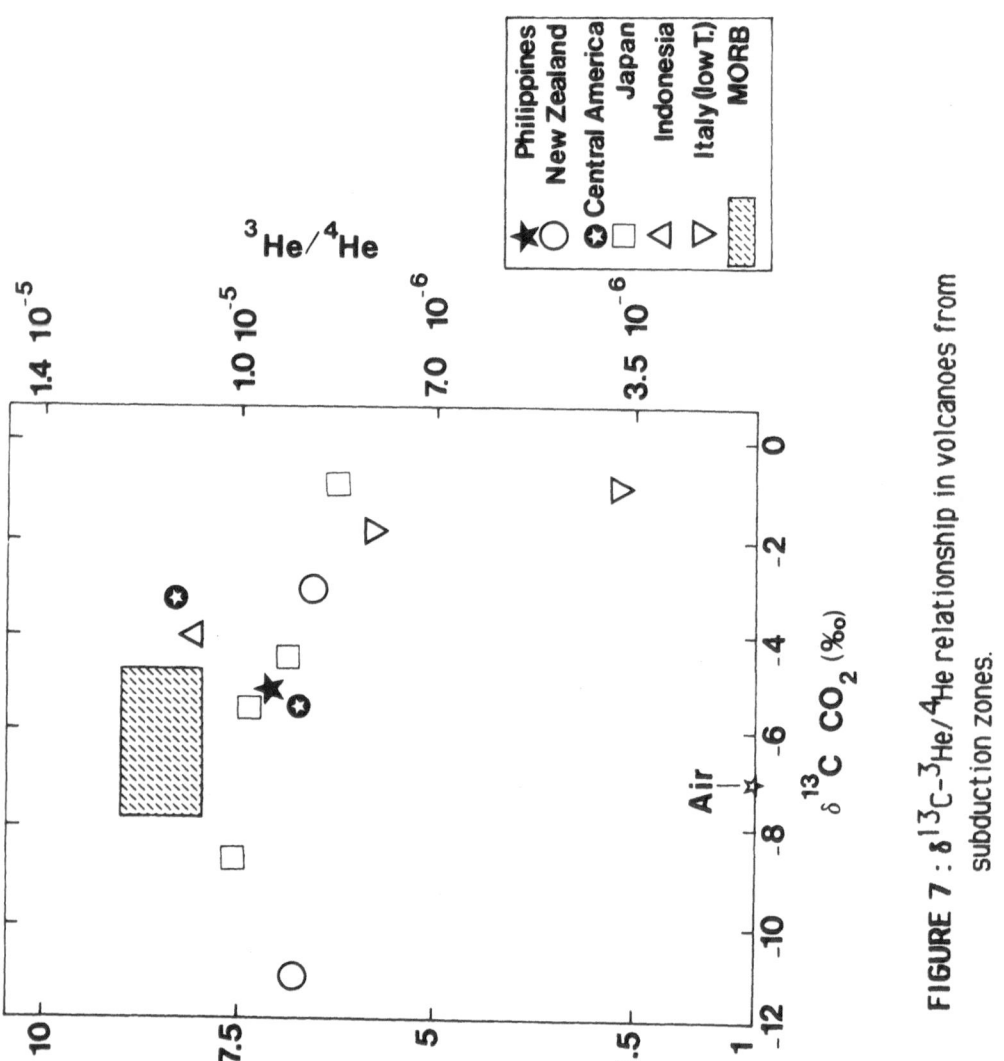

FIGURE 7 : $\delta^{13}C - {}^3He/{}^4He$ relationship in volcanoes from subduction zones.

interesting way by MARTY & JAMBON (44). Considering like us that the $C/^3He$ ratio in mantle products is aroung 10^9 (2×10^9 from our own precise estimate) and similar to that found in a number of chondrite classes; although with more uncertainty, they compare it with the $C/^3He$ ratio which would have been built in a closed atmosphere. they come up with a ratio of 10^7 ($4.24 \ 10^7$ if we take the ^3He number of ALLEGRE et al. From what they conclude that carbon has been much less degassed than He because reducing conditions prevailed in the mantle so that C behaved like a compatible element under graphite or diamond form, at least in the past when large helium outgassing occured. This would fit very nicely the idea of a very reduced primary material like the enstatite chondrites (20). Now, on the contrary, we see that carbon and helium are outgassed with a $C/^3He$ ratio fo $2 \ 10^9$ and the authors attribute that fact to the building of more oxidizing conditions. However there are serious logical flaws in that model :

There is no objection to the idea that $C/^3He$ should be similar in chondrites and in pristine mantle materials although these ratios cannot probably be very well defined in the former, due to the rather large uncertainty on the amount of ^3He formed by spallation (in certain cases the "planetary" component calculated from exposure ages correspond to negative concentrations). It can help to calculate the carbon content of the earth from the rare gas data.

However if carbon has not outgassed while ^3He has outgassed to more than 99 % from the upper levels of the mantle, then the present $C/^3He$ in the outgassed mantle is very different from the initial one and the fact that it is similar to that of chondrites is purely fortuitous. From the value of this ratio given by MARTY & JAMBON and the concentrations of ALLEGRE et al (42) for rare gases we calculate a value of 10^{10}. Then there is a fractionation of about 5 between C and ^3He during magma formation. The corresponding carbon concentration in the mantle is 1170 ppm.

Then the only possible cases are as follows :

i) carbon was never a compatible element so that no fractionation ever occurred during outgassing and the present supergene ratio of $4 \ 10^7$ is the proof that strong recycling occurred throughout the earth history (mean of $6 \ 10^{14}$ g/year).

ii) carbon behaved really like a compatible element during most of the outgassing history so that its present mantle concentration is close to 1200 ppm. Since, as claimed reasonably by MARTY & JAMBON the redox conditions now correspond to the incompatibility of carbon (no CO_2-He fractionation), then, with an average degree of partial melting of 10 %, the average initial concentration of carbon in MORB magmas is about 1.2 %, which implies that the present stock of supergene carbon can be built in about $2 \ 10^8$ years.

These conclusions are presented voluntarily in a provocative way but the starting point of MARTY & JAMBON is indeed a very serious one and their estimate of the supergene $C/^3He$ cannot seriously be questioned to more than 50 %.

This brings about the problem of carbon recycling, whose existence can be hindered by the presence of the so called "volatile subduction barrier". There is really a problem but we have shown that from an isotopic point of view this carbon is very similar to that of MORB and HOT SPOTS with the possibility of contribution of sedimentary carbon of the same average isotopic composition to a level of about 50 % (3300 ppm C in subduction zones magmas instead of 2200 ppm in MORB). If we consider the $C/^3He$ ratio, we measure values of 2.5 to 7 10^{10}, similar to those reported in (44). This is more than a factor 10 higher than the MORB ratio but this does not result from a strong radiogenic Helium component since the ^{3}He/^{4}He ratios are not lowered to less than 50 % of the MORB value (Fig. 7). Hence the volatile cannot originate from a single source unless it is different from all the known sources. This peculiar source can very well be the mantle wedge depleted in He. As an alternative model we can also propose the aged oceanic crust which as the right C/He and ∂^{13}C values. In order to build the correct ^{3}He/^{4}He ratio we have mix MORB AND RADIOGENIC HE IN PROPORTIONS FROM 8,:2 TO 5:5

REFERENCES

1 - MUEHLENBACHS K. and CLAYTON R.N., *Can. J. Earth Sci.,* **69**, 172-184 (1972).

2 - PINEAU F., JAVOY M., HAWKINS J.W. and CRAIG H., *Earth Planet. Sci. Lett.,* **28**, 299-307 (1976).

3 - KYSER T.K., O'NEIL J.R. and CARMICHAEL I.S.E., *Contri. Min. Petrol.,* **81**, 88-102 (1982).

4 - KYSER T.K., O'NEIL J.R. and CARMICHAEL I.S.E., *Contri. Min. Petrol.,* **93**, 120-123 (1986).

5 - JAVOY M., *Orogenic Mafic Ultramafic Associations Int. Coll. C.N.R.S. Grenoble,* 279-287 (1980).

6 - BOTTINGA Y. and JAVOY M., *Earth Planet. Sci. Lett.,* **20**, 250-265 (1973).

7 - GREGORY R.T. and TAYLOR H.P., *Jr Contr. Min. Petrol.,* **93**, 114-119, (1986).

8 - JAVOY M.,' Oxygene Isotopes in Magmatology', *Thesis Paris,* (1970).

9 - GREGORY R.T. and TAYLOR H.T., *Jr. Contrib. Min. Petrol.,* **86,** 2737-2755, (1981).

10 - COCKER J.D., GRIFFIN B.J. and MUEHLENBACHS K., *Earth Planet. Sci. Lett.,* **61**, 112-122 (1982).

11 - AGRINIER P., JAVOY M. and GIRARDEAU J., *Submitted to Chem. Geol.,* (1987).

12 - JAVOY M., *C.R. Acad. Sci. Paris,* **273,** 2414-2417 (1971).

13 - AGRINIER P., JAVOY M., SMITH D.C. and PINEAU F., *Isotope Geoscience,* **25**, 145-162, (1985).

14 - PINEAU F., BOTTINGA Y. and JAVOY M., *Earth Plan. Sci. Lett.,* **29**, 413-421 (1976).

15 - PINEAU F. and JAVOY M., *Earth Planet. Sci. Lett.,* **62**, 239-257 (1983).

16 - DES MARAIS D.J. and MOORE J.G., *Earth Plan. Sci. Lett.,* **69**, 43-57 (1984).

17 - MATTEY D.P., CARR R.H., WRIGHT I.P. and PILLINGER C.T.,*Earth Planet. Sci. Lett.,* **70**, 196-206 (1984).

18 - MATHEZ E.A., *Submitted to Geochim. and Cosmochim. Acta,* (1987).

19 - JAVOY M., PINEAU F. and DELORME H., *Chem. Geol.,* **57**, 41-62 (1986).

20 - EXLEY R.A., MATTEY D.P., CLAGUE D.A. and PILLINGER C.T., *Earth Planet. Sci. Lett.,* **78,** 189-199 (1986).

21 - ALLARD P., *Thesis Paris,* (1986).

22 - DELORME H., JAVOY M., CHEMINEE J-L. and PINEAU F., *Terra Cognita Spec. Issue,* **79,** (1983).

23 - FINE G. and STOLPER E., *Contrib. Mineral Petrol.,* **91**, 105-121 (1985).

24 - JAVOY M. and FOUILLAC A-M. , *D.S.D.P., Init. Rep. Legs 51, 52, 53/2,* 1153-1157 (1979).

25- NADEAU S., PINEAU F., JAVOY M. and FRANCIS D., *E.O.S. - A.G.U. Spring Metting,* (1987).

26 - MATTEY D.P., MENZIES M.A. and PILLINGER C.T., *Terra Cognita,* **5**, 147 (1986).

27 - PINEAU F., JAVOY M. and KORNPROBST J-J., *Petrol.* in press (1987).

28 - CRAIG H., *Nuclear geology in geothermal areas 53-92 Spo leto* (1963).

29 - JAVOY M., PINEAU F. and DEMAIFFE D., *Earth Planet. Sci. Lett.,* **68**, 399-412 (1984).

30 - JAVOY M. and PINEAU F., *Meteoritics,* 320-321 (1983).

138

31 -   JAVOY M. and PINEAU F., *T erra Cognita* **6,** (2), 103 (1983).

32 -   EXLEY R.A., BOYD S.R., MATTEY D.P. and PILLINGER C.T., *Earth Planet. Sci. Lett.,* **81**, 163-174, (1987).

33 -   BOYD S.R., CARR L.P., SEAL M., MILLEDGE H.J., MENDELSSOHN M., WOOD P.J., MATTEY D.P. and PILLINGER C.T., *Terra Cognita,* **5,** (2-3), 146 (1985).

34 -   THIEMENS M.H. and CLAYTON R.N., *Earth Planet. Sci. Lett.,* **62**, 165-168 (1983).

35 -   CRAIG H. and LUPTON J.E., *Earth Planet. Sci. Lett.,* **31**, 369-385 (1976).

36 -   KYSER T.K. and O'NEIL J.R., *Geochim. Cosmochim. Acta,* **48**, 2123-2133, (1984).

37-   POREDA R., SCHILLING J.G. and CRAIG H., *Earth Planet. Sci. Lett.,* **78**, 1-17 (1986).

38 -   PINEAU F. , in préparation.

39 -   PINEAU F. and JAVOY M., *Terra Cognita,* ***6 (2)***, 191, (1986).

40 -   JAVOY M., PINEAU F., and PALENZUELA P., *Terra Cognita E.U.G. meeting* (1987).

41 -   HAMILTON D.C., WAYNE BURNHAM C. and OSBORN E.F., *J. PetroL,* **5(1),** 21-39, (1964).

42 -   ALLEGRE C.J., STAUDACHER T. and SARDA P., *Earth Planet. sci. Lett.* **81,** 127-150 (1987).

43 -   JAVOY M., PINEAU F. and ALLEGRE C.J., *Nature,* **300**, 171-173 (1982).

44 -   MARTY B. and JAMBON A., *Earth Planet. Sci. Lett. in press* (1987).

MAGMA GENERATION IN SUBDUCTION ZONES

I. SELWYN SACKS, HIROKI SATO, and JULIE MORRIS
Carnegie Inst. of Washington, Dept. Terrestrial Magnetism,
5241 Broad Branch Rd., N.W.,
Washington, DC 20015

ABSTRACT. There are two general classes of models for magma generation in subduction zones; melting of the subducted slab, or magma
derivation from the asthenosphere wedge overlying the subducted
slab. Lack of knowledge of the thermal structure of island arc regions has hindered our understanding. By comparing laboratory results with seismic and heat flow observations in island arc regions
we determine that temperatures are generally below the solidus.

Recent results on the anelasticity and velocity of a spinel
lherzolite at pressures and temperatures appropriate to upper mantle conditions (Sato et al., 1987) enable this determination, Figure 1. They found that at temperatures above about 80% of the
solidus, the seismic energy absorption process is dominantly thermally activated, that is measurements made at different temperatures and pressures have a single trend when plotted as a function
of homologous temperature. Figure 2 shows these results. That
the mechanical behaviour of peridotite under upper mantle conditions is dominated by the grain boundary is suggested by the low
bulk modulus of the activation volume of the attenuation process,
(2-8 as against more than 100 GPa for a crystal) as well as by the
rather small decrease in solidus temperature at increased pressure
as found by Takahashi (1986). It is of particular importance that
Q (low Q indicates high absorption) decreases well below solidus
temperatures (due to grain boundary softening) and also that the
solidus crossing is not particularly marked. Our concept that observed low Q in the asthenosphere can only be due to partial melt
does not seem to be justified. Note that since lab measurements
are made at far higher frequencies than those used in earth studies, it is necessary to determine a calibration factor. Observed
heat flow may be used to calculate the temperature at the base of
the lithosphere (e.g. Chapman and Pollack, 1977) and this temperature compared with that derived from the lab results and the observed upper asthenosphere Q. The Japan sea lithosphere thickness
was determined to be 40 km by Evans et al. (1978). From the observed heat flow of 100 mW/m^2, the temperature at the base of the
lithosphere, (or top of the asthenosphere) is calculated to be
1100°C. The Q_p in this region was determined to be 90 by Umino and
Hasegawa (1984), from which may we derive a temperature of 1040°C.
The difference between these two temperatures (60°C) is the cor-

S. R. Hart and L. Gülen (eds.), Crust/Mantle Recycling at Convergence Zones, 139–144.
© 1989 by Kluwer Academic Publishers.

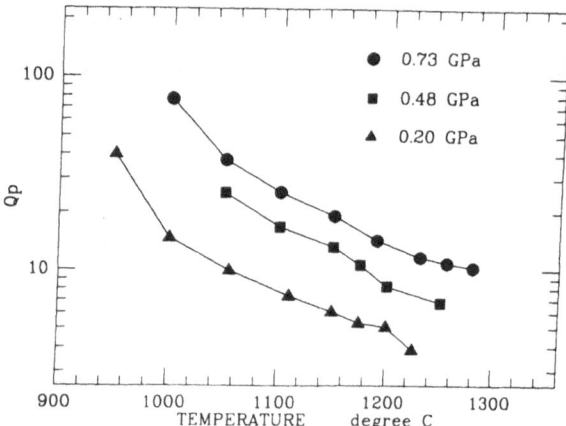

Figure 1. Attenuation of compressional waves transmitted through a peridotite sample in a Yoder-type gas-media, high-pressure apparatus. (High Q indicates low attenuation). Q_p decreases rapidly with increasing temperature even at subsolidus conditions. Q_p increases with pressure.

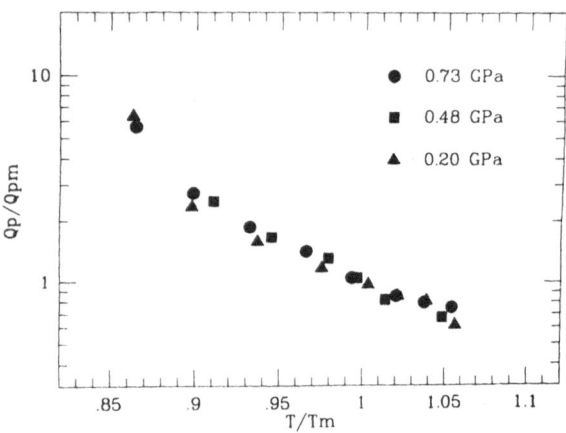

Figure 2. Q_p normalized by Q_{pm} (Q at the solidus) as a function of normalized temperature, T/T_m (T_m is the solidus). The data at the three different pressures have a single trend, indicating a dominant relationship between the melting temperature and the anelastic properties of peridotite. The close overlap of the data allows us to extrapolate to higher pressures with some confidence.

rection which is applied to all Q-derived temperatures in this arc region.

In the sub-volcano asthenosphere wedge beneath Japan, Q_p values (Sacks, 1975, Hasegawa et al., 1979) of about 200 at depths in the 80-100 km range are too high for hyper solidus material. The results from two independent Q determinations are shown in Figure 3. While direct path lengths in the asthenosphere wedge trenchward of the volcanic front are 50 km or less, the use of broad band seismic data (up to 10 Hz frequency response) and carefully chosen ray paths enable reliable anelasticity determinations in this region. The Q structure reveals a low Q asthenosphere, with values similar to those found in sub-oceanic asthenosphere. However, trenchward of the volcanic front the Q is significantly higher than that expected from normal asthenosphere.

This region is also characterized by low heat flow, 40 mW/m^2 which is about half that of Japan west of the volcanos. Assuming that the lithosphere thickness is about the same on each side of the volcanic front, the heat flow suggests that the temperature at the top of the asthenosphere east (trenchward) of the volcanos is about half of that to the west. Near its trenchward apex, the asthenosphere wedge is cooled by the overlying (Japan) lithosphere as well as from below by the upper surface of the cold subducting plate. The circulation of hot asthenosphere material is inhibited in this region but is helped by the shear heating which is most intense here because of the constriction. (Hsui and Toksoz, 1979 have modelled the thermal structure for various amounts of shear heating.) Low heat flow is therefore expected, but the Q values, though higher than those west of the volcanic front, are not as high as would be expected from such a drastic decrease in temperature. We emphasize that the lab measurements were performed on a dry peridotite and that the calibration procedure corrects both for frequency dependance as mentioned earlier, and for water (or other volatiles) content. This is because Q values are proportional to the ratio of the actual temperature to the solidus, and the solidus is affected (reduced) by volatiles. Our interpretation of the intermediate Q values near the apex of the asthenosphere wedge in conjunction with the low heat flow, is that the material there must be water saturated.

Beneath the volcanos, and to the west, the Q values are consistent with temperatures somewhat below solidus, about 0.9 T_m. Note that the spacial resolution of the Q determinations is inadequate to detect magma chambers, diapirs and the like. However, magma is generated in this region (Kushiro, 1987). Fluxing of the asthenosphere with water released from the sediments would trigger the melting. Water is released from the subducted sediments continuously from the shallowest depths with a probable peak at a depth of about 100 km where the amphibole breaks down. Tatsumi (1986) has described the various hydrous phases.

The combination of the asthenosphere convection forced by the subduction, and the cooling in the wedge apex causes a region (intermediate Q) near the apex where temperatures are too low to al-

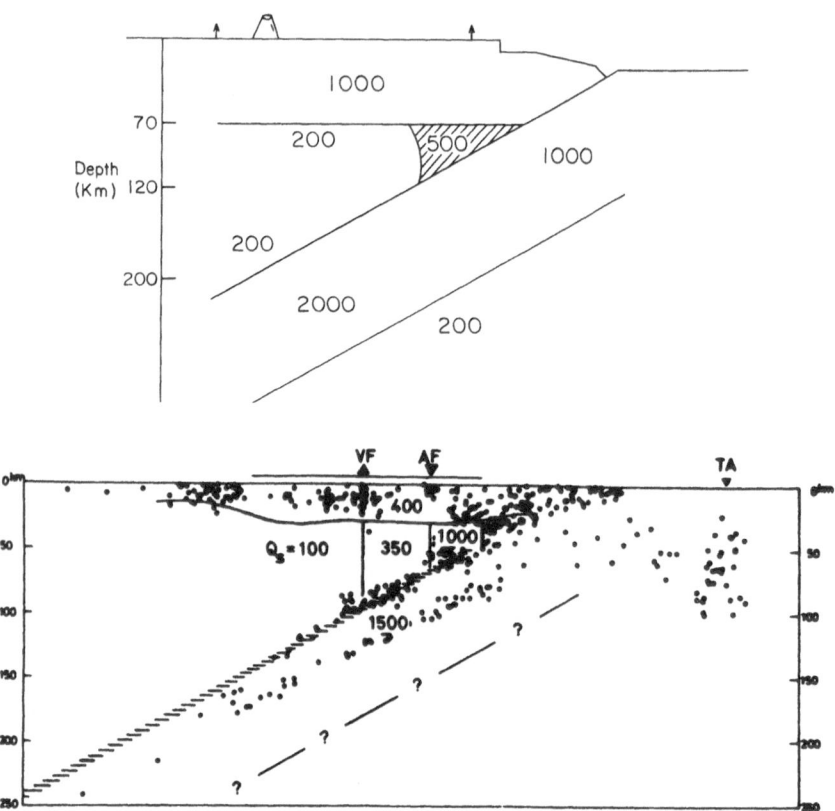

Figure 3.   Anelasticity structure from two independant studies of
the asthenosphere in the Japan region.   The upper figure shows
the results from Hasegawa et al. (1979) for Q_s.   Dots indicate
earthquakes, VF is the volcanic front, AF is the aseismic front.
Many ray paths from earthquakes on the subducted plate to numerous (>30) seismographs on land were used in the inversion.

The lower figure shows results from Sacks (1975) using data from
two broad frequency range seismographs.   Ray paths from subducted plate earthquakes over a depth range of 50-200 km were observed at seismographs indicated by arrows.   The wide spacing
of the stations used does not allow resolution of the Q_p = 200-
500 boundary.   Note Q_p shown on this figure must be divided by 2
(approx.) to be compared with the Q_s shown in the upper figure.

Below and westward of the volcanic front, Q values are low, and of
fairly common asthenosphere values.   Trenchward of the front, however, Q_p's of 500-700 are significantly higher than normal asthenosphere values.   This region is also characterized by low heat flow.

low melt even under water saturated conditions. Further from
the trench, temperatures are sufficiently high (though below dry
solidus) that addition of water will flux melting. The model is
illustrated in Figure 5.

One conclusion from this model is that melting of the slab or
even of the sediments is not required for magma generation. This
is consistent with the observations of [10]Be from a third rank volcano, Oshima Oshima, in the Japan sea (Tera et al., 1986). If the
uppermost part of the slab were significantly melted by the time
it reached the position of the first rank volcano, Esan, on the Pacific coast of Hokkaido, it is unlikely that any [10]Be-bearing sediments would survive further mantle penetration of more than 100 km.
The thermal modelling by Hsui and Toksöz, 1979, suggests fairly low
temperatures at the surface of the slab. Geochemical constraints
are required to determine if the sediment or oceanic crust is significantly melted in any other subduction zone.

A further conclusion is that the sediment and crust surviving
down to below the magma-genetic zone (250 km or so), may be released into the upper mantle away from the immediate vicinity of
the island arc region.

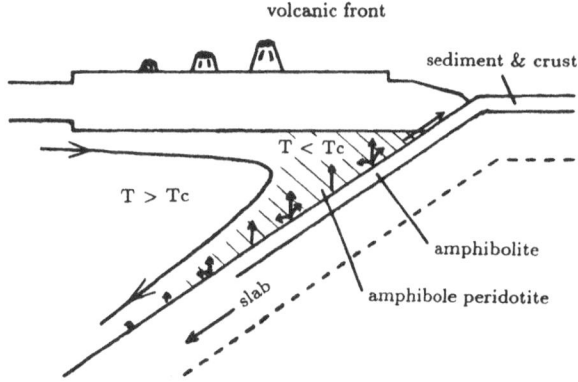

Figure 4. Model of magma generation. Water (arrows) is released
from the sediments (and eventually also from the amphibole and
later phlogopite breakdown), large quantities near the trench decreasing with depth. Where asthenosphere temperatures are below
the wet solidus, no magma can be generated and the apex of the asthenosphere wedge is water saturated. Where the temperature is
above the wet solidus, water from the slab will flux melting even
though the material may be below the dry solidus. Certainly in
Japan, and probably in many arcs, existence of [10]Be in a back arc
volcano and the Q structure, suggests that the slab and even the
sediments may not melt till greater depths.

References

Chapman, D.S., and H.N. Pollack, 'Regional geotherms and lithospheric thickness', *Geology*, **5**, 265-268, 1977.

Evans, J.R., K. Suyehiro, and I.S. Sacks, 'Mantle structure beneath the Japan sea - a re-examination', *Geophys. Res. Lett.*, **5**, 487-490, 1978.

Hasegawa, A., N. Umino, A. Takagi, and Z. Suzuki, 'Double-planed deep seismic zone and anomalous structure in the upper mantle beneath northeastern Honshu (Japan)', *Tectonophys.*, **57**, 1-6, 1979.

Hsui, A.T., and M.N. Toksöz, 'The evolution of thermal structures beneath a subduction zone', *Tectonophys.*, **60**, 43-60, 1979.

Kushiro, I., 'A petrological model of the mantle wedge and lower crust in the Japanese island arcs', in *Magmatic Processes: Physiochemical Principles*, (B.O. Mysen, Ed.), Geochem. Soc., Special Publ. No. 1, pp. 165-181, 1987.

Sacks, I.S., 'Anomalous island arc asthenosphere and continental growth', *Carnegie Inst. Wash. Year Book*, **74**, 256-266, 1975.

Sato, H., I.S. Sacks, T. Murase, G. Muncill, and H. Fukuyama, 'Qp - melting temperature relation on high pressure-temperature anelastic properties of peridotite: Attenuation mechanism and implications for the mechanical properties of the upper mantle', *J. Geophys. Res.*, submitted for publication, 1987.

Takahashi, E., 'Melting of a dry peridotite KLB-1 up to 14 GPa: Implications on the origin of peridotitic upper mantle', *J. Geophys. Res.*, **91**, 9367-9382, 1986.

Tatsumi, Y., 'Formation of the volcanic front in subduction zones', *Geophys. Res. Lett.*, **13**, 717-720, 1986.

Tera, F., L. Brown, J. Morris, I.S. Sacks, J. Klein, and R. Middleton, 'Sediment incorporation in island arc magmas: Inferences from [10]Be', *Geochim. Cosmochim. Acta*, **50**, 535-550, 1986.

Umino, N., and A. Hasegawa, 'Three-dimensional Qs structure in the northeastern Japan Arc', *Zishin*, **37**, 217-228, 1984.

RECYCLED CONTINENTAL CRUSTAL COMPONENTS IN ALEUTIAN ARC MAGMAS:
IMPLICATIONS FOR CRUSTAL GROWTH AND MANTLE HETEROGENEITY

Kay, R.W.[1,2] and S. Mahlburg Kay[1]
[1]INSTOC
Snee Hall
Cornell University
Ithaca, NY   14853
[2]Department of Geological Sciences
Snee Hall
Cornell University
Ithaca, NY   14853

ABSTRACT.   Mantle-derived magmatic rocks of the Aleutian island arc
are isotopically and geochemically distinct from their counterparts in
the ocean basins immediately to the south and north.   Aleutian magmas
contain crustal components from subducted oceanic sediment, subducted
hydrothermally altered mid-oceanic ridge basalt (MORB), and oceanic
crust that is imbedded within the Aleutian arc crust.   Olivine
tholeiite is the commonest magma type that rises from the mantle to
the crust-mantle boundary, but it is observed as a lava only
infrequently, due to crustal assimilation and fractionation.   Aleutian
olivine tholeiites originate by partial melting of upwelling
peridotite rendered buoyant by addition of excess (XS) elements from
the subducting plate.   At present, operation of the subduction process
(S-process) at arcs like the Aleutians is the main mechanism for both
crustal growth and for recycling continental crustal material into the
mantle is a second recycling mechanism.   Mantle heterogeneities (as
sampled by OIB melts) were largely created in the past by the same
recycling   processes.   Lower   crustal   delamination   accompanying
accretion of arc terranes at continental margins.

1.   INTRODUCTION

After a quarter of a century of debate, the idea that continental
crust is recycled at convergent plate margins, once thought physically
unreasonable, has gained respectability.   Coats (1962) was perhaps the
first to depict the process, now called sediment subduction at a
convergent plate margin, as he envisioned it to operate in the
Aleutian Island arc.   In the intervening years, the Aleutians have
figured prominently in the recycling debate.   This progress report
focuses on the Aleutian case and recognizes several pivotal statements
that are, in order of appearance, subject to increasing difficulty of
experimental and observational verification.   The statements, and some
alternatives to them, all obey the constraints of mass balance,
chemical equilibrium, and physics (in particular fluid dynamics).

S. R. Hart and L. Gülen (eds.), Crust/Mantle Recycling at Convergence Zones, 145–161.
© 1989 by Kluwer Academic Publishers.

2. MANTLE-DERIVED MAGMATIC ROCKS OF THE ALEUTIAN ISLAND ARC ARE ISOTOPICALLY AND GEOCHEMICALLY DISTINCT FROM THEIR COUNTERPARTS IN THE OCEAN BASINS IMMEDIATELY TO THE SOUTH AND NORTH

2.1 Excess (XS) Components: Trace Elements, Isotopes

This statement involves a comparison of mass tracers that leads to the identification of "excess" (XS) components in Aleutian arc magmas. The statement had better be true, or the observational basis for recycling is absent. Arbitrarily distinguished are two tectonic levels: crustal and subcrustal (see Figure 1). The XS components can

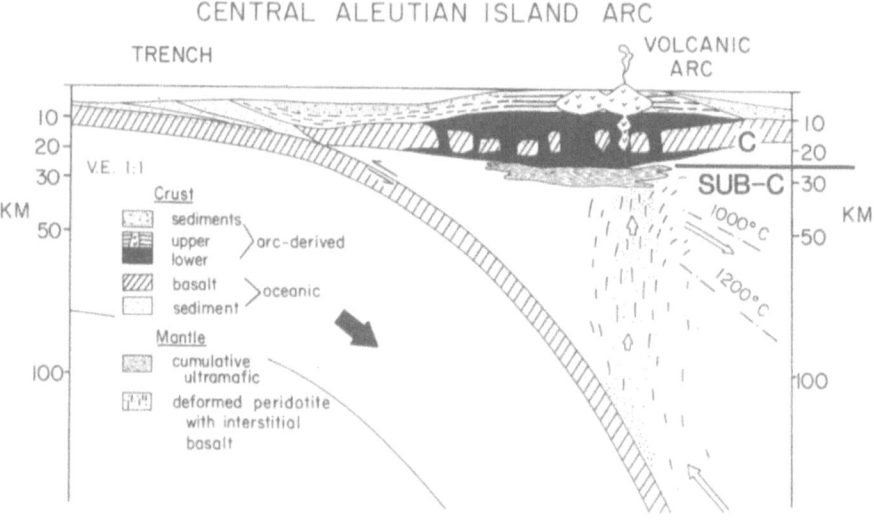

Figure 1. Cross-section (no vertical exaggeration) showing crustal formation process in the central Aleutian arc (modified after Kay and Kay 1985, 1988; and Kay et al. 1986). Schematic geotherms refer to temperatures obtained on ultramafic tectonized xenoliths (DeBari et al. 1987). Melt segregates at shallow mantle depth; crystal fractionation of ultramafic minerals create a magmatic Moho. Further magmatic fractionation and assimtation forms a basic lower crust and intermediate composition upper crust.

be added at either level or at both. In the present context, shallow-level, intracrustal recycling ("contamination") is important only because it obscures observations and confounds interpretations of the deeper recycling process. Intracrustal recycling should be recognizable as progressive geochemical change accompanying crustal-level magmatic differentiation.

To test models for subcrustal (subducting plate level) recycling, ideally one must consider only mantle-derived magmas. In both oceanic and island arc settings, these mantle-derived magmas, in equilibrium with peridotite at mantle depths, are olivine normative basalts (e.g. Nye and Reid, 1986; Gust and Perfit, 1987). Lacking Mg-rich olivine basalts in many arc volcanoes, as in the case of the Aleutians where the predominant magma type is hi-Al basalt or andesite that are not at equilibrium with either peridotite or eclogite at mantle depths (e.g. Kay et al., 1982), comparisons with oceanic rocks must be made by using element ratios and isotope ratios whose fractionation-independence has been determined from analyses of members of Aleutian lava series.

Lavas spanning a continuum of magmatic fractionation trends, the extremes of which are clearly distinguished as calc-alkaline and tholeiitic, occur in volcanoes from the main volcanic front in the Aleutian arc. In the central parts of the island arc, calc-alkaline fractionation trends are found in volcanic rocks from volcanoes within coherent segments in the arc (Figure 2), while volcanic rocks with

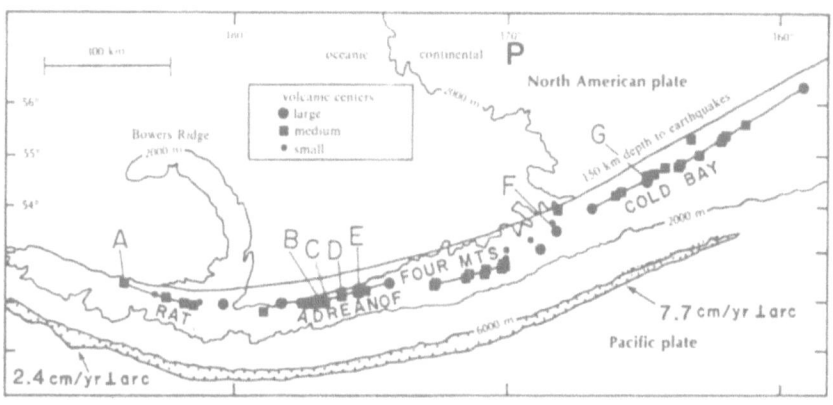

Figure 2. Map showing the four coherent volcanic segments of the insular Aleutian arc (after Kay et al. 1982) and the locations of key volcanic centers. The calc-alkaline volcanoes are Buldir (A), Moffett (B), and Adagdak (C) on Adak Island, Great Sitkin (D), and Kasatochi (E), and the tholeiitic volcanoes are Okmok on Umak Island (F), and Westdahl (G) on Unimak Island. The Pribilof islands (P) are the site of recent intraplate volcanism.

tholeiitic trends occur in volcanoes at the ends of segments or where segments cannot be easily defined. There are substantial differences in the trace element signatures between the evolved rocks of the two series (Kay, 1977; Kay et al., 1982).

148

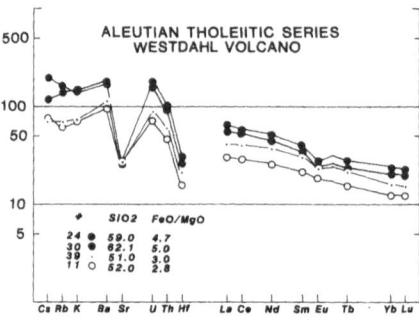 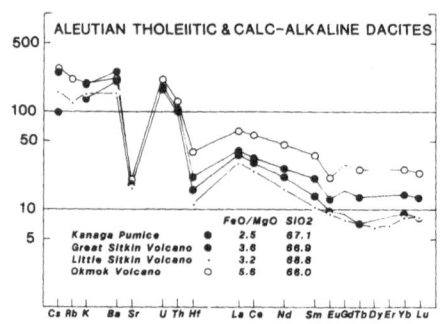

Figure 3. Trace element patterns of tholeiitic series basalts and andesites from Westdahl volcano normalized to chondrites and to ocean ridge basalts (alkali metals only -- see Kay et al. 1982) showing fractionation invarience of all element ratios (except ratios with Sr).

Figure 4. Trace element patterns (plotted as in Figure 3) for dacites from tholeiitic and calc-alkaline series. Note the lower REE contents and Eu anomalies, and higer La/Yb ratios of the calc-alkaline samples (Kanaga, Little Sitkin).

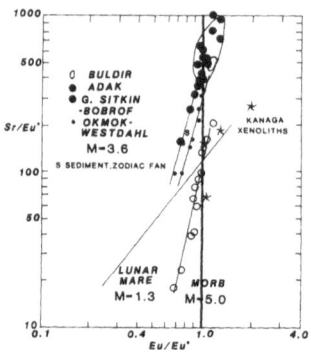 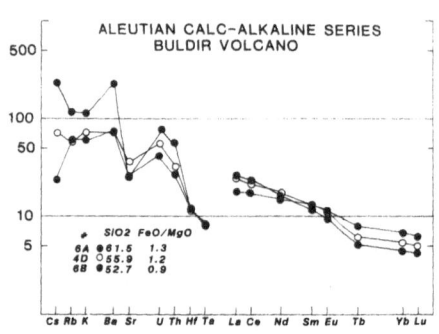

Figure 5. Plote of Sr/Eu* versus Eu/Eu* (Eu* is interpolated from REE pattern) for Aleutian volcanic rocks, xenoliths (from Kanaga Tertiary sill, stars), and sediments (S) and for MORB (closed squares) and lunar Mare (line represents trend). Slopes (M) of lines through points are suggested to be proportional to the relative fO_2 of the series, that is; Mare < MORB ≤ Aleutians. The plot also shows the relatively larger Eu anomalies and lower Sr contents in the tholeiitic (Okmok-Westdahl - open circles) and extreme transitional calc-alkaline lavas (Great Sitkin - open squares) compared to the more calc-alkaline lavas (circled field, Moffett - solid circles and Buldir - solid field). Figure from Kay and Kay (1988).

Figure 6. Trace element patterns of calc-alkaline series basalt and andesites from Buldir volcano (plotted as in Figure 3) showing variation of REE pattern (e.g. La/Yb ratio) with fractionation, an effect attributed to amphibole fractionation. Figure from Kay and Kay (1988).

Tholeiitic and calc-alkaline series rocks in the Aleutian arc are distinguished on the basis of their FeO/MgO ratios at a given SiO_2 content. The magma series show distinct mineralogical characteristics and fractionation histories that account for some trace element differences.

Incompatible trace element patterns (Figure 3) for extreme tholeiitic series lavas from Westdahl Volcano illustrate that trace elements increase systematically with increasing SiO_2 from basalts through andesites and dacites (see pattern for Okmok dacite, Figure 4). Minerals involved in the fractionation are plagioclase, pyroxene and olivine - minerals that do not fractionate most of these trace elements. The increasingly negative Eu anomalies that correlate with a decrease in the ratio of Sr to Rare Earth Elements (REE) (Figure 5) are consistent with plagioclase fractionation.

Incompatible trace element patterns (Figure 6) for extreme calc-alkaline series lavas from Buldir Volcano (this volcano has among the lowest FeO/MgO ratios of any volcano in the arc) illustrate that REE patterns become increasingly steeper with increasing SiO_2 content. Fractionation of amphibole, which occurs as an important phase in these rocks, has played a role in shaping the REE patterns. U and Th increase with differentiation but Ta and Hf are constant, reflecting the high D (distribution coefficient between mineral and host melt) value in amphibole (in general, Ta varies much more than Hf).

Figure 4 shows the contrast between high-Si andesites and dacites in the tholeiitic and calc-alkaline series. Incompatible trace elements are most enriched in the tholeiitic sample (Okmok) which has a relatively flat REE pattern and large Eu anomaly. Calc-alkaline rocks like those from Kanaga and Little Sitkin have much steeper REE patterns, lower incompatible trace element concentrations and convex downwards patterns consistent with the fractionation of amphibole and apatite. The differences in behavior of the REE relative to Th is well shown by the dacites in Figure 4, as the Th concentrations in these samples are all quite similar.

Differences in Nd and Sr isotopic composition of Aleutian volcanic rocks encompass a small range. In detail in some centers where several analyses are available, the more silicic samples are more radiogenic, while in others the opposite occurs (Figure 7). In

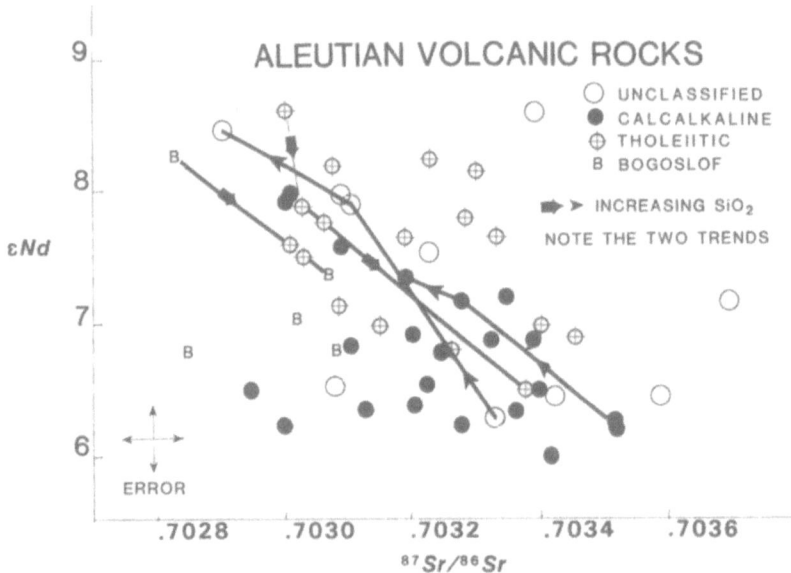

Figure 7. Plot of ϵ_{Nd} versus $^{87}Sr/^{86}Sr$ in recent Aleutian lavas. The range of values is small, especially excluding Adak Mg-andesites (Kay 1978), which plot off-scale ($^{87}Sr/^{86}Sr = 0.7028$, $\epsilon_{Nd} = 12.2$). Arrows on lines connecting points indicate disparate directions of isotopic change with increasing SiO_2 content. Data from sources listed in Kay and Kay (1988).

some centers, no pattern emerges. Rather limited existing data on temporal variation offer no substantiation to the theory of Myers et al. (1985) that isotopic variability (toward less radiogenic Sr and more radiogenic Nd) decreases with time over the life of a volcano. Okmok volcano (12 analyses) is the most completely studied to date and shows no such progression.

Neither Nd nor Sr isotope ratios show spatial variation along the arc (Figure 8). Similarly, ^{10}Be (a 1.6 my half life nuclide

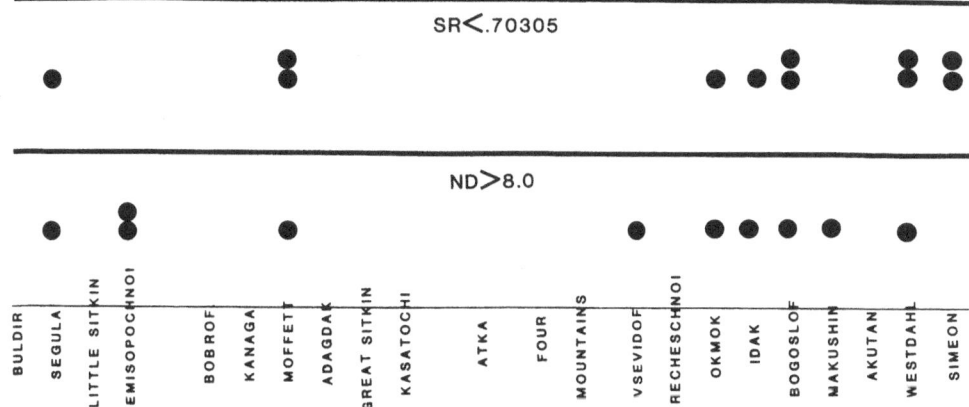

Figure 8. Occurrence of extreme values of Sr and Nd isotope ratios in Aleutian volcanic rocks showing lack of correlation with position in the arc.

produced in the earth's atmosphere) is found in equal levels in equally differentiated volcanic rocks from both eastern and western parts of the arc (Tera et al., 1986). In contrast, Figure 5 shows that the Sr/Eu* ratio at no Eu anomaly is higher for lavas from Adak, Bobrof and Great Sitkin (a group of volcanoes in the central Aleutians) than for Buldir, Okmok and Westdahl (in the western and eastern Aleutians, respectively). This appears to be related to location in the arc (Figure 2), not to calc-alkaline or tholeiitic differentiation (primitive olivine tholeiite lavas from Okmok and Adak show this difference). Ba/La, Th/Hf and Ba/Hf (and perhaps $^{206}Pb/^{204}Pb$) ratios are higher for the high Sr/Eu* lavas. These spatial variations in chemistry along the arc are of 100-500 km scale, and as they are present in all members of differentiation series, are thought to be present in the mantle-derived magmas (Kay and Kay 1988).

For the comparison of Aleutian arc magmas and oceanic magmas, three classes of oceanic data have been used: worldwide, regional, and local. The Pb isotope comparison may be used as an example. Among oceanic basalts, those from mid-oceanic ridges MORBs have relatively non-radiogenic lead, which has relatively low variability. In contrast, oceanic islands basalt (OlB) leads are more radiogenic and more variable. However, values for individual islands and islands within some rather large regions (e.g. North Pacific) have more restricted range than worldwide OIB. For this reason, rather than using worldwide OIB for comparison with Aleutian arc Pb, only North Pacific OIB and MORB have been used. Figure 9 shows the limited

152

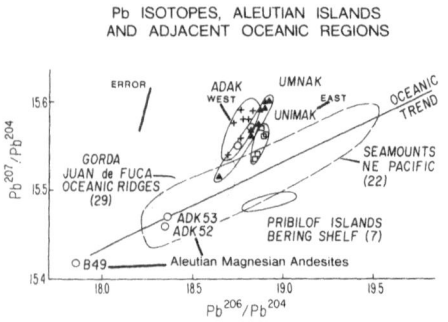

Pb ISOTOPES, ALEUTIAN ISLANDS
AND ADJACENT OCEANIC REGIONS

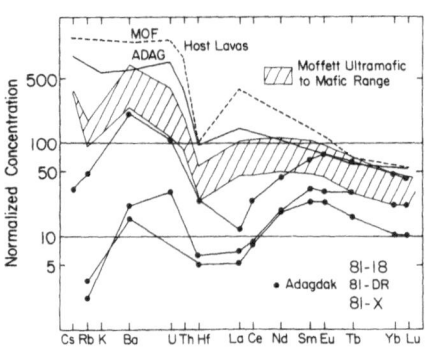

Figure 9. Pb isotope ratios in Aleutian volcanic rocks showing limited range of values and differences in $^{206}Pb/^{204}Pb$ values between three volcanic centers, with magnesian andesites from Moffett volcano (Adak) plotted separately from other Adak data. The $^{207}Pb/^{204}Pb$ values of some Aleutian volcanic rocks are higher than those in adjacent oceanic igneous rocks to the south, and in Pribilof islands intraplate volcanoes to the north, and trend toward $^{207}Pb/^{20}Pb$ - rich sediment (not shown, but located at top of Aleutian data fields). Figure modified after Kay et al. (1978).

Figure 10. Normalized (as in Figure 3) trace element concentrations of rocks from Adak volcanoes: Moffett mafic and ultramafic xenoliths (ruled field: range of three cumulate xenoliths described by Conrad and Kay, 1984), Adagdak ultramafic xenoliths, and Moffett (andesite, 56% SiO_2) and Adagdak (Table 1, col. 1) host lavas. The Moffett xenoliths have parallel trace element patterns that reflect variable mixtures of cumulus minerals (amphibole, clinopyroxene, plagioclase, and olivine) with little intercumulus melt. The ultramafic sample MM102) has the highest trace element levels and the highest proportion of amphibole (the phase richest in these trace elements). Among the Adagdak xenoliths, amphibolite 82-18 has the highest trace element level. A clinopyroxene-rich portion of xenolith 81-DR has the pattern of a clinopyroxenite at the equilibrium with its host lava. The pattern of xenolith 81-X is consistent with less clinopyroxene and more olivine than 81-DR. Figure after DeBari, et al. (1987).

range of the Aleutian Pb values and illustrates that many are outside the range of North Pacific oceanic basalts. Comparison of "Aleutian" and "oceanic" data on a local basis has not been done: no samples of basaltic crust are available from immediately south of the insular part of the Aleutian arc. A rather uncommon Aleutian magma type, called Mg-andesite, has Pb that is within the more restricted, and less radiogenic range of North Pacific MORB; trace element ratios for the Mg-andesite are also in the oceanic basalt range (Kay, 1978).

The ranges of Ba/La ratios of Aleutian arc and oceanic lavas (MORB and OIB) are largely noncoincident (e.g. Kay, 1980; Morris and Hart, 1983; Perfit and Kay, 1986), as are various ratios of alkali metals, rare earths, and high field strength elements. This statement is true even for the least-differented Aleutian lavas, and for cumulate xenoliths of a wide compositional range (from mafic to ultramafic, Figure 10).

In summary, although both geochemical and isotopic ratios vary with crustal differentiation in the Aleutians, and although regional differences occur along the arc, the overwhelming proportion of magmas that enter the Aleutian crust lie outside the range of North Pacific-Bering Sea oceanic basalts (for the latter, basalts from the Pribilof Islands and the Kamchatka Basin are used for comparison for several key ratios). Sampling bias for the oceanic basalts must be regarded as the limiting factor for establishing the truth of this statement. For the Aleutians, the proportion of seamounts (largely unsampled) is low on the Pacific crust south of the Aleutian trench so that we may argue that MORB-type oceanic crust with well-defined chemistry and isotopic composition predominates in the subducted plate. For comparison, this contrasts with the rough oceanic crust being subducted at the Marianas trench.

Definition of excess (XS) elements, those higher in the sources of arc basalts than in the source of oceanic basalts, is model-dependent, (e.g. Kay, 1980). Recent estimate (Kay and Kay, 1987) are listed in Table 1.

Table 1   EXCESS ELEMENTS IN ALEUTIAN BASALTS

	Aleutian Mg-rich Basalts		Oceanic Basalts		Excess Components[1] Relative to V25-T9	
			Hawaii	Mid-Atlantic Ridge		
	ADAG HOST	OK4	BHVO	V25-T9	ADHG HOST	OK4
La	5.85	4.29	14.9	2.9	3.7	2.3
Ce	13.3	11.03	37.4	9.82	6.1	4.5
Nd	9.0	6.4	25.0	9.30	2.1	0.2
Sm	2.21	2.00	6.15	3.00	0	0
Eu	0.720	0.696	2.08	1.08	-	-
Tb	0.370	0.374	0.89	0.77	-	-
Yb	1.31	1.29	1.90	2.96	-	-
Lu	0.201	0.199	0.265	0.43	-	-
Ba	222	164	148	9.	213	1.58
Cs	1.05	0.32	0.05	0.008	1.04	0.315
U	1.08	0.43	0.379	2.20	1.04	0.38
Th	1.99	0.915	1.21	0.09	1.92	0.825
Hf	2.09	1.04	4.30	2.20	0.48	
Ta	0.084	0.08	1.1	-	-	-
K	8900	4500	4500	650	8400	4100

[1]Excess components calculated by assuming that Sm is derived entirely from MORB - like source and that elements with higher ratios to Sm than in MORB V25-T9, are derived from an "excess component" which is listed in the table.

3. ALEUTIAN ARC MAGMAS CONTAIN CRUSTAL COMPONENTS FROM SUBDUCTED OCEANIC SEDIMENT, SUBDUCTED HYDROTHERMALLY ALTERED MID-OCEANIC RIDGE BASALT (MORB), AND OCEANIC CRUST THAT IS IMBEDDED WITHIN THE ALEUTIAN ARC CRUST

3.1. Sources of Excess (XS) Components

For Aleutian magmas, as for those of other island arcs, there are several potential sources of XS components. The aim of a geochemical tracer study is identification of the sources by matching components in magmas and sources. Generally, sampling of the likely sources is impossible, and analysis of proxies must substitute.
     The XS components of greatest interest in studies of mantle evolution and crustal growth are those that are recycled at mantle depth. Crustal or lithospheric contamination by arc basement is a separate conceptual category. In an oceanic arc like the Aleutians a confusion between arc basement and subducted plate is bound to occur: they are both oceanic crust with sediment.
     Oceanic crust was the sole occupant of the Aleutian island arc construction site. Over the past 60 m.y. the arc crust has thickened,

but appears not to have rifted. The original oceanic crust should lie under (and perhaps also over) gradually accumulating arc volcanic and plutonic rocks. Exposures of this trapped oceanic crust have not been found, although back-arc basin crust is exposed on the volcanically inactive Komandorski islands, and on the western-most Aleutian islands (e.g. Attu, Kay et al., 1986). The trapped oceanic crust (including sediment) is a possible crustal source for trace elements and Sr, Nd, and Pb isotopes in Aleutian arc magmas. Tectonized mafic and ultramafic xenoliths in Tertiary basalts from Kanaga island that are located several kilometers south of the present volcanic line, are possible samples of the buried crust. Their trace element compositions have the required "XS" elements (Figure 11), and their

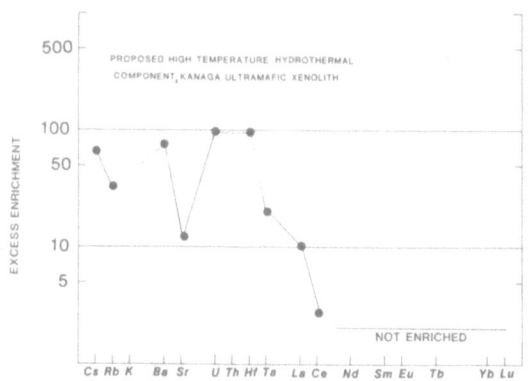

Figure 11. Excess components in ultramafic xenolith from Kanaga island (calculated as in Table 1). It is proposed that observed crustal recrystallization of the sample during deformation at high temperature was accompanied by the introduction of the excess elements in hydrothermal fluids.

high epsilon Nd values (9.6 in one case) and high Sr isotope ratios are all consistent with identification of them as altered oceanic crust. Note that the buried oceanic crust is within the Aleutian arc lithosphere and is a non-renewable contamination resource. Therefore, it should be progressively less important as a source of elements in arc magmas.

A constantly renewable source of continental crustal components is carried with the descending Pacific lithospheric plate as sediment and as hydrothermally metasomatized igneous crust. Unfortunately, little is known of the chemical and isotopic characteristics of specific sediment relevant to the Aleutian recycling hypothesis. Working with generalities, clay-rich sediment has high Pb, alkali, and alkaline earth concentrations relative to rare earths, high

$^{207}Pb/^{204}Pb$, low ϵ_{Nd}, and high $^{87}Sr/^{86}Sr$. The top of the sediment pile has high ^{10}Be, and high Ba reflecting regionally important siliceous sedimentation. One may choose an idealized, perhaps even realistic single or multiple component to represent recycled sediment (e.g. Kay, 1980), but something much more specific (local) is necessary. To explain the observed regional differences in trace element and Pb isotopes ratios in Aleutian volcanic rocks by differences in sediment inputs, resolution on the 100 km scale is obviously required, but is not available. On the scale of the North Pacific, the geochemistry of various common sediment types have been surveyed, and some shortcomings of the idealized sediment models are revealed (see Figure 12).

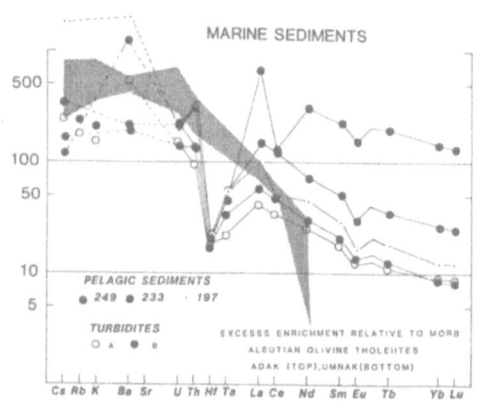

Figure 12. Normalized trace element contents (as in Figure 3) of Pacific pelagic clay samples 249, 233 and 197 (DSDP Sites 76, 307, and 38, respectively) and two composite turbidite samples from DSDP Site 183 (south of Aleutian trench 161°W, 53°N). Note negative Ce anomaly in sample 249, and highly variable ratios of alkali metals, Ba, U and Th. Excess (XS) components calculated in Table 1 are also shown. Figure from Kay and Kay (1987).

    · Some basal pelagic sediments have low Ba/La and Th/La ratios, and large negative Ce anomalies that make them unsuitable mixing end members.
    · Some siliceous turbiditic sediments have very high Ba/K and Ba/Cs ratios that make them unsuitable end members.
    As a general observation, the XS alkali metal, Ba, U, and Th component in arc magmas has far greater homogeneity than do various North Pacific sediments. Interelement ratios in the XS component do not match those in specific sediment types or in mixtures of sediment types (here, we may not have enough data). It does little good to match one element (say, Ba) and have another mismatch (say, Cs).
    The case of K should be noted, for as a major element, K is limited by clay stoichiometry to a maximum of ~8% by weight.

Therefore, K/Ba ratios in sediments often vary because of Ba variation (for example, due to a siliceous component) and a high Ba sediment used as a mixing end member to explain high Ba/La, will not explain the XS K in arc magmas (Kay, 1980).

At present, hydrothermally altered oceanic crust of early Tertiary age is thrust under the Aleutian arc. The hydrothermal fluid was sea water that contained elements derived from continental weathering. No samples of this altered oceanic crust are available, so conjucture will substitute. Undoubtedly, some of the continent-derived elements in the altered oceanic crust are the same XS elements found in Aleutian arc magmas (e.g. K, U), but equally important, some XS elements are missing (e.g. Th, which is insoluble in seawater). In a general way, the oceanic sediment and altered oceanic crust complement each other as mixing components necessary to explain XS components in Aleutian magmas.

The search for appropriate mixing components to match the XS components in Aleutian magmas may yet fail for a good reason: the XS elements are those extracted from the mantle, and need not coincide with the entire inventory of elements that are subducted (Kay, 1985). Element fractionation during subduction (informally called the S-process) may occur. In this case recycled elements in arc magmas complement recycled elements that are returned to deeper mantle levels, to reside there for long expected residence times (billions of years). The proposition that much of the trace element differences between arc and oceanic magmas are due to S-process element fractionation (in particular by accessory phases) has been argued for by Morris and Hart (1983) and discussed critically by Kay (1980).

4. ALEUTIAN OLIVINE THOLEIITES ORIGINATE BY PARTIAL MELTING OF UPWELLING PERIDOTITE RENDERED BUOYANT BY ADDITION OF XS ELEMENTS FROM THE SUBDUCTING PLATE

4.1 Relation of Tectonics and the Magma Generation Process

Figure 1 (Kay and Kay, 1985, 1988; DeBari et al. 1987) is cross section depicting a model for the magma generation process, that operates under a central Aleutian calc-alkaline volcano. A physical basis for the model as a whole, especially one that can be used in a predictive mode, is not available at present. Instead various plausible physical processes that have been studied in isolation from the larger system have been deduced to apply, and have been plugged in when needed. Deduction presupposes premises -- boundary conditions in what is an unobserved and largely unconstrained region of the earth. Several key boundary conditions can be listed:

The pattern of seismicity, travel time anomalies of seismic waves, and gravity data are the basis for an accurate depiction of the general configuration of crust, mantle, and subducting plate (e.g. Engdahl and Billington, 1986).

· The flowing region (asthenosphere) reaches shallowest depths in the hottest and most fluid-rich region. This is under the volcanic line. The region is also the one with the lowest seismic Q, has slightly lower seismic velocities, and is aseismic.

· Phase equilibrium studies indicate that olivine tholeiite magmas that rise to the mantle-crust boundary are at equilibrium with spinel lherzolite (peridotite) at depths as shallow as about 30 km -- only slightly greater than the crustal thickness. The source of the magmas is peridotite mantle. This mantle is renewable (indicating covection) and it contains a renewable inventory of XS elements, which originate at the subducting plate, located some 70 km deeper. The overall mass percentage of XS elements in the source is very low -- less than 1% as constrained by the close similarity of delta ^{18}O values of arc magmas and oceanic basalts (Nye and Reid, 1986; Kay and Kay, 1988).

The physical process that establishes the connection between the source of XS elements at 100 km and the source of the main mass of the magmas at 30 km is a rising convection limb within which there is porous flow of basalt through peridotite (which is unable to hold melt fractions in excess of several percent).

The similarities and differences between pressure - release melting at arcs and mid-oceanic ridges, resulting in olivine tholeiites with similar major element chemistry, has been examined by Kay and Kay (1988). The magmatic production rates are similar at the slowest-spreading mid-oceanic ridges (e.g. SW Indian Ridge) and at the Aleutian arc (averaged over ~60 m.y.). The driving force for convection is different in the arc case, being due to buoyancy created by volatiles released by the subducted plate. The observation that volcanic spacing along both arcs and mid-oceanic ridges is about the same length scale supports a speculative model that the characteristic volcanic spacing in both results from transverse convective rolls. In this model, following the release of the basaltic fraction from the upwelling convecting limb, residual mantle largely descends between volcanic centers (not shown in Figure 1).

5.   AT PRESENT, OPERATION OF THE SUBDUCTION PROCESS (S-PROCESS) RESULTS IN BOTH CRUSTAL GROWTH AND RECYCLING OF CONTINENTAL CRUSTAL MATERIAL INTO THE MANTLE. MANTLE HETEROGENEITIES (AS SAMPLED BY OIB MELTS) WERE CREATED IN THE PAST BY THE SAME MECHANISM, AND ALSO BY LOWER CRUSTAL DELAMINATION ACCOMPANYING ACCRETION OF ARC TERRANES AT CONTINENTAL MARGINS

5.1 Crustal Growth and Mantle Evolution: an Integrated View

Long term crustal recycling to the mantle at arcs (e.g. Armstrong, 1981) requires inefficient removal of continental components from the subducted plate in the S-process. The plate itself is recycled, but are the hydrothermally introduced elements purged and the sediment scraped off or melted out as the plate plunges into the mantle? Several observations are relevant to this question:

· Much less than a total thickness of altered oceanic MORB covered by sediment is required to supply the flux of recycled elements by subduction (e.g. Karig and Kay, 1981). If the oceanic crust is not underplated then, permissively, it may be returned to the mantle, where it represents a zero age mantle heterogeneity.

· Among Oceanic Island Basalts (OIB) some have mantle sources that resemble aged recycled crust of the Aleutian type (e.g. Kay, 1984).

· To find recycled continental crustal components from the subducted plate in Aleutian lavas is encouraging for those who favor the S-process to create mantle heterogeneity (OIB Sources: Hofmann and White, 1982). It proves that the crustal component has made it 100 km into the mantle. To fail to recognize deeply recycled continental crustal in arc magmas, (e.g. Morris and Hart, 1983), is circumstantially unfavorable to the arc-OIB connection.

The inefficient removal of subducted crustal components may occur in three places during the S-process a). during extraction of crustal components from the subducted plate, b). during extraction of primitive basaltic melt from peridotite, and c). by fractionation of basalt in the mantle or more likely at the Moho. Returned to the mantle in all three localities are mafic or ultramafic residues, or fractionated liquidus minerals that are at equilibrium with arc magmas and therefore share their isotopic signatures. Some trace element ratios in these arc-related mantle rocks are dominated by the recycled component. The production rate of these residues is large - perhaps one fifth of the production rate of MORB residues (the same as the ratio of magma production rates, Kay 1980). The return mechanism of the residues to deeper levels to the mantle is obvious for the subducted plate, but less apparent for the shallower level residues and the cummulate crystals that occur immediately below the Moho. The residues are part of the convecting region (Figure 1), therefore, they flow away; a lower lithospheric delamination mechanism may apply to the cumulates. Kay and Kay (1986, 1987) envision this to occur not during subduction, but during suturing of oceanic island arcs to continental margins.

Delamination of the mafic-ultramafic lower lithosphere of oceanic island arcs is driven by deformation and volatile induced low temperature-high pressure phase transformations that accompany arc suturing. These reactions produce dense garnet-bearing rocks (garnet granulites, eclogites) from the light feldspathic gabbros that comprise the lower crust of island arcs (e.g. Kay and Kay, 1985). The result is a density inversion, and the former lithosphere sinks into the mantle. In this view, the new crust that ends up accreted to continental nuclei is mid-crust to upper crust only, less mafic, and

160

more like mean continental crustal composition than the crust of oceanic island arcs. The recycling into the mantle of young, mafic lower crust together with its continental crustal components acquired during the S-process may be as important as the S-process itself in creating mantle heterogeneity, and steering the course of mantle evolution.

ACKNOWLEDGEMENTS

Support from the National Science Foundation is acknowledged. The Antalya NATO conference was a model of its genre; we thank the organizers and the participants.

REFERENCES

Armstrong, R.L., 1981, Radiogenic isotopes: the case for crustal recycling on a near-steady-state no-continental-growth Earth, Phil. Trans. R. Soc. Lond. A301, 443-472.

Coats, R.R., 1962, Magma type and crustal structure in the Aleutian arc, Am. Geoph. Union, Geophysical Monograph 6, 92-109.

Conrad, W.K. and R.W. Kay, 1984, Ultramafic and mafic inclusions from Adak Island: Crystallization history and implications for the nature of primary magmas and crustal evolution in the Aleutian arc, J. Petrol., 25, 88-125.

DeBari, S., S. Mahlburg Kay, and R.W. Kay, 1987, Ultramafic xenoliths from Adagdak Volcano, Adak, Aleutian Islands, Alaska: Deformed igneous cumulates from the Moho of an island arc, J. Geol., 95, 329-341.

Engdahl, E.R. and S. Billington, 1986, Focal depth determination of central Aleutian earthquakes, Bull. Seis. Soc. Amer., 76, 77-93.

Gust, D.A. and M.R. Perfit, 1987, Phase relations of high-Mg basalt from the Aleutian Island Arc: Implications for primary island arc basalts and high-Al basalts, Contrib. Mineral. Petrol., (in press).

Hofmann, A. and W. White, 1982, Mantle plumes from ancient oceanic crust, Earth. Planet Sci. Letters 57, 421-436.

Karig, D.E. and R.W. Kay, 1981, Fate of sediments on the descending plate at convergent plate margins, Phil. Trans. Roy. Soc. Lond., A301, 233-351.

Kay, R.W., 1977, Geochemical constraints on the origin of Aleutian magmas, in Island Arcs, Deep Sea Trenches and Back-Arc Basins, ed. Talwani and Pitman, Am. Gephys. Union, M. Ewing Series, 1, 229-242.

Kay, R.W., 1978, Aleutian magnesian andesites: melts from subducted Pacific ocean crust, J. Vol. Geotherm. Res., 4, 117-132.

Kay, R.W., 1980, Volcanic arc magmas genesis: Implications for element recycling in the crust-upper mantle system, J. Geol., 88, 497-522.

Kay, R.W., 1984, Elemental abundances relevant to identificaion of magma sources, Phil. Trans. Roy. Soc. Lond., A310, 535-547.

Kay, R.W., 1985, Island arc processes relevant to crustal and mantle evolution, Tectonophys., 112, 1-16.

Kay, R.W. and S. Mahlburg Kay, 1986, Petrology and geochemistry of the lower continental crust: An overview, Geol. Soc. Lond. Spec. Publ., 24, 147-159.

Kay, R.W. and S. Mahlburg Kay, 1987, Crustal Recycling and the Aleutian Arc, Geochim. and Cosmochim. Acta, submitted.

Kay, R.W., J.L. Rubenstone and S. Mahlburg Kay, 1986, Aleutian terranes from Nd isotopes, Nature, 322, 605-609.

Kay, R.W., S.-S. Sun, and C.-N. Lee-Hu, 1978, Pb and Sr isotopes in volcanic rocks from the Aleutian Islands and Priblof Islands, Alaska, Geochim. Cosmochim. Acta, 42, 263-273.

Kay, S. Mahlburg and R.W. Kay, 1985, Role of crystal cumulates and the oceanic crust in the formation of the lower crust of the Aleutian arc, Geology, 13, 461-464.

Kay, S. Mahlburg and R.W. Kay, 1988, Aleutian Magmas in space and time, Decade of North American Geology, Geol. Soc. Amer. (in press).

Kay, S. Mahlburg, R.W. Kay, and G.P. Citron, 1982, Tectonic controls of Aleutian arc tholeiitic and calc-alkaline magmatism, J. Geophys. Res. 87, 4051-4072.

Myers, J.D., B.D. Marsh and A.K. Sinha, 1985, Strontium isotopic and selected trace element variations between two Aleutian volcanic centers (Adak and Atka): implications for the development of arc volcanic plumbing systems, Contrib. Min. Pet., 91, 221-234.

Morris, J.D. and S.R. Hart, 1983, Isotopic and incompatible element constraints on the genesis of island arc volcanics from Cold Bay and Amak Island, Aleutians, and implications for mantle structure, Geochim. Cosmochim. Acta, 47, 2015-2030.

Nye, C.J. and M.R. Reid, 1986, Geochemistry of primary and least fractionated lavas from Okmok Volcano, Central Aleutians: implications for arc magmasgenesis, J. Geophys. Res., 91, 10271-10287.

Perfit, M.R. and R.W. Kay, 1986, Comment on "Isotopic and incompatible element constraints on the genesis of island arc volcanics from Cold Bay and Amak island, Aleutians, and implications for mantle structure" by J.D. Morris and S.R. Hart, Geochim. Cosmochim. Acta, 50, 477-481.

Tera, F., L. Brown, J. Morris, S. Sacks, J. Klein, and R. Middleton, 1986, Sediment incorporation in island arc magmas: inferences from [10]Be: Geochim. Cosmochim. Acta, 50, 535-550.

TOMOGRAPHIC MAPPING OF THE UPPER MANTLE STRUCTURE BENEATH THE ALPINE
COLLISION BELT

W. Spakman
Department of Geophysics
Institute of Earth Sciences
P.O. Box 80.021
3508 TA Utrecht
The Netherlands

Extended abstract

1. Introduction

The application of iterative imaging techniques has enabled a detailed
mapping of the P-wave velocity heterogeneity in large parts of the
Upper Mantle beneath Central Europe, the Mediterranean and the Middle
East (Spakman, 1985). More than half a million ISC P delay times are
simultanuously inverted for estimates of seismic wave velocity
anomalies, earthquake relocation parameters and near station velocity
anomalies. In principle the employed tomographic method is based on a
division of the Upper Mantle area in a large number of cells and a
linearization of the travel time integral with respect to a reference
velocity model.

Two different types of imaging algorithms are separately applied
to solve the huge set of linear equations involved. The use of iterative algorithms in seismic tomography has made the inversion of $O(10^6)$
data in $O(10^5)$ model parameters feasible, however, at present a drawback
of these methods is that no proper error analysis is available. We
applied inversions with synthetic data computed from synthetic velocity heterogeneity models to obtain estimates for the accuracy of the
tomographic mapping.

In this paper we present a brief outline of this research and we
will dwell upon a few remarkable results.

2. Data and method

The ISC-bulletin tapes (years 1964-1982) were selected for earthquakes
that occurred in the Mediterranean and the Middle East. An event was
selected if the reported number of first arrival observations (up to
90°) was at least 10. This resulted in about 30,000 mainly weak and
poorly determined events. Taking an absolute delay time threshold of 5

S. R. Hart and L. Gülen (eds.), Crust/Mantle Recycling at Convergence Zones, 163–172.

seconds, the ISC reports 550,500 delay times for these events which were observed by 937 regional and teleseismic stations up to 90° of epicentral distance.

The Upper Mantle cell model which is used to parameterize the slowness heterogeneities is shown in figure 1. It consists of 9 layers, which are subdivided in 52×20 approximately $1° \times 1°$ cells. The layer thickness increases with depth from 33 km for the first layer to 130 km for the last one, which reaches a depth of 670 km.

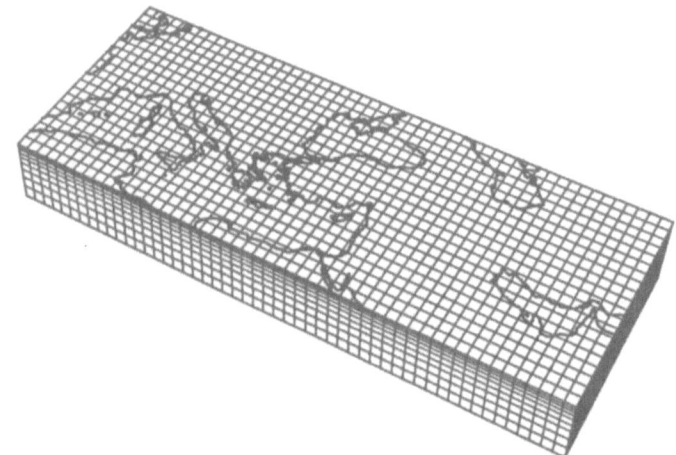

Figure 1. The cell parameterization of the Upper mantle beneath Central Europe, the Mediterranean and the Middle East. The size of the model is approximately 5700×2200×670 km. The Upper Mantle is divided into 9 layers of increasing thickness. Each layer consists of 1040 $1^0 \times 1^0$ cells.

The inclusion in the inversion procedure of 30,000 event relocations would lead to the estimation of 120,000 unknown hypocentre coordinates, which are not expected to be well resolved. Therefore events were grouped in $50 \times 50 \times 33$ km cells and instead of relocating each single event a cluster of events is relocated. 2626 event clusters could be identified which results in a contribution of 10.484 unknowns to the problem for the relocation part of the inversion alone.

Station corrections are estimated only for those stations outside the cell model for which no station delay time is reported by Dziewonski and Anderson (1983), which is the case for 226 stations. A 80 km thick cell is placed beneath these stations to account for the 'near' station slowness anomaly. In this way the total number of parameters amounted to 20,070.

The P-delay data can be related to the model parametrization by a large set of linear equations. A smoothing procedure is incorporated in the inversion. Only smoothing within the horizontal layers (retaining optimal depth resolution) is performed, in a way that the slowness heterogeneity in a cell is required to correlate by 20% of its amplitude with the solution in every adjacent cell. The ray paths are computed from the Jeffreys-Bullen reference model for P-waves, using the ISC hypocenter locations. Finally, a data averaging operation is applied by averaging the data belonging to rays coming from a cluster of events and arriving at the same station. This reduced the number of data to 292,451. Hence the tomographic problem is described by a linear system with dimensions $292,451 \times 20,070$.

This system of equations is solved using the conjugate gradient algorithm *LSQR* (Paige and Saunders 1982). A solution of the inversion problem is also computed using a member of the *SIRT* family of backprojection techniques. The tomographic method and the characteristics of both imaging algorithms are described at length by Spakman and Nolet (1987).

3. Results

3.1 The Aegean Upper Mantle

Figure 2 shows a cross section trough the Aegean Upper Mantle displaying the structure in terms of the seismic velocity heterogeneity for P-waves. In figure 2, the dipping high velocity zone (cross-hatched) is recognized as the blurred image of the Aegean (Hellenic) slab (Spakman 1986a). It is due to the cooling of oceanic lithosphere prior to subduction and possible chemical differences with the surrounding Upper Mantle, that the slab exhibits higher seismic velocities and hence it can be made 'visible' by tomographic mapping.

The low velocity anomaly lying between the slab and the high velocity lid may find its origin in frictional heating and partial melting, processes which are linked to the subduction and which are most likely the cause of Pliocene-Quaternary Aegean volcanism. It is unknown how large the source region for subduction linked volcanism can be. Considering the extent of the low velocity anomaly, we also like to point at the possibility that the anomaly is the trace of former locations of frictional heating and partial melt processes and hence of the former location of the slab trench-arc region. An indication for this is that Tertiary to Recent volcanism migrated southward in time from the northern Aegean to its present location (Fytikas et al 1984).

Cross sections taken trough the Upper Mantle beneath Greece display the same subduction process. However, there it seems that the lower part of the slab is detached. The detachment is imaged around depths of 200 km.

Also indicated in figure 2 are the hypocenters of those events that have been reported for this region by the PDE in the years 1964-1984. The geometry of the slab, as indicated by the seismicity pattern has played an important role in tectonic/kinematic models of the evolution of the Aegean area. Including other observables (e.g. Pliocene-Quaternary volcanism, late Neogene geology, tectonics) estimates for the initiation of subduction of the Eastern Mediterranean seafloor beneath the Aegean have been derived which range from 5-10 Ma (McKenzie, 1978, Mercier, 1981) to 13 Ma (Le Pichon and Angelier, 1979, 1981). In these studies it is not clearly recognized that the seismicity accompanying the subduction process depends on the age of the lithosphere involved (Vlaar and Wortel 1976). In particular, Wortel (1980, 1982) demonstrates that the occurence of the deepest

seismicity is a strong function of the age of the subducted lithosphere at the time it arrived at the trench. Reheating of the downgoing slab affects its rheological properties, eventually to the extent
that no earthquakes can be generated. For instance, 100 km old oceanic
lithosphere (i.e. at the onset of being subducted) appears to be
seismically active for at most 12 ± 3 Ma. Wortel derived these estimates from available data on the major subduction zones, therefore we

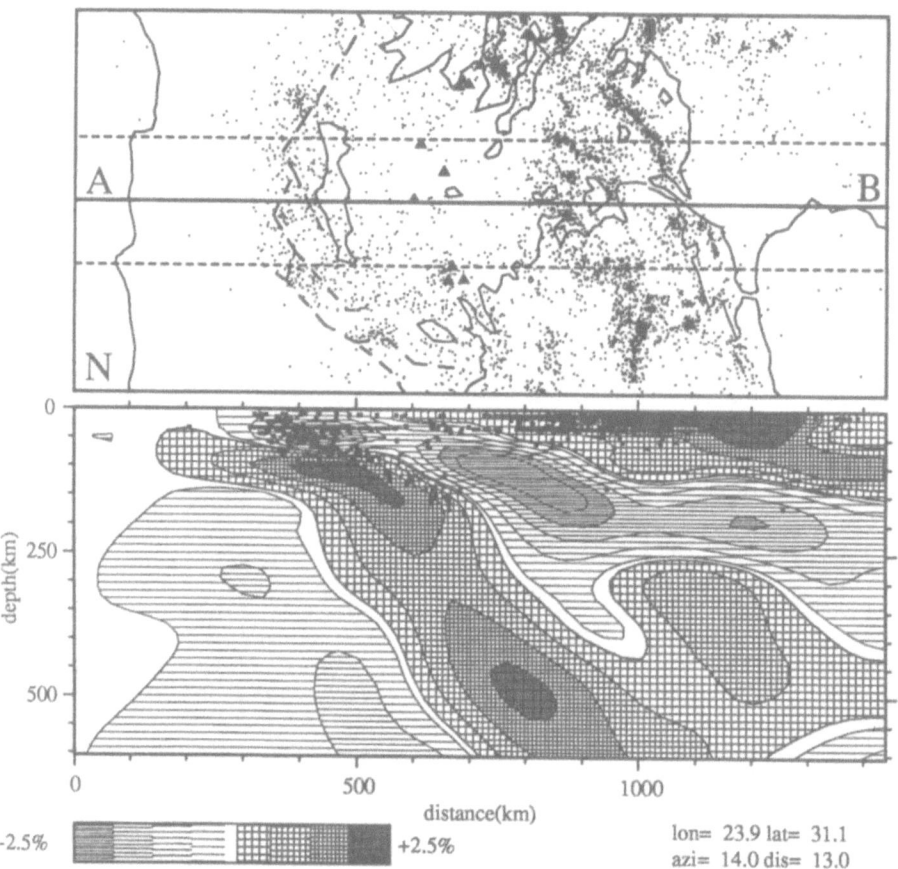

*Figure 2. A cross section through the Aegean Upper Mantle which is taken along the great circle segment
A–B. A corresponds to 23.9°E and 31.1°N. The great circle segment has an azimuth of N 14° and a
length of 13°. The upper panel displays the geographical region. Dots indicate 8139 earthquake epicenters which have been reported by the PDE in the years 1964-1985. The long-dashed line indicates the
location of the Hellenic Trench system. Black triangles denote the calc-alkaline Pliocene-Quaternary volcanic arc. A northward directed arrow is displayed in the lower left corner for orientation. The lower
panel shows a tomographic image of the Aegean Upper Mantle in cross section. The hatching indicates
the inferred P-wave velocity anomalies measured in percentages relative to the surrounding Upper Mantle reference velocity, i.e the Jeffreys-Bullen model. The scale is given at the lower left. Dots indicate the
projection on the cross section plane of those hypocenters which are located between the two dashed
lines shown in the upper panel.*

cannot directly apply this to the somewhat anomalous Hellenic subduction zone. But, it does indicate that we have to be cautious in deriving slab penetration depth estimates from the depth of the deepest earthquakes.

This inference and the tomographic results on the Aegean Upper Mantle structure lead Spakman et al. (1988) to reconsider the geodynamic evolution of the Aegean area. They arrive at a *minimum* duration time for the imaged Aegean subduction of 26 Ma, i.e. the deepest part of the slab in Figure 2 was subducted at least about 26 Ma ago.

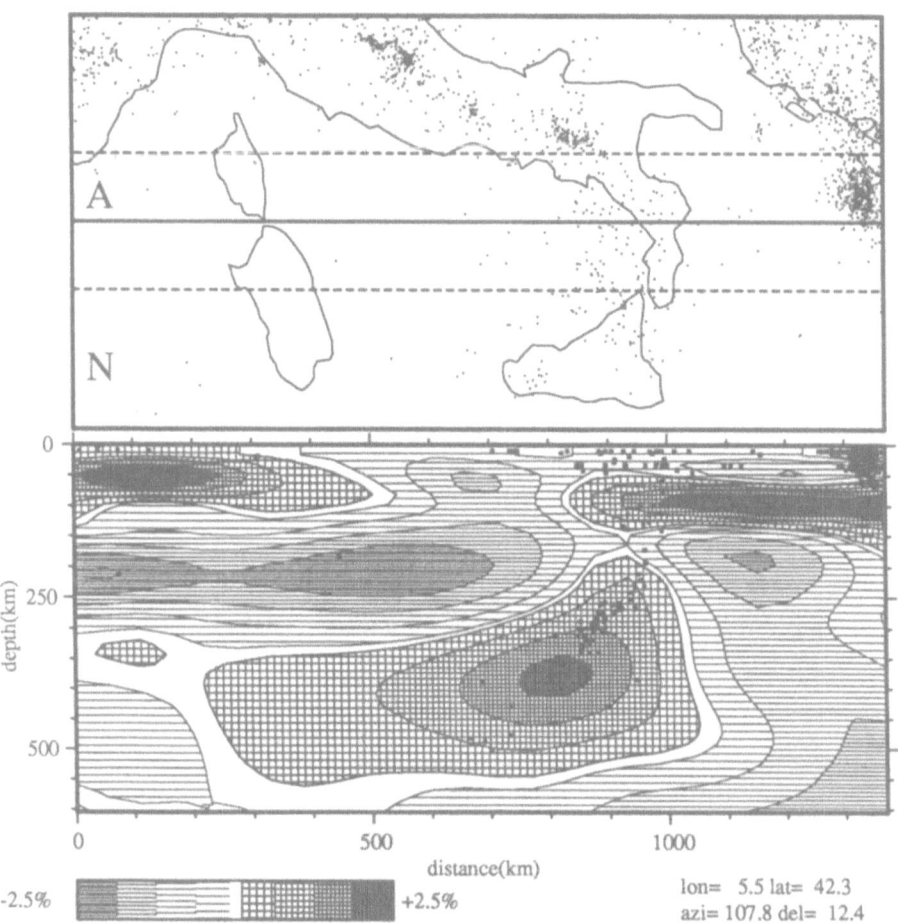

Figure 3. A cross section through the Tyrrhenian Upper Mantle which is taken along the great circle segment A−B. A corresponds to 5.5°E and 42.3°N. The great circle segment has an azimuth of N 107.8° and a length of 12.4°. The upper panel displays the geographical region. Dots indicate 3819 earthquake epicenters which have been reported by the PDE in the years 1964-1985. The lower panel shows a tomographic image of the Tyrrhenean Upper Mantle in See also caption figure 2.

The block-rotational movements in this area, as proposed by Kissel and Laj (1987) must be closely linked to spatial and temporal evolution the subduction process in which the likely slab detachment beneath western Greece and the withdrawel of the Central Aegean slab (figure 2) more or less to the south, probably have played a leading role. The Aegean Upper Mantle structure and its geodynamic implications are discussed at greater length by Spakman et al. (1988).

3.2 The Tyrrhenean Upper Mantle

Figure 3 shows a cross section through the Upper Mantle in the Tyrrhenean region. The large high velocity anomaly is the image of the Tyrrhenian slab. Between depths of 150-200 km the slab seems to be detached from the Ionian lithosphere (upper right). We have not yet succeeded in finding evidence for this disruption of the slab from resolution analysis. Moreover, in a cross section taken about 50 km to the SW and and parallel to the one shown here, the slab seems to be continuous. In a recent paper Anderson and Jackson (1987) carefully re-examined the deeper Tyrrhenean seismicity. They arrive at the conclusion that no major gaps exist in the depth distribution of earthquake foci between 100 - 200 km, indicating that the slab might not be completely disrupted yet. Also in this region much more oceanic lithosphere seems to be subducted than can be estimated from the lenght of the Benioff Zone (600 km). Again, the absence of seismicity in the deepest part of the slab may be explained by the thermal assimilation of the slab by the Upper Mantle (Wortel 1980,1982). Although the slab temperature may be too high for earthquakes to occur at these depths, thermal anomalies of several hundred degrees can still exist, which allows tomographic mapping of the silent zone of the subducted slab.

3.3 Other results

Many other structures are mapped by the tomographic inversion of the Upper Mantle heterogeneity beneath the area displayed in Figure 1. Here we mention some results given by Spakman (1986a,b).
    Beneath Southern Spain and the Western Mediterranean a northward dipping slab is discovered which may be related to an oceanic basin that once separated Spain and Africa but is now completely destroyed. The existence of this slab may provide an explanation for the peculiar 640 km deep seismicity beneath southern Spain.
    The deeper structure of the Alps displays a higher velocity lithospheric root of 150-200 km depth beneath the Western to Central Alps. Cross sections taken trough the Alps suggest the collision of two high velocity slabs of which the southern slab overtrusts the northern slab. In contrast to the higher velocities imaged beneath the Alps the Apennines are underlain by a low velocity anomaly.
    The 'plate boundary' between the Eurasian and Adriatic plates can be followed from the Alps to the Aegean area at depths between 33 - 170 km. On the whole the large scale heterogeneity of the European-Mediterranean lithosphere reflects strong correlation with the gross geological structure at the surface.
    Arguments are presented by Spakman(1986a) for a Mesozoic-Tethyan

origin of the Eastern Mediterranean basins and the Aegean and Tyr-
rhenean slabs. A dominant higher velocity anomaly in the deeper Upper
Mantle (>300 km) that roughly follows the Alpine Belt from Spain to
Iran, can be explained as (parts of) the subducted Mesozoic Tethys.
The slabs beneath southern Spain and the Western Mediterranean, Tyr-
rhenean and Aegean basins all are part of this zone.

In large parts of the model low velocity anomalies are imaged for
the P-velocity heterogeneity at depths between 100 – 240 km, for
instance, beneath Central Europe. With respect to the origin of these
velocity lows we tentatively propose an explanation in terms of the
models of lithospheric doubling and shifting of Vlaar (1983). Vlaar's
models involve obduction of old, mechanically strong lithosphere over
young (<20 Ma) oceanic lithosphere including the oceanic ridge segments. As a result of the positive buoyancy of the latter the overriding lithosphere is underplated by the young oceanic lithosphere. The
obduction of the oceanic Upper Mantle region results in a chemical and
thermal perturbation with respect to the former situation beneath the
lid and may result in magmatic penetration of the overriding plate.
The blanketing effect of the overlying plate results in long lasting
(>50-100 Ma) thermal anomalies. Between Africa and Eurasia young oceanic areas have been destroyed as a result of the convergence between
the two plates (e.g. Dercourt et al., 1986). As yet, only circumstantial evidence exists (Vlaar, 1988), but the models of Vlaar provide a
new and provocative explanation for the imaged low velocity regions.
An application of the concept of lithospheric doubling to Alpine orogeny is given by Vlaar and Cloetingh (1984).

4. Accuracy

An estimation of the upper limit of the reliability of the mapped
anomalies can be obtained from sensitivity test with synthetic 'harmonic' and 'cell-spike' velocity models (Spakman and Nolet 1987). A
sensitivity test result is shown in figure 3. The top panel displays a
'harmonic' synthetic anomaly velocity pattern in the same cross section as displayed in Figure 2. The lower panel demonstrates how well
we can resolve this pattern using the same tomographic inversion
scheme as in the actual data case. At first sight one might be disap-
pointed in the resolving power when comparing the synthetic model with
the tomographically mapped one. A more closer view indicates that,
although the amplitudes are systematically underestimated, the 'harmonic' patterns are reasonably well mapped in two thirds of the cross
section.

Within the slab, amplitude contrast of 2.5% can be resolved. The
blurring in the tomographic imaging is primarily caused by the imperfect illumination of the Aegean Upper Mantle by seismic rays. However,
we cannot rule out the possibility of cooling of the surrounding Upper
Mantle by the slab. In reality we expect the actual width of the slab
to be smaller. On the basis of this and other sensitivity tests Spakman and Nolet argue that the slab anomaly is possibly resolved with a
spatial error that varies between 50-100 km. The Upper Mantle region
north of the slab is resolved with a comparable accuracy. To the south

the resolution is much poorer.

We have restricted ourselves to studying how well synthetic velocity models can be resolved. This only gives an estimate of the upper limit of resolution and accuracy. Model errors, like the important effects of an inadequate reference velocity model from which the employed ray geometries are computed, can not be determined in this way. The tomographic results depend of coarse heavily on the illumination of the Earth by seismic rays.

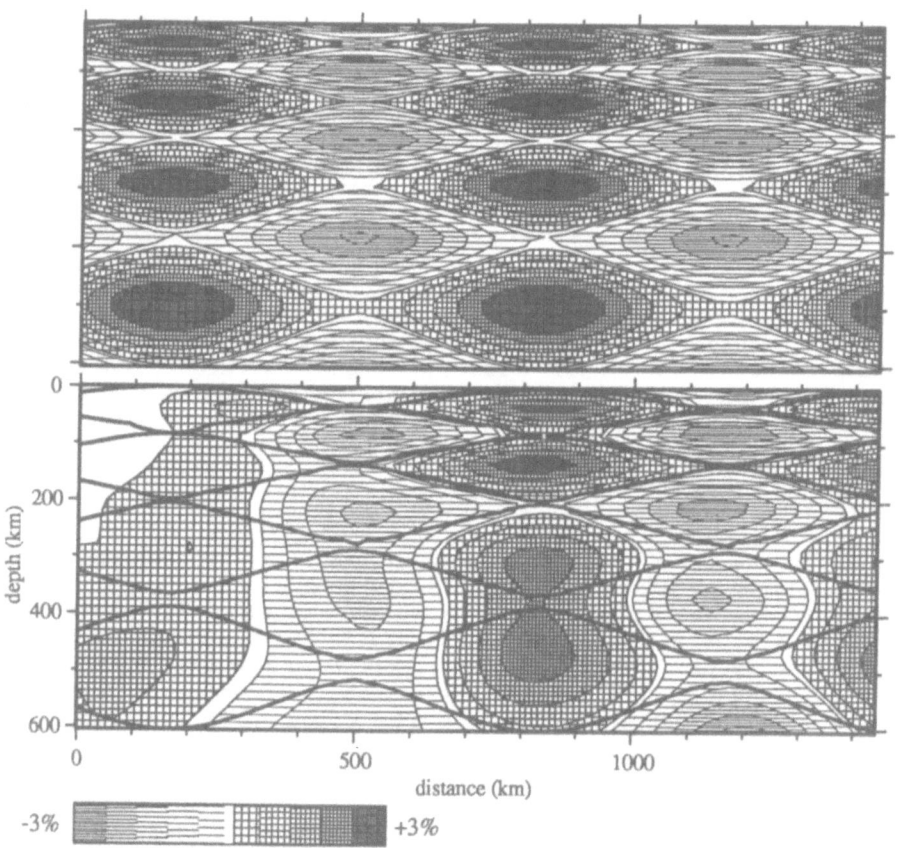

Figure 4. A sensitivity test result. The upper panel displays a synthetic 'harmonic' anomaly pattern with amplitudes between -3% and 3% relative to the surrounding Upper Mantle reference velocity. From this model (in 3D) synthetic data are computed using the same seismic ray geometries as are used in the real data case. A substantial amount of Gaussian noise is added to the data (87 % of the variance in the data is due to the addition of noise). The next step is to invert the synthetic data for the 'harmonic' model. This result is shown in the lower panel. The differences between the actual synthetic model and the inversion response must be attributed to the combined influence of the algorithm performance, the errors in the data and the imperfect illumunation of the Upper Mantle by the employed ray set. In the lower panel, the thick contour lines indicate the zero amplitude level of the actual synthetic velocity model (top panel).

An improvement of the reliability can be obtained by applying 3D ray tracing techniques. However, when using $O(10^6)$ data, the CPU time demands of existing methods have so far discouraged us from doing so.

One also needs to pay attention to the influence of the applied numerical method of solution on the results, i.e. on the accuracy of amplitudes and on the spatial resolution of the inferred anomalies. A comparison of the *SIRT* and *LSQR* algorithms in solving the same tomographic problem (i.e. the one discussed in here), shows important similarities in the shape of the anomalies in regions where the spatial resolution is expected to be relatively good, but important differences in regions of modest or poor resolution. This comparison indicates that one has to be very cautious with the interpretation of tomographic images, since it is very difficult to separate the image of real heterogeneities from the ones that are induced by errors in the data, lack of spacial resolution and the influence of the employed algorithm. This topic is studied in more detail by Spakman and Nolet (1987). They favour the *LSQR* algorithm for its superior convergence properties and its greater flexibility in use. One can obtain *SIRT* type solutions using the *LSQR* algorithm but not visa versa. The reason for this is that the *SIRT* algorithm employs an implicit (including an unwanted and unphysical) scaling of the tomographic equations. Van der Sluis and Van der Vorst (1987) show that the implicit scaling leads to an estimate of the minimum norm least squares solution, but measured in a norm which in general differs significantly from the Euclidian norm. Simply stated this means that the *SIRT* method does not solve the set of equations that are supplied to the algorithm but instead a completely reweighted system of equations. The *LSQR* method always leads to an estimate of the minimum norm least squares solution in an Euclidian sense of the equations that are supplied to the algorithm. Independent results of Nolet (1985) and Van der Sluis and Van der Vorst (1987) regarding the characteristics of the *LSQR* and *SIRT* type algorithms are all in favor of the superior convergence properties of the *LSQR* algorithm in tomographic problems.

Acknowledgements. I am grateful to G. Nolet for his advise and support in the development of the tomographic inversion scheme. I benefitted much from many discussions with M.J.R. Wortel and N.J. Vlaar on Upper Mantle processes. This research was financially supported by the Netherlands Organisation for Advancement of Pure Research (ZWO).

References:

Anderson, H. & J. Jackson, 1987, The deep seismicity of the Tyrrhenian sea, *Geophys. J.R. Astron. Soc.*, in press.
Dercourt, J., L.P. Zonenshain, L.E. Ricou et al., 1986, Geological evolution of the Tethys belt from the Atlantic to the Pamirs since the Lias, *Tectonophysics*, **123**, 241-315.
Dziewonski, A.M. and D.L. Anderson, 1983. Travel times and station corrections for P waves at teleseismic distances, *J. Geophys. Res.*, **88**, 3295-3314.

172

Fytikas, M., F. Innocenti, P. Manetti, R. Mazzoulin, A. Peccerillo & L. Villary, 1984, Tertiary to Quaternary evolution of volcanism in the Aegean region, In: J.E. Dixon & A.H.F. Robertson (Eds), *The geological evolution of the Eastern Mediterranean*, Blackwell (Oxford): 824 pp.

Kissel, C., C. Laj & A. Poisson, 1987, A paleomagnetic overview of the tertiary geodynamical evolution of the Aegean Arc, *Terra Cognita*, **7**, 108.

Le Pichon, X. & J. Angelier 1979 The Hellenic Arc and trench system: a key to the neotectonic evolution of the Eastern Mediterranean, *Tectonophysics*, **60**, 1-42.

Le Pichon, X. & J. Angelier 1981 The Aegean sea, *Phil. Trans. R. Soc. Lond.*, **A300**, 357-372.

McKenzie, D.P., Active tectonics of the Alpine-Himalayan belt: the Aegean Sea and surrounding regions, *Geophys. J. Royal Astron. Soc.*, **55**, 217-254, 1978.

Mercier, J.L., Extensional-compressional tectonics associated with the Aegean arc: comparison with the Andean Cordillera of south Peru - north Bolivia, *Phil. Trans. Royal Soc. Lond.*, **A300**, 337-355, 1981.

Nolet, G. 1985 Solving or resolving inadequate and noisy tomographic systems. *J. of Comp. Phys.*, **61**, 463-482.

Spakman, W. 1985 A Tomographic Image of the Upper Mantle in the Eurasian-African-Arabian Collision Zone, *EOS*, Vol. **66**, No. 46, 975.

Spakman, W. 1986a Subduction beneath Eurasia in connection with the Mesozoic Tethys, *Geol. Mijnbouw*, **65**, 145-153.

Spakman, W. 1986b The upper mantle structure in the Central European-Mediterranean region, In: R. Freeman, St. Mueller & P. Giese (eds), *European Geotraverse (EGT) Project, the central segment*, European Science Foundation, Strassbourg, 215-222.

Spakman, W. & G. Nolet 1987 Imaging algorithms, accuracy and resolution in delay time tomography, In: N.J. Vlaar, G. Nolet, M.J.R. Wortel & S.A.P.L. Cloetingh (Eds), *Mathematical Geophysics: a survey of recent developments in seismology and geodynamics*, Reidel, Dordrecht, 155-188.

Spakman, W., M.J.R. Wortel and N.J. Vlaar, 1988, The Hellenic subduction zone: a tomographic image and its geodynamic implications, *Geoph. Res. Lett.*, in press.

Van der Sluis, A. & H.A. van de Vorst, 1987, Numerical solution of large, sparse linear systems arising from tomographic problems, In: G. Nolet (Ed), *Seismic tomography*, Reidel, Dordrecht, 1987, 53-87.

Vlaar, N.J. 1983 Thermal anomalies and magmatism due to lithospheric doubling and shifting, *Earth Planet. Sci. Lett.*, **65**, 322-330.

Vlaar, N.J. 1988 Subduction of young lithosphere: lithospheric doubling as a possible scenario, *Proceedings NATO Advanced Workshop on Crust-Mantle Recycling at Convergence Zones*, Antalya, Turkey.

Vlaar, N.J. and S.A.P.L. Cloetingh 1984 Orogony and ophiolites: Plate tectonics revisited with reference to the Alps. In: H.J. Zwart, P. Hartman & A.C. Tobi: Ophiolites and ultramafic rocks. *Geol. Mijnbouw* 63, 159-164.

Wortel, M.J.R 1980, Age-dependent subduction of oceanic lithosphere, *Phd. Thesis*, University of Utrecht, 147 pp.

Wortel, R., Seismicity and rheology of subducted slabs, *Nature*, **296**, 553-556, 1982.

FROM PLATE TECTONICS TO GLOBAL DOMAIN TECTONICS

Levent Gülen
Center for Geoalchemy
Department of Earth, Atmospheric and
Planetary Sciences
Massachusetts Institute of Technology

The assessment of crust/mantle recycling processes requires a detailed knowledge of the time-dependent, three-dimensional dynamic behavior of the earth as a whole system including physical and chemical properties of the components involved. Unfortunately, plate tectonics has so far failed to provide us with such a dynamic picture. Plate tectonics may be described as a kinematical description of the thin spherical outer shell of the earth. Even the driving mechanism of the plates still remains an unanswered question. As repeatedly pointed out, plate tectonics also cannot satisfactorily account for continental deformation.

A "Global Domain Tectonics" model will be presented for the sake of initiating, hopefully, an interdisciplinary discussion. This model is based on a striking scale-independent similarity and mirror-symmetry that have been recently observed for the pattern of continental deformation along the Alpine-Himalayan orogenic belt (Gülen, 1987). The Sistan Geo-suture (Figure 1) forms an axis of symmetry dividing this collisional belt into two parts: eastern Mediterranean in the west (from Greece to Afghanistan) and Asia in the east (from Afghanistan to Indonesia). The following analogies of regional geological structures, which may be referred to as "deformational domains," can be made for the two collisional subsystems (Figures 2, 3):

Eastern Mediterranean	Asia
Anatolian "plate"	Tibet
Taurides	Himalayas
E. Anatolia	Hindu Kush-Pamir
Greater Caucasuses	Tien Shan
Aegean Arc	Java-Banda Arc
Aegea	Indonesia
Black Sea	Tarim Basin
African-Arabian Plate	Indian Plate

173

S. R. Hart and L. Gülen (eds.), Crust/Mantle Recycling at Convergence Zones, 173–179.
© 1989 by Kluwer Academic Publishers.

Figure 1: Schematic map of active tectonics in Eastern Mediterranean and Asia. Solid arrows represent major plate/block motions with respect to Eurasia. Open arrow pairs indicate extrusion-related extensional areas. Note that the Sistan Geo-suture forms an axis of "mirror symmetry" dividing the Alpine-Himalayan orogenic belt into two parts. Map compiled after Molnar and Tapponnier (1975), Tapponnier et al. (1986).

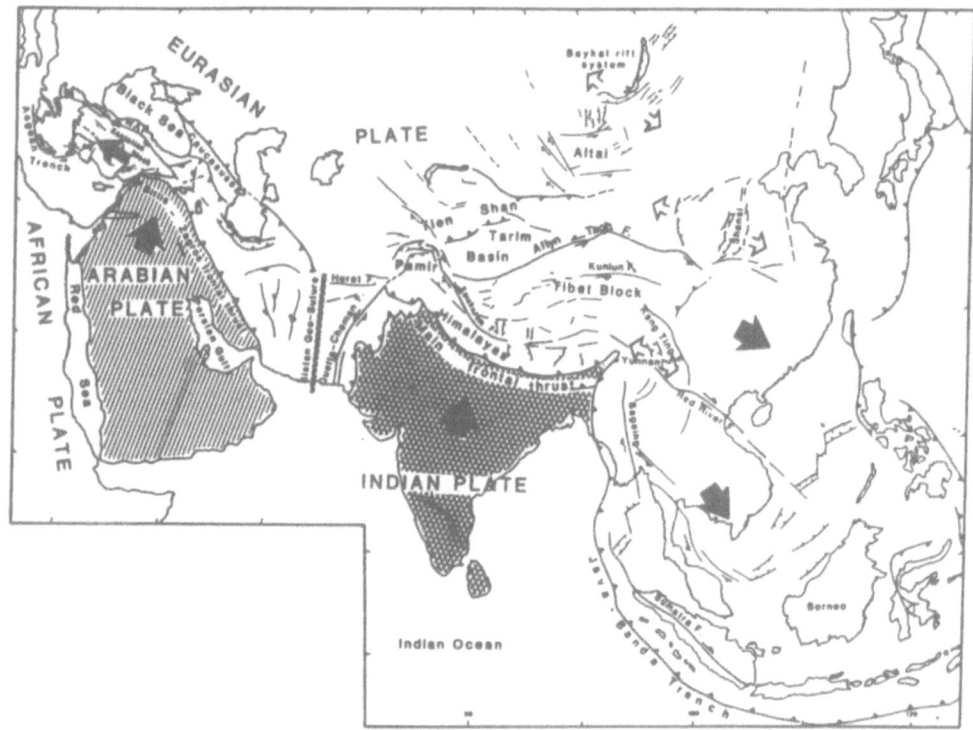

Numerous analogies can also be made on a fault-by-fault basis:

Eastern Mediterranean	Asia
North Anatolian	Altyn Tagh-Gansu
East Anatolian	Karakorum
Antalya	Sagaing
Burdur	Red River
Tabriz	Herat
Main Recent	Quetta-Chaman

Interestingly, the axis of symmetry, the Sistan Geo-Suture, that is obtained from the pattern of continental deformation coincides with the axis that separates divergent, absolute plate velocity vector fields obtained by using a hot-spot reference frame (Figure 4; Chase, 1978; Minster and Jordan, 1978). The same axis of symmetry is also applicable to the global seismic tomography domains (Masters et al., 1982; Anderson, 1984; Dziewonski, 1984, Forte and Peltier, 1987). All these observations may point out the presence of two, counter rotating, toroidal global mantle convection cells which do not correspond one-to-one to the plate structure of the earth's lithosphere. The recognition of at least two "Global Tectonic Domains:"

--Indicates "fractal" nature of the pattern of continental deformation.
--May provide the unification of "plate tectonics" and "continental tectonics"
--Allows for considerable refinement of our current understanding of the plate structure of the eastern Mediterranean.
--Suggests the importance of "mantle drag" among the plate driving forces especially for continent-bearing plates, thus supporting the "continental tectosphere" model of Jordan (1975). This, in turn, necessitates a coupling between continents and mantle convection cells.
--If these global tectonic domains are viewed as topological planes, then some of the plate kinematical problems associated with the eastern Indian Ocean may be solved, for example, by assuming that the Indian Plate may have once been attached to the Australian plate. In fact, this assumption is in perfect agreement with the model suggested by Powell et al. (1980) for a revised fit between east and west Gondwanaland before the break-up in the late Jurassic.
--Based on the fairly good similarity and symmetry of the Paleo-Tethyan suture zone on both sides of the axis, and the unidirectional northward drift of rifted-away Gondwanian continental slivers and their continued accretion to the southern margin of Eurasia

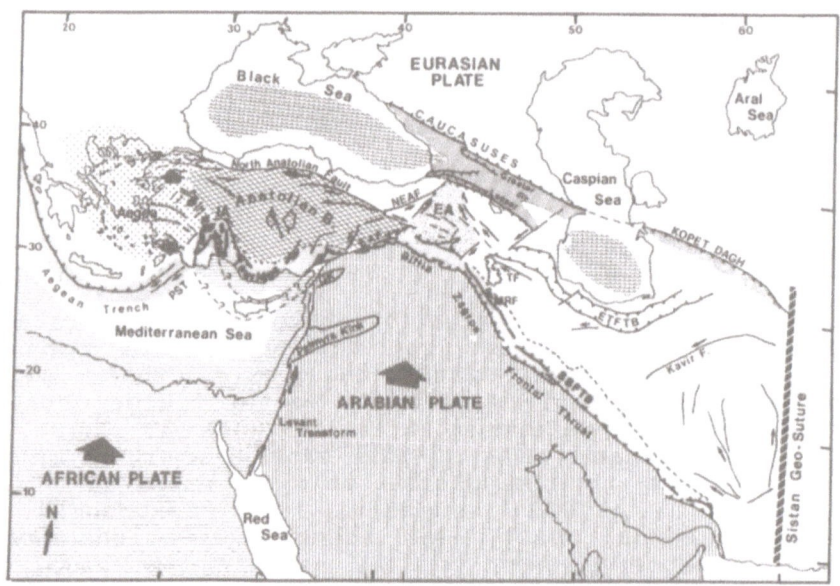

Figure 2: Schematic map showing major active tectonic elements in the Eastern Mediterranean. Deformation domains and plates are indicated by different patterns. Wavy pattern denotes the oceanic crust of the Black Sea and the southern Caspian Sea. Solid large arrows indicate the motions of African and Arabian plates with respect to Eurasia. The smaller arrow pair represents extension in the Aegea. Key to the abbreviations: IA: Isparta Angle, EA: Eastern Anatolian Deformational Domain, PST: Pliny-Strabo Trenches, NEAF: North East Anatolian Fault: EAF: East Anatolian Fault, MRF: Main Recent Fault, TF: Tabriz Fault, BF: Burdur Fault, GK: Gaziantep Kink, ETFTB: Elbroz-Talesh Fold and Thrust Belt, SSFTB: Sanandaj-Sirjan Fold and Thrust Belt.

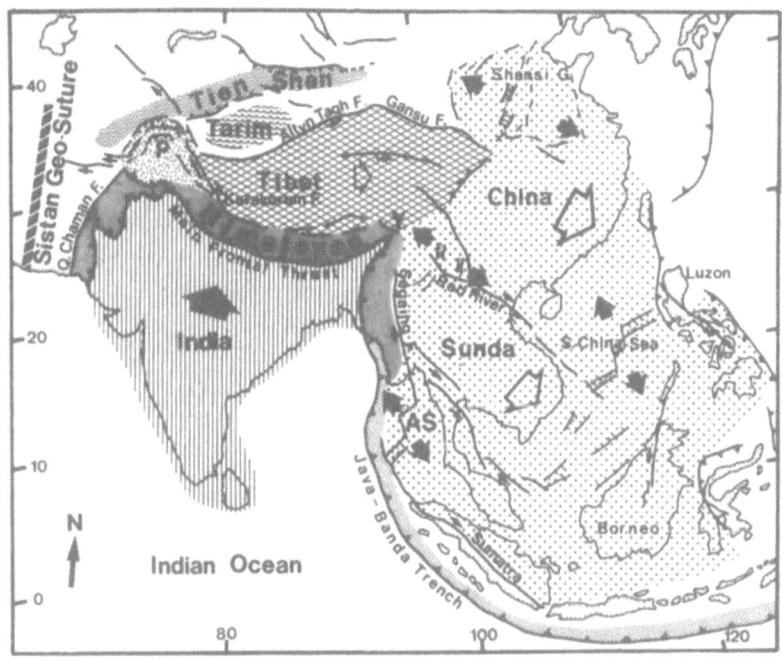

Figure 3: Schematic map of active tectonics in South eastern Asia. Deformational domains are indicated by different patterns. Solid large arrow represents the motion of Indian plate with respect to Eurasia. Open arrows indicate extrusion directions for Tibet, China and Sunda. Solid arrow pairs denote the extensional regions. Key to the abbreviations: P: Pamir Hindu Kush syntaxis, Y: Yunnan Syntaxis, AS: Andaman Sea. Map compiled after Molnar and Tapponnier (1975), Tapponnier et al. (1986).

178

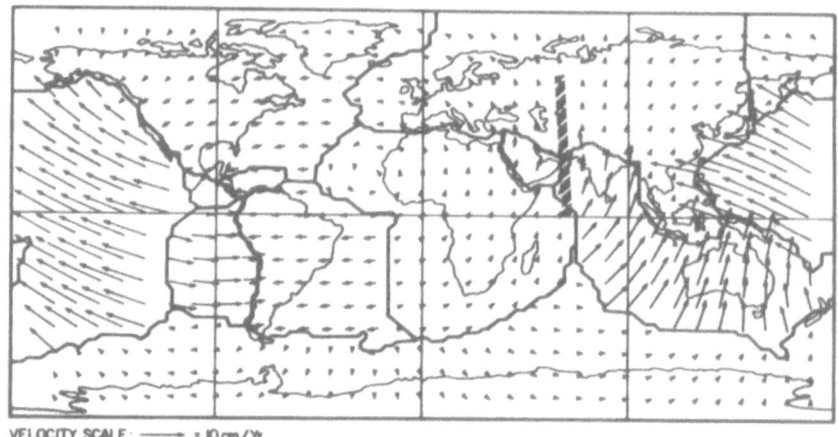

VELOCITY SCALE: ⟶ = 10 cm / Yr

Figure 4: Surface plate velocity field constructed from the absolute angular velocity vectors of Minster and Jordan (1978) in the "hot spot" reference frame after Forte and Peltier (1987). Note that the Sistan Geo-Suture forms a boundary separating counter-clockwise rotating African-Arabian-Eastern Mediterranean-European field and clockwise rotating Asian field.

since the Permian, one can infer that the existing first-order, two-cell, toroidal mantle convection has been operational at least since the Permian time.

--Although a significant amount of work needs to be done to reveal the details, the remarkable coincidence of axis of symmetries of seemingly independent data sets may suggest an intimate relationship among the pattern of continental deformation, plate driving forces, mantle structure, mantle convection, the geoid and possibly the geomagnetic field within a unified dynamic framework.

REFERENCES

Anderson, D.L., (1984) 'The earth as a planet: paradigms and paradoxes.' Science, 223, 347-355.

Chase, C., (1978) 'Plate kinematics: the Americas, East Africa, and the rest of the world.' Earth and Planet. Sci. Lett., 37, 355-368.

Dziewonski, A.M., (1986) 'Mapping the lower mantle: determination of lateral heterogeneity in P-velocity up to degree and order 6.' J. Geophys. Res. 89, 5929-5952.

Forte, A.M. and Peltier, W.R. (1987) 'Plate tectonics and a spherical earth structure: the importance of poloidal-toroidal coupling.' J. Geophys. Res. 92, 3645-3679.

Gülen, L., (1987) 'Striking similarities between active tectonics of the eastern Mediterranean and Asia.' Terra Cognita, 7, 117.

Jordan, T.H., (1975) 'The continental tectosphere.' Rev. Geophys. Space Phys., 13, 1-12.

Master, G., Jordan, T.H., Silver, P.G., and Gilbert, F., (1982) 'Aspherical earth structure from fundamental spheroidal-mode data.' Nature 298, 609-613.

Minster, J.B. and Jordan, T.H., (1978) 'Present-day plate motions.' J. Geophys. Res. 83, 5331-5354.

Molnar, P. and Tapponnier, P. (1975) 'Cenozoic Tectonics of Asia: Effects of a continental collision.' Science 189, 419-426.

Powell, C. McA., Johnson, B.D., and Yeevers, J.J., (1980) 'A revised fit of east and west Gondwanaland.' Tectonophys. 63, 13-29.

Tapponnier, P., Peltzer, G., and Armijo, R. (1986) 'On the mechanics of the collision between India and Asia.' In: Collision Tectonics, eds. M.P. Coward and A.C. Ries, Geol. Soc. Spec. Publ. London, 112-158.

ISOTOPIC AND CHEMICAL SIGNATURES OF RECYCLED OCEANIC AND CONTINENTAL CRUST

S.B. Jacobsen
Department of Earth and Planetary Sciences
Harvard University, Cambridge, MA 02138 USA

Recycling of oceanic and continental materials in subduction zones necessitates the incorporation of recycling explicitly into models for mantle-crust evolution. Isotopic and chemical constraints show that elements like Rb, U, Th and Pb will have very short residence times (0.1-1.0 Ga) in the mantle and be close to steady state. In contrast, elements like Sr, Nd, Sm, Hf and Lu will have much longer residence times (5-10 Ga) and be far from steady state.

Both recycled sediments and basaltic oceanic crust show large ranges in U/Pb, Rb/Sr, Sm/Nd and Lu/Hf ratios, that, with time, will cause large variations in Pb, Sr, Nd and Hf isotopic ratios. The major processes that contribute to these chemical heterogeneities in the oceanic crust are magmatic differentiation, hydrothermal alteration and seafloor weathering. The Rb-Sr and U-Th-Pb systems are affected in various ways by hydrothermal circulation and seafloor weathering. In contrast, the systems Sm-Nd and Lu-Hf are not affected by these processes. Studies of ophiolites ranging in age from 0.1 to 0.5 Ga provide a clue to what Pb-Nd-Sr systematics in oceanic crust of different ages might be. Comparison of ophiolite data with variations observed in the ocean island basalts suggests that some of the ocean island data arrays in Pb-Nd-Sr isotope space may be due to the remelting of recycled oceanic crust.

The chemical and isotopic variability in subducted sediments are due both to age differences between the various continental source areas of the sediments and fractionations that occur during weathering and erosion. The Pb-Nd-Sr systematics of subducted sediments have been established on the basis of data on river water suspended loads and deep sea sediments. Comparison with ocean island basalt data suggests that subducted sediments may also be an important component in some ocean island basalts. It is, however, possible that most of the subducted sediments and the most intensely altered oceanic crust are remelted in

S. R. Hart and L. Gülen (eds.), Crust/Mantle Recycling at Convergence Zones, 181–182.
© *1989 by Kluwer Academic Publishers.*

subduction zones and become part of island arc or
continental margin magmatism and are thus never remixed with
a larger volume of the mantle.

Most evidence indicates that recycling rates were
higher early in the history of the Earth. We may be
observing the evidence of much of this today as isotopic
heterogeneities in the sources of ocean island basalts.

MANTLE CONVECTION WITH ACTIVE CHEMICAL HETEROGENEITIES
- EXTENDED ABSTRACT -

Ulrich R. Christensen
Max-Planck-Institut für Chemie,
Saarstraße 23,
6500 Mainz, F.R. Germany

The main source of mantle heterogeneity is the subduction of
basaltic oceanic crust. Whether or not coherent lumps of continental
crust are recycled in significant amounts is unkown. These heterogeneities will - at least in parts - be torn into progessively longer
and thinner streaks (or schlieren, tendrils, bands, lamellae), giving
rise to a "marble cake mantle" (Allègre and Turcotte, 1986). The
question of how fast the thickness reduction proceeds to the point of
complete mixing has been studied by means of numerical modelling (Olson
et al., 1984; Hoffman and McKenzie, 1985; Gurnis 1986b). The various
results are in disagreement and seem to depend on model assumptions.
In all these models (with one exception: Gurnis, 1986a) the chemical
heterogeneity has been taken as passive, which means that differences
in physical properties, like density or viscosity, between the injected
former crust and the host rock are neglected. The purpose of this
study is to construct models of active chemical heterogeneity. It
splits into two parts. In the first part the consequence of a
viscosity difference between the streaks and their host rocks is
considered; it leads to anisotropic rheological behavior. In the second
part density differences are considered to determine the feasibility of
segregation of light and heavy material.

At moderate depth subducted oceanic crust will transform into
eclogite. The difference in viscosity between the eclogitic rock,
dominated by pyroxenes, and the peridotitic host, dominated by olivine,
is uncertain, but some orders of magnitude are likely (Avé Lallemant,
1978). In a marble cake mantle, consisting of many bands which are
nearly parallel on a local scale, this will lead to an effective
rheological anisotropy. A banded structure will deform much easier
under simple shear parallel to the banding than under normal stress
aligned with the banding (or pure shear). The viscosity will be a
tensor quantity. In a steady convective circulation the bands will
tend to align with the stream-lines. In two-dimensional numerical
convection models stream-line oriented viscous anisotropy with a
difference on the order 10-100 between "shear" and "normal" viscosity
is studied. One surprising result for bottom heated convection at
high Rayleigh-number is the formation of a large, entirely stagnant,
cell core, surrounded by a ring of rapid flows in the thermal boundary
layers (Fig. 1). This stagnant core remains a stable feature upon
various model modifications, e.g., variation of the aspect ratio, or
introduction of moderate temperature dependence of viscosity. It does,

S. R. Hart and L. Gülen (eds.), Crust/Mantle Recycling at Convergence Zones, 183–189.
© *1989 by Kluwer Academic Publishers.*

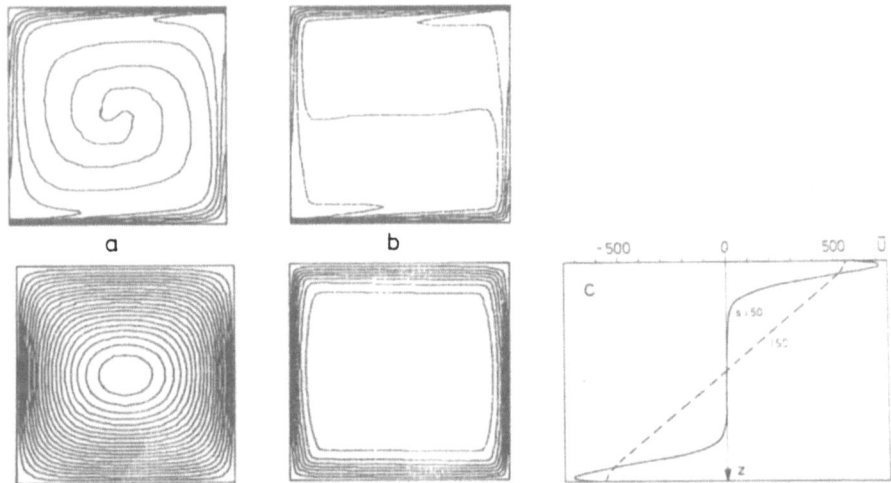

Figure 1 Bottom heated convection at a Rayleigh number of 500,000.
Isotherms and stream-lines are displayed for isotropic viscosity (a)
and for an anisotropy factor of 50 (b). In c) vertical profiles of mean
horizontal velocity for both cases.

however, disappear with significant internal heating. It may thus be
relevant for the upper mantle, being depleted in heat-producing
elements, within a two-layer model of mantle convection. Due to the
movement of plate boundaries, the circulation is certainly not steady,
and a model were the cell core serves as a long term chemical reservoir
of primitive mantle (e.g. Spohn and Schubert, 1982) does not seem
feasible. However, the temporary existence of large stagnant region
would prolong mixing times in the upper mantle.

In a strongly time-dependent flow the bands will no longer be
aligned with the instantaneous stream-lines. Rather they are
dynamically orientated by the flow pattern. Some numerical experiments
of time-dependent convection have been done for cells of large aspect
ratios. The results are somewhat preliminary. They suggest that
viscous anisotropy suppresses the onset of time-dependent boundary
layer instabilities to a certain degree (thus prolonging mixing
times). However, at sufficiently high Rayleigh number cold or hot
blobs detach from the upper and lower thermal boundary layer. This
leads to a repeated folding and refolding of the bands, and any large
scale coherence in the orientation of the bands may be quickly
destroyed. In this case the mantle would soon appear isotropic.
Further processes which would destroy the anisotropy are possible
folding or boudinage on the scale of individual bands, which is

especially expected for non-Newtonian flow laws and strong viscosity
differences between stiffer eclogitic bands and softer host rock.
Therefore, the occurence of anisotropic viscosity in the mantle must be
regarded as speculative. On the other hand, necking and boudinage
within the high-viscosity band of subducted crust could prolong the
life-time of large-sized heterogeneities very considerably. Once
broken into pieces, the eclogite blocks would be carried along in the
convecting mantle without much further stretching. A more detailed
account on the effects of anisotropic viscosity can be found in
Christensen (1987).

The eclogitic former crust has an excess density of about 200 kg m^{-3}
compared to peridotite. This contrast is likely to remain through all
the phase transitions, which take place on the way to the lower mantle
(Ringwood, 1982). The same is true - with the opposite density
contrast of about -40 kg m^{-3} - for the depleted dunite/harzburgite
layer below the oceanic crust. It has been speculated that the
subducted crust can sink under its own weight, segregate from the
mantle, and accumulate at same barrier, e.g. the core-mantle boundary
(Hofmann and White, 1982). Simple Stokes sphere estimates, however,
show that this segregation is not easy to achieve. With the standard
mantle viscosity of 10^{21} Pas a block of eclogite has to be 50 km large
to sink with 1 cm/a, a 5 km piece would make only 0.01 cm/a. This
consideration, and numerical models of constant viscosity convection
with buoyant markers, led Gurnis (1986a) to conclude that segregation
is at best marginally feasible and that mixing times would hardly be
affected by the density contrast. However, in the real mantle, due to
the temperature and pressure-dependence of the rheology, regions with a
viscosity several orders of magnitude below average are likely to
exist, namely the hot lower boundary layer, hot rising sheets or
plumes, and possibly a low-viscosity asthenosphere. In these regions
even kilometer-sized bodies may segregate. Numerical models may help
to decide if segregation is likely to occur with plausible assumptions
about the mantle rheology.

There are three different methods in use to model chemical
heterogeneity in a convecting medium, and tests have been performed to
determine their respective performance. Most reliable is the method of
marker chain, which delineates the boundary between different media.
Its disadvantage is that, as the boundary stretches, new markers have
to be inserted to maintain sufficient resolution, and soon the number
of markers becomes prohibitively large. Another method is to indicate
the volume of a chemical reservoir by the concentration of statisti-
cally distributed markers. Here the number of markers remains
constant, but care must be exercised in interpreting the results.
After sufficient straining has occured an isolated marker may stand for
an extremely long and thin schlieren stretched over an unknown region.
Even worse, when the markers are buoyant, the single marker may sink
(or rise) due to its weight, whereas the schlieren, for which it
stands, would not. Thus there is the possibility for spurious
segregation. In the third method chemical differences are represented
by a continuous function, analogous to the temperature, which is
advanced along a simple advection equation. Here the problem is

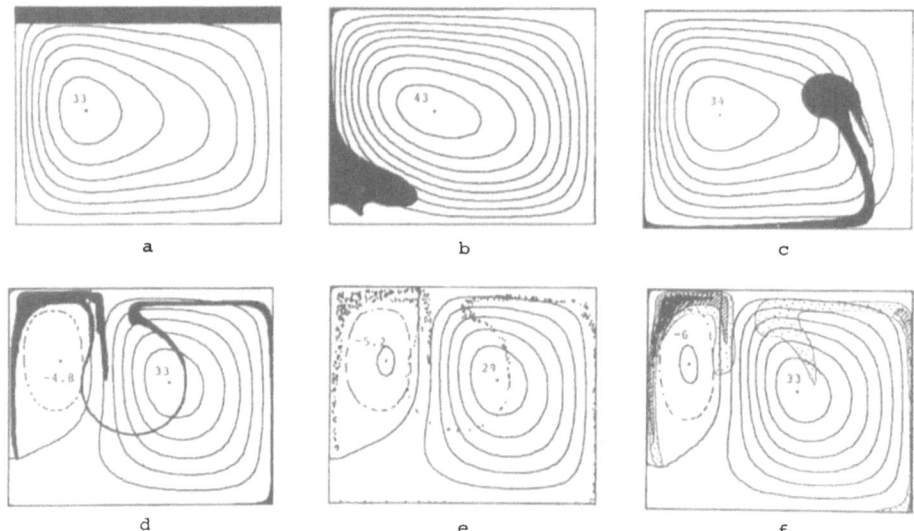

Figure 2   Convection at a Rayleigh number of 35000 with partial
internal heating. Initially a buoyant chemical layer of half the
thermal buoyancy was put on top. a)-d) Subsequent stages of the
evolution, obtained with the method of marker chain. e) Stage d
obtained with distributed markers. f) Stage d obtained by using a
continuous function to model the chemistry.

numerical diffusion, which tends to enhance mixing. In some sense the
method is complementary to the previous one, because it will under-
estimate any segregation rate. Over a time interval of moderate length
the three methods give consistent results, as shown by the model in
Fig. 2. A steady convection cell, partially heated from within, was
taken as initial state. At time zero a positively buoyant chemical
layer was placed into the upper boundary layer. The chemical buoyancy
is less than the thermal buoyancy and subduction is not inhibited.
After being subducted, the light material rises from the bottom as a
chemical plume. Next a cold thermal blob forms in the upper boundary
layer, detaches, and breaks through the layer of buoyant material. At
least until this stage all three methods give consistent results.
Although this simple model was only deviced to test numerical codes, it
shows one interesting feature: During subduction, the light material
may accumulate into a body which is thicker than the original "crust"
due to vertical compressive stresses (or dilatory strain) in the lower
part of the downgoing current. It is well known that focal mechanisms

indicate down-dip compression in subducted slabs below 400 km. This is
attributed to either a viscosity increase with depth, or perhaps a
barrier between upper and lower mantle. The subducted crust may be
squeezed into a body which is much wider than the original 6-7 km,
provided it is sufficiently deformable. This would strongly favour
segregation from the rest of the mantle.

Using the method of distributed markers, a model has been set up
which incorporates a number of relevant features. The viscosity is
taken to depend on temperature and pressure. At time zero, a double
chemical layer is placed on top with a heavy (eglocitic) crust and a
buoyant (harzburgitic) layer underneath, the latter being three times
as thick. The chemical buoyancy of the topmost layer is equal to the
maximum thermal buoyancy. Both layers combined are neutrally buoyant
in the chemical respect. Subduction is initiated in the center of the
cell and some thickening of the layers is observed (Fig. 3a). Subsequently, the flow becomes strongly time-dependent with plumes and
cold blobs detaching from the hot and cold boundary layer, respectively. Within the hot low viscosity regions in the lower boundary
layer and in the hot plumes strong segregation is indeed observed. At

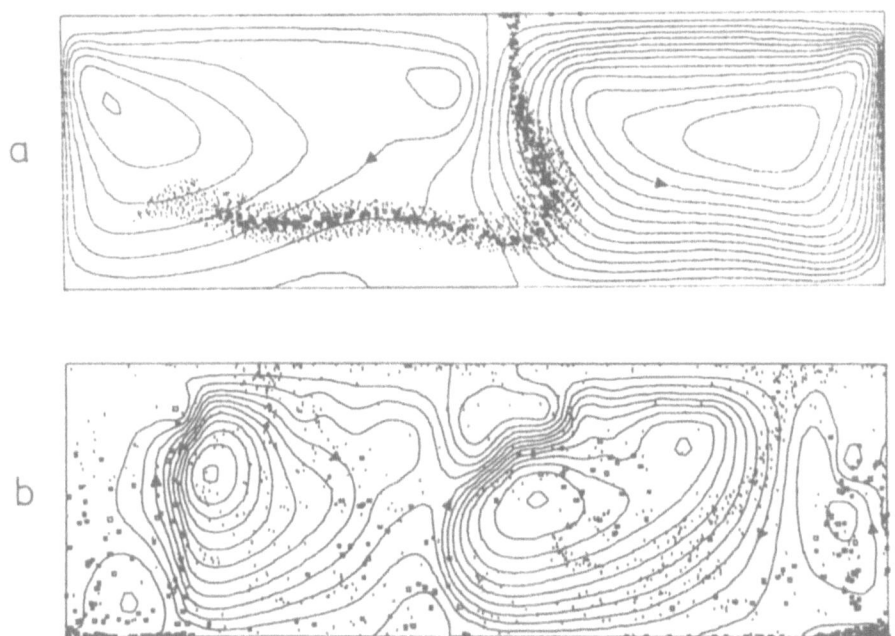

Figure 3   Convection with a Rayleigh number of 500,000. The viscosity
would vary by temperature alone by a factor of 250,000 and by pressure
alone by a factor of 500. Dots represent 'light' markers and squares
'heavy' markers.   a) Subduction of the chemical layer has completed. b)
After 8 transit times.

the end of the model calculation (Fig. 3b) part of the heavy markers
have accumulated at the bottom, whereas the light ones are more
scattered. It is not entirely clear, if spurious segregation due to
the discretization into markers plays a large role in this model. An
independent verification with another numerical method is warranted.
If the subducted crust is broken into boudins, the discretization into
markers would be quite realistic. At least in this special case
segregation must be accepted as a real feature of this model. It is
especially interesting that the segregated "mantle dreg" is not
distributed uniformly along the bottom, rather it piles up beneath
rising plumes. Although there is no wholesale rise of the heavy
material as anticipated by Hofmann and White (1982), some of it seems
to be entrained into the thermal plume. If it had been stored at depth
for a sufficient time, the entrainment may thus explain the isotopic
character of hot spot volcanism. The uneven distribution of mantle
dreg (or the existence of "inverse continents") at the core mantle
boundary is in accord with the observation of strong seismic
heterogeneity at or near the core-mantle boundary (e.g. Morelli and
Dziewonski, 1986). The tendency for a heavy "sediment" at the
core-mantle boundary to accumulate beneath rising plumes with some
entrainment into the plume is also seen in simple models starting from
a layer of uniform thickness at the bottom (Christensen, 1984).

The numerical models presented here are a first attempt to clarify
the behaviour of active chemical heterogeneities in the convecting
mantle. They draw attention to the key role, which the rheology of the
mantle may play for the questions of mixing and/or segregation. More
systematic and careful numerical modelling may enhance our
understanding of the geometry, distribution, and dynamics of chemical
reservoirs in the mantle.

References

Allègre, C. and D.L. Turcotte, 1986. Implications of a two-component
    marble-cake mantle, Nature, 323, 123-127.
Avé Lallemant, H.G., 1978. Experimental deformation of diopside and
    websterite, Tectonophysics, 48, 1-27.
Christensen, U., 1984. Instability of a hot boundary layer an
    initiation of thermo-chemical plumes, Annales Geophysicae, 2,
    311-320.
Christensen, U., 1987. Some geodynamical effects of anisotropic
    viscosity, Geophys. J. R. astr. Soc., in press.
Gurnis, M., 1986a. The effects of chemical density differences on
    convective mixing in the earth's mantle, J. Geophys. Res., 91,
    11407-11419.
Gurnis, M., 1986b. Stirring and mixing by plate-scale flow: large
    persistent blobs and long tendrils coexist, Geophys. Res. Letts.,
    13, 1474-1477.
Hoffman, N.R.A., and D.P. McKenzie, 1985. The destruction of
    geochemical heterogenities by differential fluid motions during
    mantle convection, Geophys. J. R. astr. Soc., 82, 163-206.

Hofmann, A.W. and W.M. White, 1982. Mantle plumes from ancient oceanic crust, Earth Planet. Sci. Letts., 57, 421-436.

Morelli, A., and A.M. Dziewonski, 1986. Topography of the core-mantle boundary determined with reflected and refracted waves; Abstract; EOS Trans. AGU, 67, 1099-1100.

Olson, P., Yuen, D, and D. Balsiger, 1984. Mixing of passive heterogeneities by mantle convection, J. Geophys. Res., 89, 425-436.

Ringwood, A.E., 1982. Phase transformations and differentiation in subducted lithosphere: Implications for mantle dynamics, basalt petrogenesis, and crustal evolution, J. Geol., 90, 611-643.

Spohn, T., and Schubert, G., 1982. Modes of mantle convection and the removal of heat from the Earth's interior, J. Geophys. Res., 87, 4682-4696.

CRUSTAL RECYCLING AND SUBDUCTION: SYSTEMS PERSPECTIVE

Jän Veizer
Derry Laboratory, Department of Geology
University of Ottawa and Ottawa-Carleton Geoscience Center
Ottawa, Ontario K1N 6N5
Canada

ABSTRACT. From the planetary point of view, the Earth can be imagined as a system comprised of intertwined natural populations propagated through time via recycling. This recycling, or "birth/death" process, imposes age patterns on natural populations of the solid earth, hydrosphere, atmosphere and living entities. Mathematically, the concept is analogous to that of population dynamics in living systems. The populations of the Earth system form a hierarchical structure. The hierarchy of crustal tectonic realms contains populations of $\leqslant 10^{24}$ to 10^{26} grams in size, with half-lives in the 10^7-10^9 years time range. The approximate parameters for major constituents of the oceans are $\leqslant 10^{24}$ grams and 10^2-10^7 years, for the atmosphere $\leqslant 10^{21}$ grams and 10^{-2}-10^7 years, and for living systems $\leqslant 10^{14}$-10^{19} grams and 10^{-3}-10^2 years, respectively.
Assuming a steady-state system, the calculated half-lives (τ_{50}) for the hierarchy of custal populations are the following: basins of active margins ~27 Ma, oceanic sediments ~51 Ma, oceanic crust ~59 Ma, basins of passive margins ~75 Ma, immature orogenic belts ~78 Ma, mature orogenic belts ~355 Ma and platforms ~361 ma. The half-lives for continental basement depend on the resistivity of the particular isotopic dating system to later resetting and range from ~673 Ma for K/Ar, through ~987 Ma for Rb/Sr and U-Th/Pb, to $\leqslant 1728$ Ma for Sm/Nd isotopic pairs. Based on these half-lives, the average probabilities of preservation for the tectonic realms over geologic time are summarized in Figure 1.
The observed oblivion ages for the above tectonic realms are usually 3.0 to 3.5 times their respective half-lives. This is less than theoretically expected for a steady-state system (Figure 2). For fast cycling populations (e.g. the oceanic crust) the measurements reflect increasing destruction rates with the ageing of the ocean floor. In contrast, for slow cycling populations (e.g. the continental basement) the deficiency of the old rocks reflects the rates of generation of continental crust in the course of the Archean and Early Proterozoic eons.

S. R. Hart and L. Gülen (eds.), Crust/Mantle Recycling at Convergence Zones, 191–195.
© 1989 by Kluwer Academic Publishers.

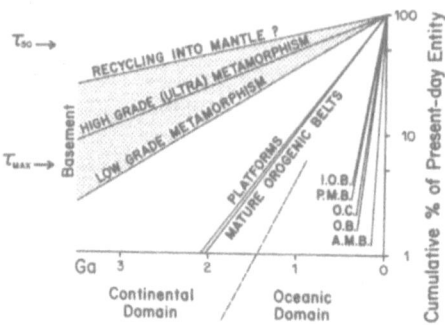

Figure 1. Preservation probabilities for major global tectonic realms calculated from recycling rates listed in Veizer and Jansen (1985). Explanations: A.M.B. - active margin basins, O.B. - oceanic intraplate basins, O.C. - oceanic crust, P.M.B. - passive margin basins, I.O.B. - immature orogenic belts. The preservation probabilities for basement were calculated on the assumption that low-temperature heating episodes reset mostly K/Ar ages, high-temperatures also Rb/Sr and U-Th/Pb, and recycling via mantle all isotopic pairs, including Sm/Nd. τ_{50} is half-life or 50th percentile and τ_{MAX} is oblivion age or 5th percentile. The latter is taken as the 95% probability of destruction and the point where the resolution of the data base approaches the background scatter. Reproduced from Veizer (1988) by permission of Pergamon Journals Inc.

The recycling of crustal tectonic realms is reflected in temporal distribution of mineral deposits. The inventories of global economic accumulations of metals (Pb, Zn, Cu, Au, Mn, Fe, Cr, Ni, U, Al and Sn) and their grouping into nine genetic categories show that the total cumulative tonnages for all categories of ores and metals decrease with geologic age. However, the genetic categories within each commodity show a consistent age pattern. In general, the rates of tonnage decrease with age are in the following succession: weathering crusts > detrital-sedimentary > hydrothermal > chemical-sedimentary > volcano-sedimentary > ultramafic > metamorphic ores (Figure 3). This progression reflects a decreasing role of recycling, and an enhanced evolutionary component, from weathering crusts to metamorphic categories of deposits.

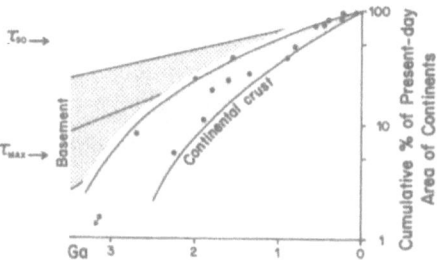

Figure 2. The observed present-day distribution of continental crust as compared to the probabilities of preservation for a steady-state system (cf. Figure 1). Reproduced from Veizer (1988) by permission of Pergamon Journals Inc.

Simulation of metallogenesis during geologic history, based on the concepts of recycling for host tectonic realms, enables differentiation of steady-state features of mineralization from its non-uniformitarian evolutionary component. The approach is designed to maximize the steady-state interpretation and the resulting picture probably understates the extent of evolution. The overall evolution of the Earth can be divided into five partially overlapping stages. These are: (1) The greenstone belt stage, dominant in the Archean and petering out ~1.8 Ga ago, with gold mineralization, Algoma type iron ores, and massive sulfides; (2) The cratonization stage, with the peak in the Late Archean and Early Proterozoic, and containing the paleo-placer gold and uranium deposits, the bulk of iron (Superior type) reserves, and related manganese deposits; (3) The rifting stage, at ~1.8 ± 0.3 Ga ago, typified by mafic-ultramafic dykes and layered complexes, and related mineralization of Cr, Ni, Cu, Au, and Fe as well as by frequent base metal deposits, hydrothermal uranium, and volcano-sedimentary manganese; (4) The stable craton phase, ~1.7 to 0.9 (or 0.6) Ga ago, with common alkalic volcanism and plutonism and a dirth of mafic to intermediate volcano-plutonic associations and metallic ores. The significant ores have been of exogenic type and confined to the cratonic sedimentary cover (unconformity uranium, chemical-sedimentary manganese, and Zambian type copper); (5) The Phanerozoic stage of continental dispersal, characterized by frequent and varied mineralization, particularly of the hydrothermal type. The fundamental control of mineralization appears to have been the

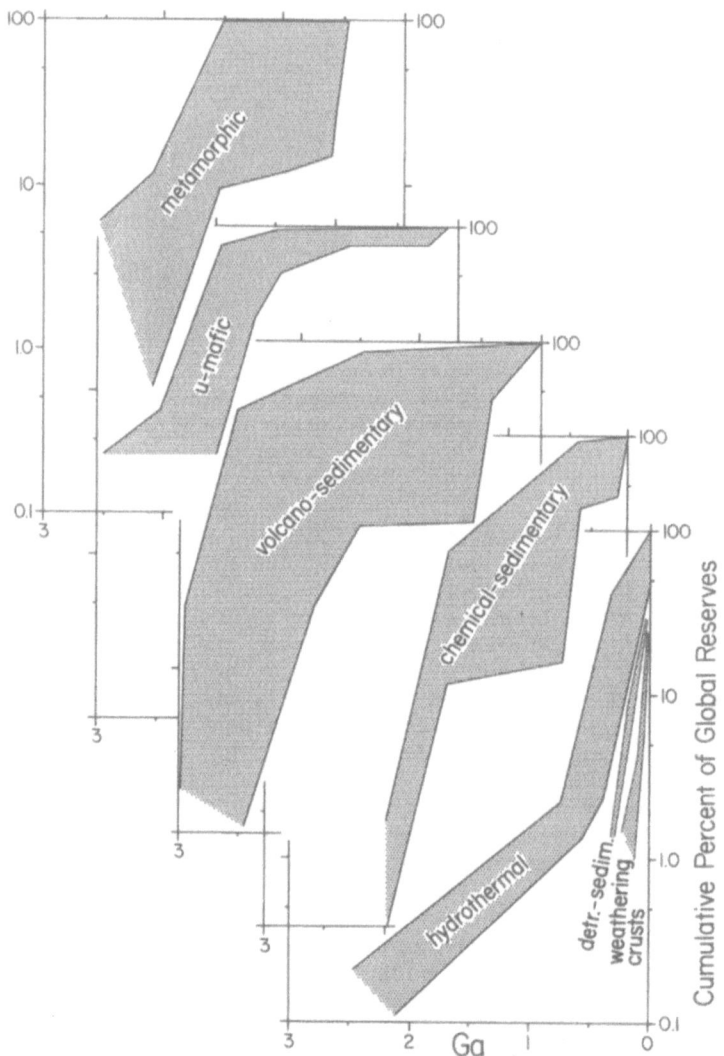

Figure 3. Generalized present-day age distribution patterns for genetic categories or ores. Reproduced from Veizer et al. (1989) by permission of American Journal of Science.

tectonic evolution of continental crust, with the stages of aggregation and cratonization (Archean-Early Proterozoic), subsequent rifting (~1.8 ± 0.3 Ga ago), thickening expressed in relative stability and intertia (Middle and Late Proterozoic), and fragmentation and dispersal of the existing (super) continent(s) during the Phanerozoic.

I believe that the above crustal evolution reflects the process of self-organization common to many natural, human, or technological systems. Such evolution proceeds via initial fast generation (and recycling) of poorly organized populations (e.g. the "permobile" Archean regimes), to be later outcompeted by slower but more organized and efficient populations (e.g. the ocean to continent autocatalytic tectonic module). In general system science theory this is sometimes referred to as the r to K transition.

In my view, the Sm/Nd systematics of sediments suggests that the global sedimentary mass has been about 90% cannibalistic since about the earliest Proterozoic. This imposes an upper limit on the possible sediment subduction into the mantle. The calculated upper limit is at least four times less than the quantities invoked from isotope systematics of the mantle. Consequently, the crust to mantle return flow must be accomplished by processes other than sediment subduction. Alternatively, the model estimates are based on faulty assumptions.

References.
Veizer, J. (1988) 'Continental growth: Comments on the "Archean-Proterozoic transition: Evidence from Guyana and Montana"' by A.K. Gibbs, C.W. Montgomery, P.A. O'Day and E.A. Erslev. Geochim. Cosmochim. Acta 52, 789-792.

Veizer, J. (1988) 'The evolving exogenic cycle'. In: Chemical cycles in the Evolution of the Earth, Eds. C.B. Gregor, R.M. Garrels, F.T. MacKenzie and J.B. Maynard, Wylie, New York, Chapter 5, 175-220.

Veizer, J. and Jansen, S.L. (1985) 'Basement and sediment recycling-2: Time dimension to global tectonics'. J. Geology 93, 625-643.

Veizer, J., Laznicka, P. and Jansen, S.L. (1989) 'Mineralization through geologic time: Recycling perspective'. Am. J. Sci., in press.

ONE-AND-A-HALF LAYER CONVECTION?...

H.-C. Nataf
École Normale Supérieure
Département de Géologie
24, rue Lhomond, 75231
Paris, Cedex 05, France

Recycling at convergence zones is very much dependent upon the fate of the subducting lithosphere in a convecting mantle. Hot spots also play a key role in the understanding of the circulation in the mantle.

In the present contribution, I review some of the seismological and geodynamical constraints on the layering of convection and on the origin of hot spots. This leads me to propose a speculative model in which convection is layered (upper mantle/lower mantle), but where narrow intense density anomalies (such as hot spots or slabs) can partly pass across the interface.

THE LAYERING OF CONVECTION IN THE MANTLE

The debate is an old one: is the circulation associated with the motion of the plates mantle-wide or restricted to the upper mantle?

Many of the arguments against layered mantle convection come from the ideas one has concerning the interface between the two layers. In particular, if heat can be carried only by conduction across the interface, one expects boundary layers to form above and below the interface. This yields a temperature increase across the boundary. Arguments have been developed to prove that this was in contradiction with observations. Davies (1983) proposed that this would reduce the viscosity of the lower mantle down to an unacceptable value. Spohn & Schubert (1982) have argued that the deduced temperature in the lower mantle would be too high. Dziewonski & Woodhouse (1987) have found that there was no evidence for boundary layers in the models of lateral heterogeneities of the transition zone obtained from seismic tomography.

I will challenge these arguments, using a revised value for the temperature increase expected across the interface. Indeed, recent progress in the understanding of convection with a temperature- and pressure-dependent viscosity (Richter et al., 1983; Morris & Canright, 1984; Christensen, 1984, 1985; Nataf, 1986) lead me to propose that the

197

S. R. Hart and L. Gülen (eds.), Crust/Mantle Recycling at Convergence Zones, 197–200.
© 1989 by Kluwer Academic Publishers.

temperature should increase by no more than 300 K above the interface and 150 K below it.

However, convincing evidence has been found by Creager & Jordan (1984, 1986) that the lower mantle was anomalously fast under several subducting slabs. If the mantle is layered, this would imply that downwelling convection currents systematically form under subducting slabs. This is not unexpected since laboratory experiments on layered convection show that "thermal coupling" is the preferred mode of coupling (Moreno & Nataf, 1985). Nevertheless, even in that case one does not expect the coupling to be systematic. Therefore, we will follow Creager & Jordan in assuming that their observations imply that some slabs do penetrate into the lower mantle. I will, however, argue that other observations suggest that this is not always the case, or else that slabs can rise up back into the upper mantle after they have warmed up. Indeed large scale heterogeneities are detected at the base of the upper mantle (Masters et al., 1982; Nataf et al., 1984; Woodhouse & Dziewonski, 1984): they are best explained as remnants of cold subducted lithosphere. In my view, the 2% (300K) anomalies observed are difficult to explain if convection is whole mantle.

THE ORIGIN OF HOT SPOTS

The fact that hot spots do not seem to move with respect to each other indicate that their origin is, at least, deeper than the base of the lithosphere. However, seismic surveys fail to detect anomalous mantle under most hot spots, except in the lithosphere. These two observations put together indicate that the "pipe" that feeds a hot spot is narrow (<100 km).

In turn, that and the fact that hot spots last for more than 100 Ma indicate that their feeding is regulated. This excludes, for example, that they form from large chunks of buoyant material, as proposed by Davies (1984), since these would tend to rise as a whole. Only in a thermal boundary layer can regulation occur: while material that has become unstable is rising up, new material replaces it that will also become unstable, and so on.

Which boundary layer do hot spots come from?

Let's suppose that they come from the lower boundary layer of the upper mantle. One needs then to explain how they can tap the lower mantle, in order to explain their "primitive" isotopic component. Laboratory experiments show that material of the lower layer can indeed be entrained by uprising currents in the upper layer (Moreno & Nataf, 1985). Quantification of this phenomenon shows that the contamination of the material could be large enough to explain the observations. It remains difficult, however, to explain why hot spots move so little. Indeed, the return flow in the upper mantle should make them move at a speed of at least 5mm/year, in contradiction to the observations.

Probably higher large-scale horizontal velocities are, in fact, needed in order to explain the relative homogeneity of MORB (Richter et al., 1982). Finally, hot spots do not seem to be associated with hot regions of the upper mantle as revealed by seismic tomography, while the lower mantle is hot where hot spots are found.

I will therefore consider that hot spots come from the thermal boundary layer at the core-mantle interface. The major problem is then to explain how they can rise across the interface between the upper and the lower mantles.

ONE-AND-A-HALF LAYER CONVECTION

The two major problems we are left with are of the same kind: how can buoyant material (hot: hot spots; cold: slabs) get across the interface between the two layers of the mantle while not disrupting the overall layering?

Our laboratory experiments indicate that when the density contrasts due to convection become as large as the intrinsic density contrast that makes the layering, the deformation of the interface becomes huge and overturn occurs. This is because large cushions of hot (cold) material get stuck below (above) the interface. In the case of narrow intense density anomalies, it is, however, possible that they could rise across the interface, _if_ they do not pile up at the interface, or if they do so only after they have lost most of their buoyancy. Thermals in a Newtonian fluid might never fulfill that condition, but I find it conceivable that the somewhat rigid slabs, and the hot spots that burn their way up, would.

Finally, I note that in such a model some material gets across the interface. Therefore, part of the heat can be advected through the interface, and the temperature increase across it is lowered.

REFERENCES

Christensen U.R., 'Heat transport by variable viscosity convection and implications for the Earth's thermal evolution,' Phys. Earth Planet. Int., 35, 264-282, 1984.

Christensen U.R., 'Heat transport by variable viscosity convection II: pressure influence, non-Newtonian rheology and decaying heat sources,' Phys. Earth Planet. Int., 37, 183-205, 1985.

Creager K.C. and T.H. Jordan, 'Slab penetration into the lower mantle,' J. Geophys. Res., 89, 3031-3044, 1984.

Creager K.C. and T.H. Jordan, 'Slab penetration into the lower mantle beneath the Mariana and other island arcs of the northwestern Pacific,' J. Geophys. Res., 91, 3573-3589, 1986.

200

Davies, G.F., 'Viscosity structure of a layered convecting mantle,' Nature, <u>301</u>, 592-594, 1983.

Davies G.F., 'Geophysical and isotopic constraints on mantle convection: an interim synthesis,' J. Geophys. Res., <u>89</u>, 6017-6040, 1984.

Dziewonski A.M. and J.H. Woodhouse, 'Global images of the Earth's interior,' Science, <u>236</u>, 37-48, 1987.

Masters G., T.H. Jordan, P.G. Silver and F. Gilbert, 'Aspherical earth structure from fundamental spheroidal-mode data,' Nature, <u>298</u>, 609-613, 1982.

Moreno S. and H.-C. Nataf, 'Laboratory models of two-layer convection,' (abstract), Terra Cognita, <u>5</u>, 144, 1985.

Morris S. and D. Canright, 'A boundary-layer analysis of Benard convection in a fluid with strongly temperature-dependent viscosity,' Phys. Earth Planet. Int., <u>356</u>, 355-373, 1984.

Nataf H.-C., 'Elements d'anatomie et de physiologie du manteau terrestre-tomographie sismique et convection experimentale,' These d'Etat, Univ. Paris-Sud, Juin, 1986.

Nataf H.-C., I. Nakanishi and D.L. Anderson, 'Anisotropy and shear-velocity heterogeneities in the upper mantle,' Geophys Res. Lett., <u>11</u>, 109-112, 1984.

Richter F.M., S.F. Daly and H.-C. Nataf, 'A parameterized model for the evolution of isotopic heterogeneities in a convecting system,' Earth Planet. Sci. Lett., <u>60</u>, 178-194, 1982.

Richter F.M., H.-C. Nataf and S.F. Daly, 'Heat transfer and horizontally averaged temperature of convection with large viscosity variations,' J. Fluid Mech., <u>129</u>, 173-192, 1983.

Spohn T. and G. Schubert, 'Modes of mantle convection and the removal of heat from the earth's interior,' J. Geophys. Res., <u>87</u>, 4682-4696, 1982.

Woodhouse J.H. and A.M. Dziewonski, 'Mapping the upper mantle: three-dimensional modeling of earth structure by inversion of seismic waveforms,' J. Geophys. Res., <u>89</u>, 5952-5986, 1984.

MINERAL PHYSICS AND MANTLE EVOLUTION

Raymond Jeanloz
Department of Geology and Geophysics
University of California
Berkeley, California 94720
USA

The properties of the upper mantle, including the presence and nature of the 400 km seismological discontinuity, are readily explained in terms of the properties of rock with a garnet periodotite composition having $Mg/(Mg + Fe) = 0.89$ ($\pm$ 0.02). Notably, the dominant minerals of the uppermost mantle, olivine (~50 percent), pyroxenes (~30 percent) and garnet (~15 percent), transform to denser phases with depth. The transformation of olivine to spinelloid (β-phase) and spinel (γ) structures at 13.4 GPa provides a natural explanation for the 400 km discontinuity. An average temperature of 1700 ($\pm$ 300) K is required at this depth according to the high P/T phase-equilibrium determinations, and this is compatible with petrological estimates of xenolith equilibration temperatures. At temperatures exceeding 1500-1800 K, the pyroxenes and garnets react to form a silica-rich garnet (majorite) over a broad pressure range, corresponding to depths of ~200 to 700 km; at lower temperatures (appropriate to subducted slabs) the majorite reacts to form silicate ilmenite.

In the lower mantle, at pressures exceeding 22 ($\pm$ 2) GPa, all of the important minerals of the upper mantle (olivine, pyroxenes and garnet) transform to an assemblage of silicate perovskite and minor coexisting oxide. As the $(Mg, Fe)SiO_3$ perovskite is stable throughout the depth (pressure) range of the lower mantle, it is the single most abundant mineral in the Earth: it makes up 70-95 percent of the lower mantle, or roughly 40 percent of the entire planet. Measurements of the melting temperature of this phase between 25 and 65 GPa, place an upper bound on the temperature of the lower mantle (to 1600 km depth) of 3000 ($\pm$ 300) K. The silicate perovskite exhibits unusual geochemical behavior, compared to the common minerals of the crust and upper mantle. For example, Fe-Si and Mg-Si coupled substitutions (with Si in > 6-fold coordination) have been documented in this mineral phase.

The density of the lower mantle is 2 ($\pm$ 1) percent larger than that of an upper-mantle composition at equivalent pressures and at temperatures exceeding 2000 K. For an upper mantle composition to satisfy the observed density of the lower mantle, a temperature of 1200 ($\pm$ 200) K would be required. This is well below the minimum estimate of 2000 K inferred for the top of the lower mantle from the petrological (phase equilibrium) constraints in the upper mantle and transition zone. Assuming an olivine to pyroxene ratio between 1:1 and 2:1, a composition of $Mg/(Mg + Fe) = 0.80$ ($\pm$ 0.02) satisfies constraints on density and temperature in the lower mantle. In addition, we find that the properties of the lower mantle can be satisfied with a silica-rich (e.g., pyroxene) composition.

Although elastic properties can not at present be reliably used to infer lower mantle composition, the density strongly suggests that the upper and lower mantle differ in bulk composition. With the evidence that all parts of the mantle are vigorously convecting, this

S. R. Hart and L. Gülen (eds.), Crust/Mantle Recycling at Convergence Zones, 201–202.
© 1989 by Kluwer Academic Publishers.

implies that the upper and lower mantle do not intermix significantly over geological time scales. Leakage between these regions (e.g., by rising plumes or sinking slabs) can plausibly occur, but global heat loss must be impeded by thermal boundary layers at ~700 km depth according to this result. Hence, temperatures in the lower mantle are likely to be close to the melting point (i.e., at ~ 2800 (± 200) K). In support of this conclusion are the high temperatures (~4700 K) currently estimated for the top of the outer core, based on high-pressure melting experiments on iron and iron alloys. Also, the lateral variations in wave velocity and the V_P/V_S ratio documented by seismic tomography can be explained in terms of partial melting in the lower mantle.

High-pressure studies of silicate melts indicate that magmas in the lower mantle are approximately neutrally buoyant relative to coexisting crystals. As a result, melts tend to be entrained by convection in the lower mantle, with little opportunity to separate from source regions. In contrast, it is the large buoyancy of melts in the upper mantle that leads to rapid heat transfer and geochemical differentiation near the surface. Thus, differentiation processes appear to be fundamentally different and much more sluggish throughout the bulk of the mantle (i.e., in the lower mantle), as compared with conventional thinking based on near-surface experience. Minor plumes of exceptional composition may violate this general conclusion and cause leakage between the upper and lower mantle, however.

Experiments with the laser-heated diamond cell demonstrate that liquid iron and iron alloys react extensively with crystalline silicates and oxides at pressures above 50 to 80 GPa. Thus, it appears inevitable that significant chemical reactions occur at the core-mantle boundary and that the composition of the core has evolved over geological time. That oxygen dissolves into the core, plausibly becoming the main alloying constituent of iron in the outer core, is supported by experiments which document the change of oxygen from lithophile to (at least partial) siderophile character. Apparently, the seismologically anomalous D" layer at the base of the mantle is the most significant metasomatic zone of the planet.

CRUST MANTLE RECYCLING: INPUTS AND OUTPUTS

Malcolm T. McCulloch
Research School of Earth Sciences
The Australian National University
Canberra, A.C.T., 2601, Australia

ABSTRACT. Geochemical and isotopic compositions of continental derived materials that are available as inputs into modern subduction zones are assessed. Oceanic sediments from fore-arc, back-arc and continental margins have a wide range of isotopic and chemical compositions reflecting a variety of source terrains. Intra-oceanic forearc sediments have compositons similar to adjacent arc volcanics while sediments deposited in the passive margin of continental shelves have Proterozoic and in some cases Archean provenances. This indicates that a wide spectrum of materials are available for incorporation into modern and also presumeably ancient subduction zones. Studies of blue-schist and associated eclogite terrains indicate sediment subduction to depths of ~20km to 60 km, but direct constraints on the quantity of continental derived sediments subducted into the deep mantle remains elusive.

An attempt is also made to identify the outputs of crust mantle recycling, that is magmas derived from mantle sources containing components of ancient recycled sediments. A select class of ultra-potassic magmas, the olivine and leucite bearing lamproites, appear to have the appropriate geochemical signatures but their source regions may be located in the lower subcontinental lithosphere rather than the upper mantle.

1. INTRODUCTION

With the possible exception of core formation, the singularly most important event in Earth history has been the differentiation and formation of the Earth's continental crust from the upper mantle. Understanding the processes which control crust-mantle interactions is an essential prerequisite in any attempt to constrain the overall growth and evolution of the continental crust as well as the upper and lower mantle. Recycling of continental crust into the upper mantle as opposed to intra-crustal recycling, can have a profound effect on the chemical composition of the mantle as well as limit the overall rate of growth of the continents.

Crust-mantle recycling occurs in three main tectonic settings (McLennan, 1987). The first and probably most important setting, is in subduction zones where sediments derived from either the upper continental crust or the arc-crust itself, are transferred into fore-arc regions and subducted together with oceanic crust. A variation of this scenerio is the tectonic erosion of continental or subcreted fore-arc crust (Karig and Kay, 1981) that can occur at the leading edges of subduction zones. Geophysical studies, in particular seismic, has shown the existence well developed horst and graben structures (eg Wortel et.al., 1985). These provide natural "traps" for sediments and hence conventional

S. R. Hart and L. Gülen (eds.), Crust/Mantle Recycling at Convergence Zones, 203–213.
© 1989 by Kluwer Academic Publishers.

arguments regarding the difficulty of subducting low density materials are no longer applicable. However what is not evident, is the flux of sediments that are ultimately delivered into the deep mantle via these traps? This is uncertain for two reasons. Firstly it is difficult to quantify the present-day sedimentary flux that is delivered to subduction zones (Karig and Kay, 1981) and is actually subducted as opposed to undergoing intra-crustal recyling. Secondly, relatively shallow (<100 km) island arc magmatic processes can result in an unknown proportion of the subducted sediment (and oceanic crust) undergoing partial melting and hence contributing to the island-arc products. This latter process is evident from the widespread presence of sedimentary components in most island arc volcanic rocks (Kay, 1980, Gill, 1981, McCulloch and Perfit, 1981, Davidson, 1986 and White and Dupre, 1986). Recycling of continental crustal materials within island arc systems may be viewed as a form of intra-crustal recycling as the arcs themselves are ultimately accreted onto and become part of the continental crust. The geochemical modification of the mantle wedge which occurs above subduction zones during arc volcanism, may however have significant longer term effects than is generally appreciated.

A second method of recycling continental crust into the mantle is via hydrothermal alteration of oceanic crust at mid-ocean ridge spreading centers. Enrichments in alkali elements and $^{87}Sr/^{86}Sr$ ratios and both depletions and enrichments of $^{18}O/^{16}O$ ratios have been documented at ocean -ridges (eg Muehlenbachs and Clayton, 1976) and inferred for the deep oceanic crust from ophiolite studies (Gregory and Taylor, 1981). Fluxes are quantifiable, being dependent mainly on the volume of spreading ridges, but as in the case of sediment subduction it is likely that a significant (but unknown) proportion of altered oceanic crust will undergo dehydration reactions and be partially or completely transferred into the mantle wedge immediately above the subduction zones and therefore is not delivered into the deep mantle.

The final mechanism for crust-mantle recycling is direct delamination of continental material from the base of continental lithosphere (McKenzie and O'Nions, 1983). The major difficulty with this hypothesis is that spinel or garnet lherzolite which is undoubtedly the major constituent of the sub-continental lithospheric mantle is significantly *less* dense than fertile peridotite (Ringwood, 1982). To overcome the lower intrinsic density of lherzolite a temperature difference of ~500^{0}C is required between the sub-continetal lithosphere and the surrounding mantle. This could possibly be achieved by double thickening of continental crust (eg. via continent-continent collision) and rapid burial of the lower crustal portion into relatively warm asthenospheric mantle (Houseman et. al., 1981). Crust-mantle delamination would then be dependent on the relative timescales for thermal re-equilibration and the rate of erosion of the upper crust versus the lowermost continental lithosphere. This process, if it occurs, is considered to be geochemically subordinate compared to recyling at subduction zones, due to the exceedingly low concentration of incompatible elements in sub-continental lithospheric materials. It should be noted however that lherzolites have a surprisingly diverse range of isotopic compositions (eg McDonough and McCulloch, 1987, Stosch and Lugmair, 1986) and if delamination does occur it may have a significant effect on mantle compositions.

From the preceding discussion it is clear that the main site for crust-mantle interactions is at subduction zones, with the major uncertainty in quantifying the extent of deep mantle (>100 km) recycling being the estimates of the geochemical fluxes of continental sediments and altered oceanic crust that are transported into deep mantle, rather than being transferred into island-arc magmas and hence ultimately being reincorporated as continental crust. To overcome this largely unresolved problem and yet still provide meaningful constraints on crust-mantle recycling, we will focus on two main aspects: the *inputs* and *outputs*. Inputs are considered to be the range of materials of continental derivation that are presently being delivered or may in the future be delivered to subduction zones of which the most important is oceanic sediments. In this paper preliminary results of systematic geochemical and isotopic studies of oceanic sediments from a variety of tectonic settings are presented. Compositions of oceanic sediments are compared with continental crustal components inferred for arc magmas as well as eclogitic xenoliths and metamorphics associated with subduction zones. This will provide a guide to the full compositional spectrum of materials that are currently being or may in the future be subducted, as well as compositions of materials that have been subducted to at least relatively shallow levels (~60 km) in the upper mantle. The second aspect that will be examined are the *outputs*, or products derived from partial melting of mantle sources dominated by or at least containing significant quantities of materials of continental derivation. If continental crustal recycling has occurred on a scale similar to that envisaged by Armstrong (1968) and DePaolo (1980) then products derived from partial melting of both shallow (ie sediment contaminated island arc magmas) and *deep* (>100 km) mantle of sources should be widespread and identifiable. The case will be made that some members of an unusual suite of igneous rocks, the ultrapotassic kimberlites and lamproites, may represent possible products of the latter process.

2. SUBDUCTION ZONE INPUTS

2.1 Oceanic sediments

Modern deep-sea sedimentary environments can be classified using plate tectonic theory into a number of catagories (Valloni and Maynard,1981). These include the fore-arc and back-arc regions of continental margin (e.g. Andes) and intra-oceanic arcs (e.g. Aleutian and Marianas), sediments deposited in trailing-edge basins produced during intercontinental rifting (e.g. Atlantic) and sediments from leading-edge strike-slip related basins (e.g. southern California). Due to the relatively short lifetime of oceanic crust (~10^7 years) compared to time scales for crust-mantle recycling (10^9 years) it is necessary to examine the composition of sediments in the complete spectrum of oceanic environments. From our knowledge of the limited longevity of oceanic crust, it is likely for example that even sediments deposited on the passive margins of continental shelves (eg. Atlantic coast of North America) will ultimately be involved in subduction processes. For this reason a study the Nd and Sr isotopic compositions (McLennan et al., 1985) of a suite of modern deep-sea sands which are representative of all of these environments has been undertaken. Preliminary results are shown in figure 1 together with additional data from the literature (McCulloch and Wasserburg, 1978, Goldstein and O'Nions, 1981, and White et al., 1985).

Deep sea sands, modern equivalents to ancient greywackes, are characterised by highly variable trace element and isotopic relationships, which can be related to particular tectonic environments. Samples from the fore-arc regions of intra-oceanic island arcs (Aleutians, Solomans and Marianas) have ε_{Nd} values (6.7 to 8.0) and in one case a LREE depleted pattern. Th/U ratios are low (1.0-2.4) reflecting the dominantly depleted mantle source of primitive island arcs. Sediments from both fore-arc and back-arc regions of more evolved arcs such as Sunda and Japan have REE patterns with substantial negative Eu-anomalies and ε_{Nd} values of from 1.5 to -14.3, consistent with a larger component of recycled upper continental crust. Continental fore-arc settings are represented by samples from the Andean subduction system, and have ε_{Nd} values ranging from 4.2 to -3.2. This range is very similar to that found in the volcanic and plutonic rocks of the Andes (Figure 3) implying that the fore-arc basins contain significant quantities of sediment eroded from the arc itself, indicating a highly

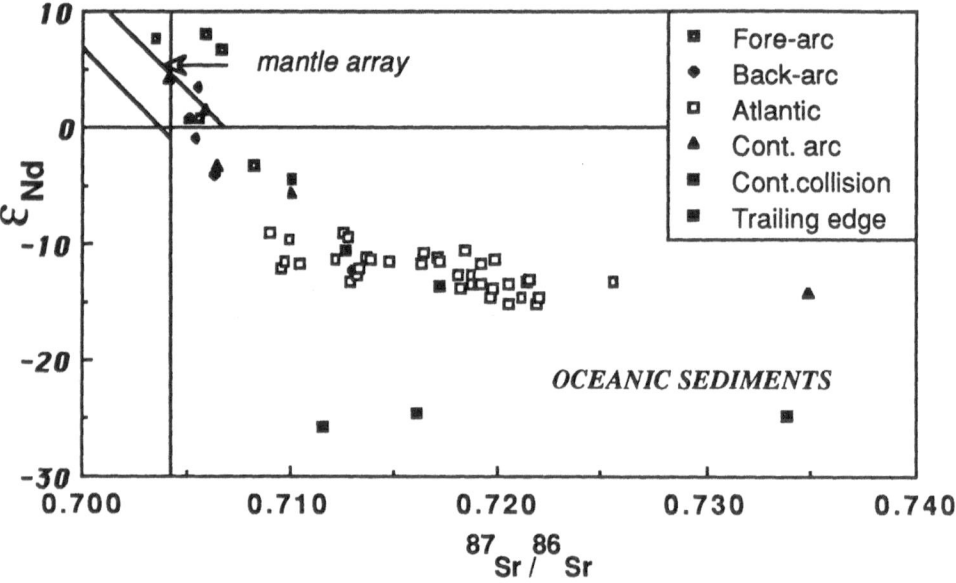

<u>Figure 1.</u> Nd and Sr isotopic compositions of modern oceanic sediments from a variety of tectonic settings (data from McLennan et. al., 1985). Atlantic sediments are from the Lesser Antilles fore-arc (White et. al.,1985). Samples with the most negative ε_{Nd} values indicate late Archean provenances, and are from trailing edge margins and a continental collision zone (Bay of Bengal).

cannibalistic system. However, probably the most significant finding of this study, is an Archean provenance for sediments from trailing edge basins. Two examples of this are sediments in the trailing edge basins adjacent to Newfoundland and Angola, (southern Africa) which have highly negative ε_{Nd} values (-24 to -25) and late Archean Nd model ages. This implies that ancient and isotopically highly evolved continental crust is in some cases deposited in ocean basins and at some stage may become available for subduction.

2.2 Eclogites and Blue-Schists

An alternative approach to constraining inputs into the crust-mantle recycling system, is the examination of subduction related eclogites and blue schist terrains. These can provide direct constraints on the types of materials that have been subducted to depths of ~20 km to 60 km. An example are the late-Cretaceous eclogites and blue-schists from the Hispaniola-Puerto Rico subduction zone. These occur along the northern coast of the Dominican Republic, immediately west of the Puerto Rico trench, as lenses and pods of eclogite enclosed in serpentine and interlayered greenschist-amphibolite marble units. Mineral compositions are similar to 'type C' eclogites with jaditic pyroxenes and almandine-grossular garnet. Calculated temperatures of equilibration range from 550°C to 750°C at pressures of 10 kb to 20 kb. The eclogites and blue-schists are LREE enriched (La_N/Yb_N=2.8-12.8) with significant Eu anomalies. Those with the highest REE contents also have enrichments in U, Th, Zr, Nb and Hf. In general the trace element

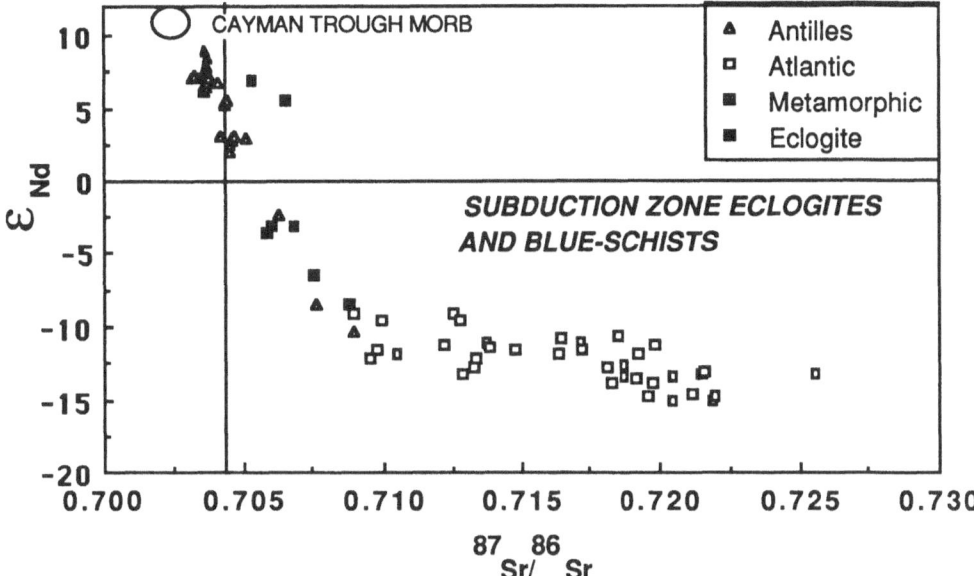

Figure 2. Nd-Sr isotopic compositions of eclogites and blue-schists from the Hispaniola-Puerto Rico subduction zone (Perfit and McCulloch, 1983). The eclogites and blue-schists have compositions consistent with mixtures of fore-arc sediments as represented by the Atlantic sedimentary data of White et. al., (1985). Compositions of arc magmas from the Greater Antilles (Perfit and McCulloch, 1983) and Lesser Antilles (Hawkesworth et. al., 1979, Davidson, 1985, 1986 and White and Dupre, 1986) overlap with the eclogites and subduction related metamorphics.

abundances are unlike those of 'Alpine' eclogites, but resembling those in the metamorphic rocks from the Puerto Rico trench. Initial Nd and Sr isotopic compositions in the eclogites range from ε_{Nd} = +6.9 to -6.4 and $^{87}Sr/^{86}Sr$ ratios vary from 0.70532 to 0.70755. As shown in figure 3 they plot along the same trend as defined by volcanic rocks from the Lesser Antilles (Hawkesworth et. al., 1979, Davidson, 1985, 1986 and White and Dupre, 1986) and the Greater Antilles (Perfit and McCulloch, 1983). The importance of the Hispaniola-Puerto Rico data to the general problem of crust-mantle recycling is that it provides a link between the sedimentary inputs and the outputs. This is indicated by the isotopic similarity between regurgitated sediment plus MORB mixtures (ie eclogites plus metamorphics) and the arc magmas. It should be noted however that there are generally significant geochemical discrepancies between the compositions of arc magmas and postulated sources (sediment-MORB mixtures) which is generally attributed to the complex role of melting and chemical transport processes that are occurring in the intervening mantle wedge (eg Kay, 1980, Gill, 1981, McCulloch and Perfit 1981 and Davidson, 1986).

3. OUTPUTS OF CRUSTAL RECYCLING

3.1 Island Arc Magmas

It is now generally recognised that in most island-arc systems there is some involvement of upper continental crust. Two examples of extreme contamination by terrestrial sediments are the Banda-Sunda arcs of Indonesia and the the Lesser Antilles arc of the Caribbean. Figure 3 illustrates the large range in Sr and Nd isotopic compositions found in a single island (Wetar) from the Banda arc. This together with sympathetic oxygen isotope (McCulloch et. al., 1984) and geochemical variations provides unequivocal evidence for the large-scale involvement of sediments. There are several questions that need to be considered. Firstly, are sedimentary signatures that are observed in arc magmas, actually those of subducted sediments incorporated at depth, or instead due to relatively shallow (within the arc crust) assimilation/partial melting of sediments accreted onto fore-arc prisms? In the central portion of the Banda arc (Wetar) thick sedimentary prisms are present, together with the geochemical evidence showing the control of relatively shallow plagioclase fractionation on assimilation and partial melting of sediments. This suggests that the upper crustal signatures are due to intra-crustal melting of the relatively thick fore-arc prisms. However in the case of the Lesser Antilles it has already been argued that subducted sediments are likely to an important component of arc magma genesis (see White and Dupre [1986] and Davidson [1986] for a detailed discussion of the Lesser Antilles). Thus no simple generalisations can be made, although it is clear that subducted sediments are an important component in some if not all arcs. The role of sediments in intra-oceanic arcs such as the Aleutian and Marianas arcs is more contentious due to their more subtle effects. Unambiguous evidence for subducted sediments appears to be provided by the existence of ^{10}Be (Tera et. al., 1986) and models requiring the addition of plume sources (Morris and Hart, 1983) are not warranted when a complete assessment of arc geochemistry is undertaken (e.g. McCulloch and Perfit, 1981, Perfit and Kay, 1986)

One aspect that has received little attention is the fate of the mantle wedge overlying the subduction zone. To account for many of the geochemical complexities of island arc magma genesis it is now generally recognised (e.g. Gill, 1982) that the arc mantle wedge has undergone significant enrichments of mobile and incompatible elements, induced for example by dehydration reactions in the downgoing slab. Upon cessation of subduction, or as a result of subduction reversal, this enriched portion of the mantle may again become part of the main convecting upper mantle. Thus enrichments produced at relatively shallow levels may be convected into deeper parts of the mantle. In the SW Pacific (Perfit et. al.,1982) and the Japan back-arc (Nakamura et.al.,1985), there is evidence that some ocean island volcanics (ie plume related volcanism) have been modified by interaction with arc contaminated upper mantle. Considering the present-day volumes of subduction enriched mantle wedges this may be an important although indirect mechanism for crust-mantle recycling. The key but still unanswered question, is what proportion of subducted sediments escape processing during island arc magmatism and are directly recycted into the deep mantle ? One means of assessing this question is to identify magmatic products derived from mantle source regions containing recycled continental crust. A case is presented for such an example in the following section.

3.2 Ultra-Potassic Magmatism

Ultra-potassic magmas (olivine and leucite lamproites), for example from the West Kimberley region of Western Australia, represent an interesting geochemical phenomenon as they possess high abundances of K_2O, (up to ~10%), incompatible elements (Rb, Ba

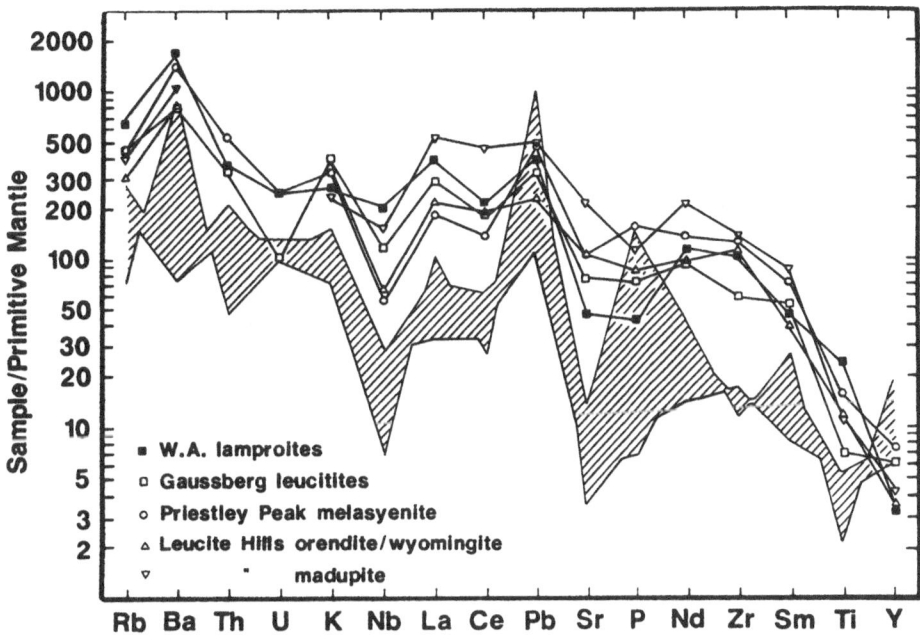

Figure 3. Normalised trace element abundances in order of increasing compatibility for ultra-potassic magmas from Western Australia (McCulloch et al.1983, Jaques et al.1984, and Nelson et al. 1986) Antarctica (Sheraton and England,1980) and Leucite Hills (Vollmer et al., 1984). Hatched area is the field for sediments (3 slates and a pelagic mud composite) from Thompson et al. (1982). Although the sediments have lower overall abundaces the relative patterns are very similar to the potassic magmas. In perticular both show enrichments in Pb and Ba and depletions in Nb. Normaliation is with respect to primitive mantle abundances (eg Nelson et al., 1986).

and LREE's) enrichments and high K_2O/Na_2O (>10), together with high MgO, Ni and Cr contents (Jaques et.al.,1984). In addition recent isotope studies (McCulloch et. al., 1983, Fraser et al., 1985 and Nelson et. al., 1986) have found Nd and Sr isotopic compositions requiring long term (>1000 Myrs) enrichments in Rb/Sr and Nd/Sm ratios. This unique combination of major and trace and isotopic characteristics is consistent with of partial melting of a depleted peridotite (garnet lherzolite or harzburgite) source that has been enriched in incompatible elements. The nature of the enrichment process is however contentious. It may be an essentially magmatic process, with enrichments being due to the accumulation of small degree partial melts from the LVZ, concomitant with incremental addition of phogopite and incompatible enriched titanates (eg Jaques et. al., 1984). Alternatively the enrichment may be due mixing of depleted mantle peridotite with subducted sediments (e.g. Nelson et.al. 1986 and Nelson and McCulloch, 1987).

210

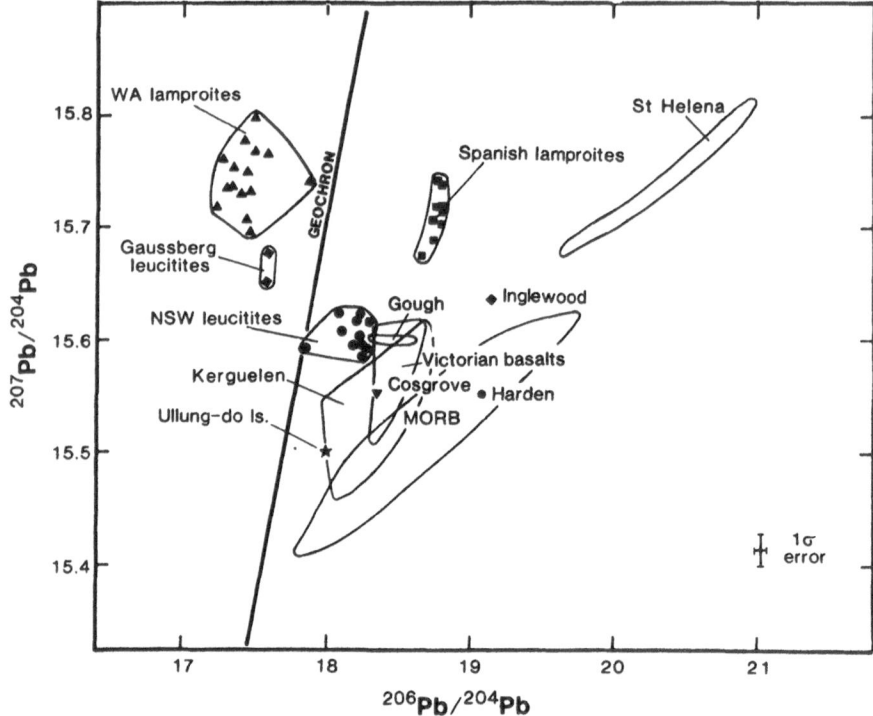

<u>Figure 4.</u> Pb isotopic compositions of ultra-potassic magmas (data from Nelson et.al., 1986). The West Australian lamproites lie significantly to the left of the geochron and have elevated $^{207}Pb/^{204}Pb$ but retarded $^{206}Pb/^{204}Pb$ ratios implying an early (>2000 Myr) high U/Pb stage followed by a low U/Pb stage. This is analogous to the Pb isotope evolution anticipated for ancient recycled sediments.

Pb isotopic studies (Fraser et. al., 1985, and Nelson et.al., 1986) suport a subduction origin, as the potassic magmas have compositions which lie to the left of the geochron (figure 4). This indicates a multistage history of U/Pb fractionation, involving an earlier high U/Pb stage to generate the high $^{207}Pb/^{204}Pb$, followed by a low U/Pb stage to account for the retarded $^{206}Pb/^{204}Pb$ compositions (figure 4). Ancient (>2000 Myr) subducted sediments would be expected to have this Pb isotopic character with the early high U/Pb stage reflecting the upper crustal sedimentary provenance and the low U/Pb ratio a result of U and Pb fractionation during weathering and sedimentation.

Due to the extensive pre-history of low U/Pb, this ancient Pb signature is markedly different from the Pb composition of modern sediments. Conversely, the Pb isotopic signature of OIB's is not consistent with involvement of ancient recycled sediments, although recycled oceanic crust appears to play an important role (White and Hofmann, 1982).

The mechanism by which sediments are incorporated into the ultra-potassic source remains speculative. The main constraint is that some ultra-potassic magmas are diamond bearing, indicating depths of melting of >100 km. It is therefore still possible that the subcontinental lithospheric mantle is the reservoir for the storage of these ancient, isotopically evolved components. If this is the case, then enrichment may have occurred during an earlier subduction event. An alternate origin is partial melting of a subducted megalith (Ringwood, 1982). This origin cannot be discounted but the close association of ultra-potassic magmas with old cratonic blocks suggests an important role for the sub-continental lithosphere.

4. CONCLUSIONS

1. Despite the still limited data base, it is clear that oceanic sediments, the main continental input into subduction zones, have highly variable isotopic compositions reflecting the heterogeneous composition of continental crust. Orogenic areas with early Proterozoic model ages appear to be the dominant present-day input but several examples of oceanic sediments having late Archean provenances have also been found.

2. Island arcs environments are the most obvious interface for crust-mantle recycling but the processes involved are complex and still poorly understood. There are difficulties in obtaining quantitative estimates of both continental crustal inputs and outputs. These arise not only from uncertainties in the esimates of volumes of sediments that are delivered to trenches, but also in distinguishing between subducted versus inter-arc crustal melting of continental components. In addition the role of the mantle wedge both as a buffer between the slab and arc magmas and its long-term effect in enriching the upper mantle is poorly constrained.

3. Specific suites of ultra-potassic magmas (e.g. Western Australian lamproites) are an example of magmas whose source region may have been enriched by deep (>100 km) recycling of continental sediments. The restriction of these magmas to old cratonic blocks, suggests however that their sources reside in the lower subcontinental lithosphere with enrichments reflecting ancient subduction related undeplating events. The Pb isotopic composition is unusual, lying to the left of the geochron and may therefore represent a complementary reservoir to MORB's and OIB's.

ACKNOWLEDGEMENTS

This synopsis has benefitted from collaborative studies and enlightning discussions with many colleagues including Dick Arculus, Wally Johnson, Linton Jaques, Eizo Nakamura, Dave Nelson, Scott McLennan, Bill McDonough and Mike Perfit. I am particularly grateful to Scott McLennan, Dave Nelson and Mike Perfit for permission to utilise unpublished data from ongoing studies.

REFERENCES

Armstrong R.L. 1968. A model for the evolution of strontium and lead isotopes in a dynamic earth. *Rev. Geophys.* 6, 175-199.

Davidson J.P. 1985. Mechanisms of contamination in Lesser Antilles island arc magmas from radiogenic and oxygen isotope relationships. *Earth Planet Sci. Lett.* 72, 163-174.

Davidson J.P. 1986. Isotopic and geochemical constraints on the petrogenesis of subduction-related lavas from Martinique, Lesser Antilles. *J. Geophys. Res.* 91, 5943-5962.

DePaolo D. 1980. Crustal growth and mantle evolution: inferences from models of element transport and Nd and Sr isotopes. *Geochim. Cosmochim. Acta.* 44, 1185-1196.

Fraser K.J., HAWKESWORTH C.J., ERLANK A.J., MITCHELL R.H. & Scott-Smith B.H. 1985. Sr, Nd and Pb isotope and minor element geochemistry of lamproites and kimberlites. *Earth Planet. Sci. Lett.* 76, 57-70.

Gill J.B. 1981. *Orogenic Andesites and Plate Tectonics.* Springer-Verlag, Berlin.

Goldstein S.L. & O'Nions R.K. 1981. Nd and Sr isotopic relationships in pelagic clays and ferromanganese deposits. *Nature* 292, 324-327.

Gregory R.T. & Taylor H.P. Jr. 1981. An oxygen isotope profile in a section of cretaceous oceanic crust, Samail Ophiolite, Oman: Evidence for $\partial^{18}O$ buffering of the oceans by deep (>5km) seawater-hydrothermal circulation at mid-ocean ridges. *J. Geophys. Res.* 86, 2737-2755.

Hawkesworth C.J., O'Nions R.K. & Arculus R.J. 1979. Nd and Sr isotope geochemistry of island arc volcanics, Grenada, Lesser Antilles. *Earth Planet Sci. Lett.* 45, 237-248.

Houseman G.A., McKenzie D.P. & Molnar P. 1981. Convective instability of a thickened boundary layer and its relevance for the thermal evolution of continental convergent belts. *J. Geophys. Res.* 86, 6115-6132.

Jaques A.L., Lewis J.D., Smith C.B., Gregory G.P., Ferguson J., Chappell B.W. & McCulloch M.T. 1984. The diamond-bearing ultrapotassic (lamproitic) rocks of the West Kimberley region, Western Australia. *In* Kornprobst J.ed. *Kimberlites II: The mantle and crust-mantle relationships.* pp 225-254. Elsevier, Amsterdam.

Karig D.E. & Kay R.W. 1981. Fate of sediments on the descending plate at convergent margins. *Philos. Trans. R. Soc. London. A.* 301, 223-251.

Kay R.W. 1980. Volcanic arc magmas: Implications of a melting-mixing model for element recycling in the crust-upper mantle system. *Journ. Geol.* 88, 497-522.

McCulloch M.T. and Wasserburg G.J. 1978. Sm-Nd and Rb-Sr chronology of continental crust formation. *Science* 200, 1003-1011.

McCulloch M.T. & Perfit M.R. 1981. $^{143}Nd/^{144}Nd$, $^{87}Sr/^{86}Sr$ and trace-element constraints on the petrogenesis of Aleutian island arc magmas. *Earth Planet . Sci. Lett.* 56, 167-179.

McCulloch M.T., Jaques A.L., Nelson D.R. & Lewis J.D. 1983. Nd and Sr isotopes in kimberlites and lamproites from Western Australia: an enriched mantle origin. *Nature* 302, 400-403.

McCulloch M.T., Compston W., Chivas A., & Abbot M.,1984. Neodymium, strontium, lead and oxygen isotopic and trace element constraints on magma genesis, on the Banda island-arc, Wetar, *27th I.G.C. (Moscow),* **11**, 344.

McDonough W.F. & McCulloch M.T. 1987. Isotopic heterogeneity in the southeast Australian subcontinental lithospheric mantle. *Earth Planet. Sci. Lett.* (in press).

McKenzie D. & O'Nions R.K. 1983. Mantle reservoirs and ocean island basalts. *Nature* 301, 229-231.

McLennan S.M., McCulloch M.T. & Taylor S.R. 1985. Trace element geochemistry of modern deep sea turbidite sands. *EOS* 66,1136.

McLennan S.M. 1987. Recycling of continental crust. *Pure and Applied Geophysics* (in press).

Morris J.D. & Hart S.R. 1983. Isotopic and incompatible element constraints on the genesis of island arc volcanics from Cold Bay and Amak Island, Aleutians, and implications for mantle structure. *Geochim. Cosmochim. Acta.* 47, 2015-2030.

Muehlenbachs K. & Clayton R.N. 1976. Oxygen isotope composition of the oceanic crust and its bearing on seawater. *J. Geophys. Res.* 81, 4365-4369.

Nakamura, E., Campbell, I.H., & Sun, S.-S. 1985. The influence of subduction processes on the geochemistry of Japanese alkaline basalts. *Nature* 316, 55-58.

Nelson D.R., McCulloch M.T. & Sun S.-S. 1986. The origins of ultrapotassic rocks as inferred from Sr, Nd and Pb isotopes. *Geochim. Cosmochim. Acta.* 50, 231-245.

Nelson D.R. & McCulloch M.T. 1987. Enriched mantle components and mantle recycling of sediments. *Proceed. 4th Int.Kim. Conf.* (in press)

Perfit M. & McCulloch M.T. 1982. Trace element, Nd-Sr isotope geochemistry of eclogites and blueschists from the Hispaniola-Peurto Rico Subduction Zone. *Transactions American Geophys. Union* 63, 1133.

Perfit M., McCulloch M.T. & Johnson R. 1983. Isotopic and trace element differences in late Cainozoic volcanic rocks from West Melanesia. *Geol. Soc. Aust. 6th Geol. Conv.* 9, 144

Perfit M. & Kay R.W. 1986. Comment on "Isotopic and incompatible element constraints on the genesis of island arc volcanics from Cold Bay and Amak Island, Aleutians, and implications for mantle structure" by J.D. Morris and S. R. Hart *Geochim. Cosmochim. Acta.* 50, 477-482.

Ringwood A.E. 1982. Phase transformations and differentiation in subducted lithosphere: implications for mantle dynamics, basalt petrogenesis, and crustal evolution. *J. Geol.* 90, 611-643.

Sherton J.W. & England R.N. 1980. Highly potassic mafic dykes from Antarctica. *J. Geol.Soc. Aust.* 30, 295-304.

Stosch H.-G. & Lugmair G.W. 1986. Trace element and Sr and Nd isotope geochemistry of peridotite xenoiths from the Eifel (West Germany) and their bearing on the evolution of the subcontinental lithosphere. *Earth Planet Sci. Lett.* 80, 281-298.

Taylor S.R. & McLennan S.M. 1985. *The Continental Crust: Its composition and evolution.* Blackwell, Oxford, U.K.

Tera F., Brown L., Morris J., Sacks I.S., Klein J. & Middleton R. 1986. Sediment incorporation in island arc magmas: Inferences from [10]Be. *Geochim. Cosmochim. Acta.* 50, 535-550.

Valloni R. & Maynard J.B. 1981. Detrital modes of recent deep-sea sands and their relation to tectonic setting: a first approximation. *Sedimentology* 28, 75.

Vollmer R., Ogden P., Schilling J.-G., Kingsley R.H. & Waggoner D.G. 1984. Nd and Sr isotopes in ultrapotassic volcanic rocks from the Leucite Hills, Wyoming. *Contrib. Mineral. Petrol.* 87, 359-368.

White W.M. & Hofmann A.W. 1982. Sr and Nd isotope geochemistry of oceanic basalts and mantle evolution. *Nature* 296, 821-825.

White W.M., Dupre B. & Vidal P. 1985. Isotope and trace-element chemistry of sediments from the Barbados Ridge-Demerara Plain region, Atlantic Ocean. *Geochim. Cosmochim. Acta.* 49, 1875-1886.

White W.M. & Dupre B. 1986. Sediment subduction and magma genesis in the Lesser Antilles: Isotopic and trace element constraints. *J. Geophys. Res.* 91, 5297-5941.

Wortel M.J.R. & Cloetingh S.A.P. 1985. Accretion and lateral variations in tectonic structure along the Peru-Chile Trench. *Tectonophysics* 112, 443-462.

THE STRUCTURE OF MANTLE FLOW AND THE STIRRING AND PERSISTENCE OF CHEMICAL HETEROGENEITIES

Geoffrey F. Davies
Research School of Earth Sciences
Australian National University
GPO Box 4, Canberra, ACT 2601
Australia

ABSTRACT.   The bathymetry of the oceans and other observations require mantle convection to comprise a plate-scale flow which penetrates deep into the mantle and plumes rising from great depth. Material injected into this regime can persist for billions of years before being sampled at a spreading center or a hotspot.

1. INTRODUCTION

The occurrence of ancient chemical heterogeneities in rocks derived from the mantle, combined with the direct inference from plate tectonics that the mantle is convecting, has led many people to conclude that the mantle must be chemically layered in order to preserve distinct source types for the required billions of years. The viability of this conclusion has depended on two things: a lack of definitive geophysical evidence for or against such layering, and the assumption that mantle convection would cause chemical heterogeneities to be homogenised in much less than a billion years.

Here, a new and quite strong geophysical argument is presented which weighs heavily against a chemical boundary in the upper part of the mantle, and quantitative results are presented which show that some heterogeneities can indeed survive mantle convection for billions of years, contrary to most people's intuition (including mine) and some conclusions drawn from recent analyses.

Although mantle convection is commonly regarded as poorly constrained, there are in fact several well-established sets of obervations which potentially provide quite strong constraints, including the earth's gravity field, geoid, heat flux and topography, deep earthquakes, seismic tomography and chemical and isotopic variations in mantle-derived rocks. The best-established of these, the earth's topography, provides a direct and robust constraint on the flux of hot, buoyant material to the base of the lithosphere, and this limits the location of any chemical boundary in the mantle to aproximately the lower quarter of the mantle. This is demonstrated in section 2 below

215

S. R. Hart and L. Gülen (eds.), Crust/Mantle Recycling at Convergence Zones, 215–225.
© 1989 by Kluwer Academic Publishers.

using two-dimensional, numerical convection models and then using more general arguments which allow the approximate quantification of other forms of topography, including hotspot swells.

Layering is only one aspect of the structure of mantle convection which is relevant to stirring of heterogeneities. There may be modes of convection other than the "large scale" flow associated with the moving plates, and plumes, upper mantle cells and upper boundary layer instabilities have all been suggested. Topography again constrains each of these to be minor, plumes being the only "small-scale" mode for which there is clear evidence. Small-scale modes are considered in section 3.

If plate-scale, deep mantle flow is dominant, then residence times are several billion years and survival times of heterogeneities range up to more than 1 billion years. The critical recent finding is that a significant fraction of a heterogeneity remains compact for a very long time, even though much of it gets strung out into tendrils which increase in length and decrease in thickness exponentially with time. With moderate increases in viscosity with depth, which are indicated by some independent lines of evidence, survival times of several billion years are readily attainable.

2. OCEAN BATHYMETRY AND MANTLE LAYERING

If mantle convection were layered, with a boundary at 670 km depth, then the upper mantle would have to be heated mainly from below, because upper mantle radioactivity is sufficient to account for only a few per cent of the surface heat flow (Jochum et al., 1983). This means that there would be a hot boundary layer at the base of the upper mantle through which most of this heat would flux, and from which buoyant material would rise.

Numerical models show that the topography generated when such material approaches the surface is comparable to that due to the cooling and subsiding top thermal boundary layer. Figure 1 shows an example where a spreading center is migrating over a fluid heated from below. The hot fluid rising from the base has not kept up with the spreading center, and is swept further away to the left as it approaches the surface. The result is a gross asymmetry of topography, as shown in Figure 2: the spreading center has migrated right off the topographic high. Perturbations to the gravity and geoid, not shown here, also show large asymmetries (80 mgal and 30 m, respectively). Even where the spreading center is stationary, large positive anomalies in gravity (50 mgal) and the geoid (15 m) occur, along with deviations from best-fit square-root-of-age subsidence by hundreds of meters near the spreading center.

The existence of a hotspot swell like Hawaii's depends on new topography being elevated as the plate passes over the hotspot source. If the swell width is w, its mean height is h and the plate's velocity over the source is v, then we can define a swell magnitude, S, as

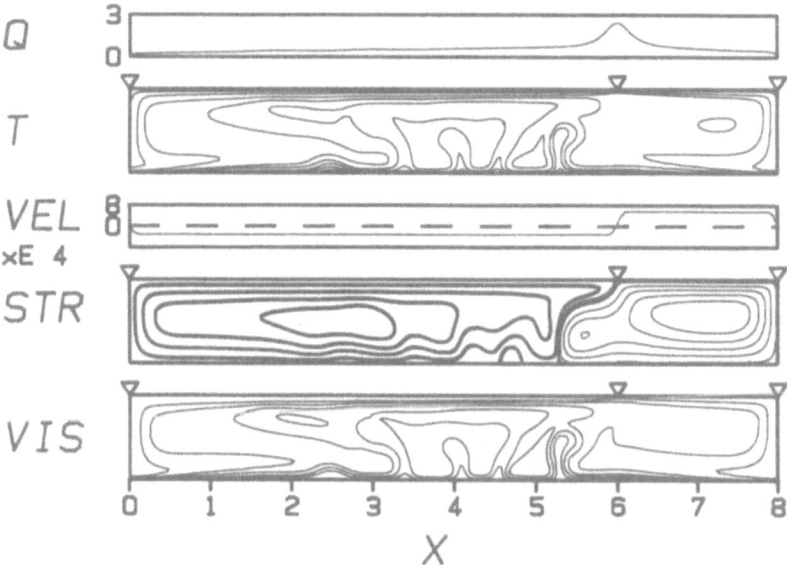

Fig. 1. Convection under a migrating spreading center and with base heating. Horizontal velocity (VEL) on the top surface is imposed to simulate diverging plates. The spreading center started at the middle and its present location is marked by the triangle. Other boxes show surface heat flux (Q), isotherms (T), streamlines (STR) and temperature-dependent viscosity (VIS).

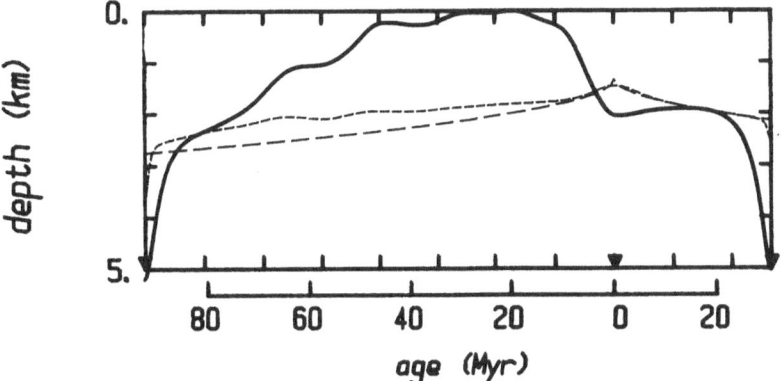

Fig. 2. Topography (depth below maximum) of the model in Figure 1 (solid curve). Also shown is the topography predicted by a cooling halfspace (long-dashed) and from the thickness of the top thermal boundary layer (short-dashed). From Davies (1987a).

$$S = hwv \qquad\qquad (1)$$

which, by isostasy, is equal to the buoyancy flux of material to the base of the lithosphere. It is also a measure of the heat flux, Q, carried to the base of the lithosphere:

$$Q = (\rho C_p/\alpha)S \qquad\qquad (2)$$

For the Hawaiian swell, with a width of about 1,000 km, a mean height of about 900 m and $v = 96$ mm/yr, $S = 2.7$ m^3/s and $Q = 2.2 \times 10^{11}$ W, compared with a total heat flux out of the earth of about 4.2×10^{13} W. For the Cape Verde swell, $v < 5$ mm/yr, $h \approx 1400$ m, $S < 0.22$ m^3/s and $Q = 1.8 \times 10^{10}$ W. This makes it clear that the Cape Verde swell, and others under slow plates, are large because the buoyant material rising under them tends to accumulate, while the Pacific hospot swells are large because the underlying plume sources are large. Davies (1987a) estimates a total plume buoyancy flux of no more than about 30 m^3/s, transporting no more than about 2.5×10^{12} W, or only about 6% of the earth's total heat flux.

It is clear from these numbers that plumes cannot account for the flux of buoyant material which would have to be rising from a 670 km boundary layer. This quantifies the observation that hotspot swells are not major features of the earth's topography, comparable with the mid-ocean ridge system.

Mid-ocean ridges and hotspot swells are the only evident candidates for being the topographic expressions of buoyant material rising from a 670 km boundary layer, and it seems that they can tolerate only about 10% of the necessary flux. This argues strongly against the existence of such a boundary layer in the upper part of the mantle.

The plume flux is actually consistent with estimates of the heat flux out of the core, which supports the idea that plumes originate from a thermal boundary layer at the core-mantle boundary.

3. SMALL-SCALE FLOW

Two other types of mantle flow have been postulated: upper mantle cells or rolls (Richter, 1973) and boundary-layer instabilities at the base of the mantle (Parsons and McKenzie, 1978). Arguments and models like those given above demonstrate that any upper mantle mode, such as rolls with a width comparable to the depth of the upper mantle, would generate large topographic anomalies (1 km or more, peak to trough) if it transported a large fraction of the earth's heat flux (Fleitout and Yuen, 1984; Craig and McKenzie, 1986; Davies, 1987b). Such anomalies have never been distinguished from hotspot tracks, and such modes probably account for only a minor heat flux if they exist at all.

Boundary-layer instabilities in the lower oceanic lithosphere have been postulated as the reason that old oceanic lithosphere tends to be shallower than predicted by the cooling halfspace model (Parsons and McKenzie, 1978). O'Connell and Hager (1980) gave a simple argument that such an instability would not in fact slow the subsidence, and might actually accelerate it. In Figure 3, one such boundary layer instability is seen occurring in a numerical

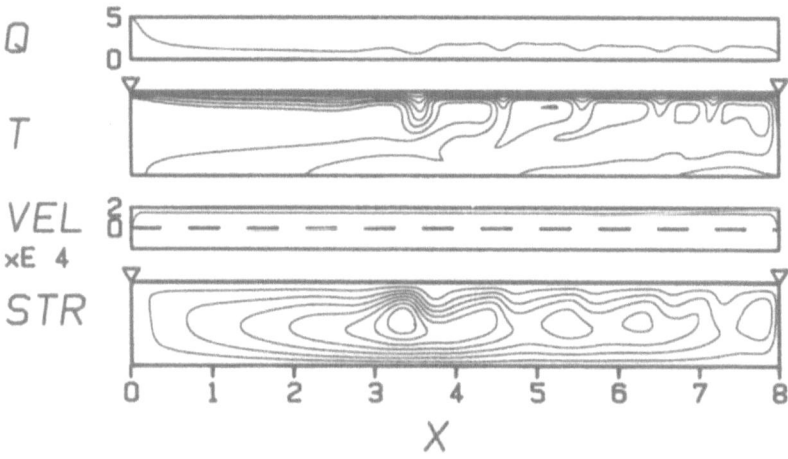

Fig. 3. A boundary-layer instability in a single, long convection cell. A uniform horizontal velocity (VEL) is imposed on the top, simulating a single plate moving to the right. Format is the same as in Figure 1, except that here the viscosity is uniform.

convection model. The topography resulting from it is shown in Figure 4. The main effect on the topography is a series of large depressions where cool fluid is sinking away from the boundary. A secondary effect is a small temporary elevation at the onset of the instability. The overall subsidence is not stopped. This instability happens to be transverse to the plate direction, but the longitudinal mode usually considered would have analogous effects. Thus the "flattening" bathymetry of old lithosphere is not evidence for a boundary-layer instability, and if such a mode exists, it must be minor. The best evidence is a series of near-ridge, very small-scale geoid lineations (Haxby and Weissel, 1986). The deviation of the old sea floor from the simple square-root-of-age subsidence seems to be quite well explained by the greater density of hotspot tracks there (Heestand and Crough, 1981; Schroeder, 1984).

220

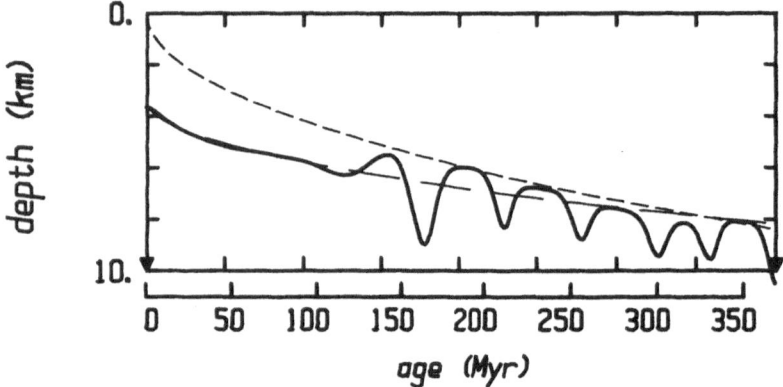

Fig. 4. Topography calculated from the model in Figure 3 (solid curve). The subsidence predicted from the cooling halfspace model is shown short-dashed, while the long-dashed curve is the best-fitting square-root-of-age curve. From Davies (1987b).

4. THE STRUCTURE OF MANTLE FLOW

A fairly simple picture of mantle convection emerges from the above results. There is no boundary at 670 km depth. Plate-scale flow penetrates deep into the mantle and accounts for a large fraction, roughly 90%, of heat transport through the mantle. Plumes originate deep in the mantle, probably at the core-mantle boundary, and account for about 10% of the mantle heat flux. No other modes of mantle convection have been distinguished (except possibly under some young lithosphere), and if they exist they must be minor, or their topographic expression would be evident.

These results have important implications for stirring of chemical heterogeneities by mantle convection: whereas multiple scales of flow could stir fairly rapidly, large-scale flow, with only narrow axial plumes perturbing it, allows some heterogeneities to survive for long periods, as will now be demonstrated.

5. STIRRING AND PERSISTENCE OF HETEROGENEITIES

Widely different conclusions have been drawn from recent studies of stirring by convection. On the one hand, Hoffman and McKenzie (1985) and Kellogg and Turcotte (1986) conclude that the upper mantle will be homogenised within a few hundred million years, while on the other hand Olson et al. (1984) and Gurnis and Davies (1986) conclude that traces of subducted lithosphere would survive for 1 - 2 Gyr.

The resolution of this disagreement has been provided by Gurnis (1986a), who has shown that Hoffman and McKenzie and Kellogg and Turcotte overestimated the stirring rate by focussing on the rate at which the **average** thickness of a stirred heterogeneity decreases, rather than on the **greatest** thickness. The average thickness tends to decrease exponentially with time. This occurs when the body passes near a stagnation point of the flow, but only a small fraction of the volume of the body is involved in the "tendril" which results from stagnation-point stretching of the body. Much of the body is thinned at a rate proportional to

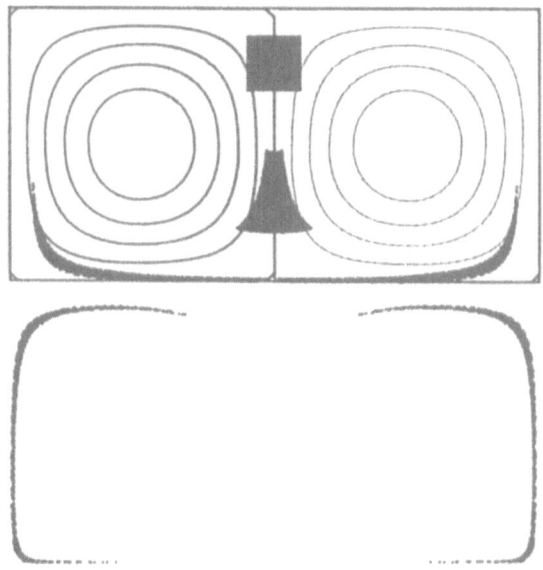

Fig. 5. Stretching of a passive heterogeneity at a stagnation point. A block of 2500 passive tracers (top square) is advected by the steady flow shown by the streamlines, and shown at several successive times. At the last time, the tracers are shown below the box so they can be distinguished from the box margin. The original block then comprises two relatively thick sections joined by a tendril so thin that no tracers have plotted within it.

1/time. The process is illustrated in Figure 5. With unsteady flow, where a body can be transferred to a neighboring cell, there is partial unmixing. The result is that although the average thickness of the body (defined as half the perimeter divided by the initial volume) decreases exponentially, some of the body remains relatively thick for a very long time (Figure 6).

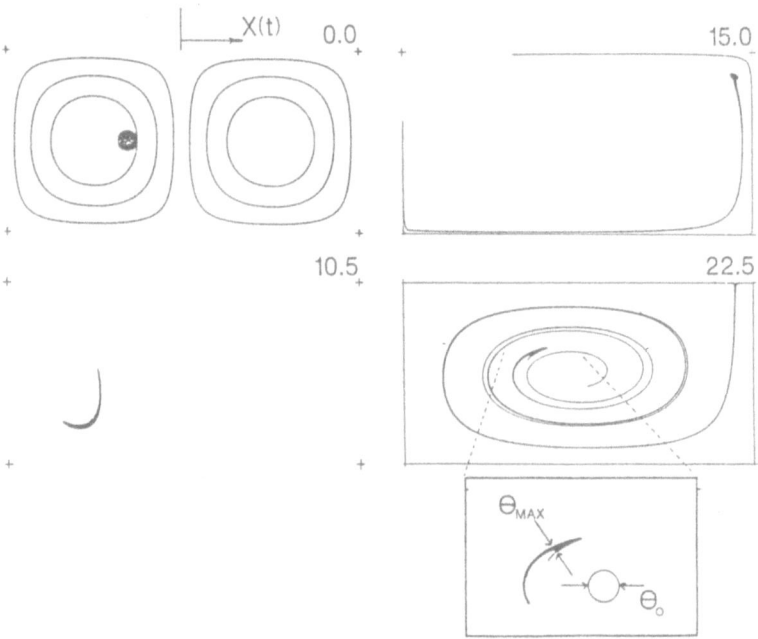

Fig. 6. Stirring of a passive heterogeneity by time-dependent
flow. The boundary between the cells oscillates left and
right, but streanlines are shown only in the first frame. The
original blob (black circle, upper left) is shown at several
later times (denoted by the number of transit times); only
the corners of the box are shown, so as not to obscure the
tracers. The blob is stretched into extremely thin tendrils,
but the thickest part of it is still a significant fraction of
its original size (inset). From Gurnis (1986a).

Gurnis (1986a) finds that the average time required to reduce the
thickest part of the body to 10% of its original thickness is about 200
transit times, where a transit time is the time taken for the fluid to
traverse the depth of the fluid layer at a typical fluid velocity. In
the whole mantle, this corresponds to about 3 Gyr. Thus, the occurrence
of "turbulent" stirring (Hoffman and McKenzie, 1985; Kellogg and
Turcotte, 1986) does not preclude the occurrence of a significant number
of ancient mantle heterogeneities (Gurnis and Davies, 1986).

6. RESIDENCE TIMES OF RECYCLED MATERIAL

As well as the time for which a heterogeneity can survive stirring
by mantle flow, we must also consider the average time it will reside in
the mantle before being sampled at a spreading center: the residence

time. A simple calculation taking the total spreading rate of about 3 km²/yr, assuming sampling to a depth of (say) 50 km under a spreading center and the mass (4×10^{24} kg) and mean density (4500 kg/m³) of the mantle yields a time of about 6 Gyr to sample the mass of the mantle at present rates. Numerical models by Gurnis and Davies (1986), using passive tracers in convection models, confirm that the mean residence time of recycled mantle material is of this order, given present rates. If the likelihood of faster rates in the past, due to the cooling of the earth's interior with time, is considered, the mean residence time is still between 1 and 2 Gyr, depending mainly on assumptions about the variation of the depth of sampling under spreading centers with time.

The possibility of a moderate increase in viscosity with depth in the mantle is indicated by some recent studies of post-glacial rebound, rotation of the earth and the geoid. This results in lower convection velocities at depth, less vigorous stirring and longer residence times. Gurnis (1986b) finds that for viscosity increases of about a factor of 100 through the mantle, residence times increase by about 50% (the effect of past variation of rates is included in this result). Thus, mean residence times in the range 1.5-3 Gyr are to be expected. This range includes the apparent mean age of MORB source (1.6-1.8 Gyr) and the apparent ages of oceanic islands (1-3 Gyr).

7. CONCLUSIONS

The observed bathymetry of oceans places strong constraints on the form of mantle convection. The absence of major topographic expression of rising buoyant mantle precludes the existence of a boundary at 670 km depth separating upper mantle and lower mantle convection layers. The only clearly recognisable topography due to buoyant mantle material is the hotspot swells, and these can account for only about 10% or less of the heat transported by the mantle. If other modes of mantle convection exist with length scales less than the plates, they must be minor. Thus mantle convection comprises plate-scale flows which penetrate deep into the mantle, with narrow plumes of buoyant material, probably arising from the core-mantle boundary, rising through this flow.

Intuitive estimates of the efficiency of stirring and homogenisation by mantle convection have been completely misleading. Even some quantitative estimates have been very misleading because they have characterised the stirring process in ways which are very insensitive to the presence of sizeable heterogeneities. When quantitative models take proper account of heterogeneities, it is found that some heterogeneities persist for very long times (hundreds of transit times or billions of years). Furthermore, the average time before sampling at a spreading center (the residence time) is found to be comparable to observed mean ages of MORB and OIB sources (about 2 Gyr) if account is taken of the probable increase of mantle viscosity with depth by about two orders of magnitude and of faster convection rates in the past. Further quantification and testing of the stirring process will be necessary, but at this stage there is no reason to appeal to layering to explain the observations of ancient mantle heterogeneities.

The above conclusions on stirring and survival of heterogeneities apply to the plate-scale flow. The flow associated with plumes probably does not greatly affect the stirring, since they are narrow structures, the material flux is relatively small, and the flow rates induced outside the plumes themselves are very small. Other small-scale modes, such as boundary-layer instability rolls, would substantially increase stirring rates, but these modes must be minor if they are present at all. This possibility needs to be better quantified, but the present indications are that small-scale modes do not greatly affect stirring.

REFERENCES

Craig, C.H. and D. McKenzie, The existence of a thin low-viscosity layer beneath the lithosphere, Earth Planet. Sci. Lett., 78, 420-426 (1986).
Davies, G.F., Ocean bathymetry and mantle convection I. Large-scale flow and hotspots, submitted for publication (1987a).
Davies, G.F., Ocean bathymetry and mantle convection II. Small-scale flow, in preparation (1987b).
Fleitout, L. and D.A. Yuen, Secondary convection and the growth of the oceanic lithosphere, Phys. Earth Planet. Interiors, 36, 181-212 (1984).
Gurnis, M., Stirring and mixing in the mantle by plate-scale flow: large persistent blobs and long tendrils coexist, Geophys. Res. Lett., 13, 1474-1477 (1986a).
Gurnis, M., Convective Mixing in the Earht's Mantle, Ph.D. Thesis, Australian National University (1986b).
Gurnis, M. and G.F. Davies, Mixing in numerical models of mantle convection incorporating plate kinematics, J. Geophys. Res., 91, 6375-6395 (1986).
Haxby, W. F. and J. K. Weissel, Evidence for small-scale mantle convection from Seasat altimeter data, J. Geophys. Res., 91, 3507-3520 (1986).
Heestand, R.L. and S.T. Crough, The effect of hotspots on the oceanic age-depth relation, J. Geophys. Res., 86, 6107-6114 (1981).
Hoffman, N.R.A and D.P. McKenzie, The destruction of chemical heterogeneities by differential fluid motions during mantle convection, Geophys. J. Roy. Astr. Soc., 82, 163-206 (1985).
Jochum, K.P., A.W. Hofmann, E. Ito, H.M. Seufert and W.M. White, K, U and Th in mid-ocean ridge basalt glasses and heat production, K/U and K/Rb in the mantle, Nature, 306, 431-436 (1983).
Kellogg, L.H. and D.L. Turcotte, Mantle homogenization (abstract), Eos, Trans. Amer. Geophys. Union, 67, 367 (1986).
O'Connell, R.J. and B.H. Hager, On the thermal state of the earth, in Physics of the Earth's Interior, Proc. Enrico Fermi International School of Physics (Course LXXVIII), edited by A. Dziewonski and E. Boschi, North Holland, p. 270-317 (1980)
Olson, P., D.A. Yuen and D. Balsiger, Mixing of passive heterogeneities

by mantle convection, J. Geophys. Res., **89**, 425-436 (1984).

Parsons, B. and D. McKenzie, Mantle convection and the thermal structure of the plates, J. Geophys. Res., **83**, 4485-4496 (1978).

Richter, F. M., Convection and the large-scale circulation of the mantle, J. Geophys. Res., **78**, 8735-8745 (1973).

Schroeder, W., The empirical age-depth relation and depth anomalies in the Pacific ocean basin, J. Geophys. Res., **89**, 9873-9884 (1984).

INTRA-MANTLE FRACTIONATION VS. LITHOSPHERE RECYCLING: EVIDENCE FROM THE SUB-CONTINENTAL MANTLE

C.J. Hawkesworth, P.D. Kempton, D.P. Mattey, Z.A. Palacz, N.W. Rogers, and P.W.C. van Calsteren
Department of Earth Sciences
The Open University
Walton Hall
Milton Keynes, MK7 6AA, U.K.

Radiogenic isotope ratios have an increasingly important role in attempts to describe and model processes in the upper mantle, but their significance is still controversial. Isotope compositions which used to be regarded as diagnostic of the continental crust are increasingly observed in fragments of the sub-continental mantle, 'mantle isochrons' now tend to be attributed to semi-continuous mixing processes rather than discrete mantle events, and metasomatism has arguably emerged from a period in the conceptual doghouse to regain its position as a distinctive, albeit shallow level mantle process. Of particular current interest is the extent to which the radiogenic isotope ratios of mantle derived rocks reflect intra-mantle fractionation processes, rather than the recycling of material from the oceanic and/or continental lithosphere.

Isotope variations in oceanic basalts indicate that chemical heterogeneities have existed for 2 Ga (Chase, 1981), whereas in continental areas diamond inclusions have yielded ages of 3.2-3.3 Ga (Richardson et al., 1984) and mafic igneous rocks have inferred source ages of up to 2.5 Ga (Fraser et al., 1985). Critical to models for earth evolution and mantle convection systems is therefore whether material can age sufficiently in the upper mantle, and if so, where. A common solution is to argue that the lithosphere is rigid and hence preserves a record of the chemical variations generated at its formation and in subsequent enrichment/depletion events. Discussions of recycled lithosphere thus tend to invoke either 'old' lithosphere as a source of distinctive isotope ratios (McKenzie & O'Nions, 1983), or crust formation and subduction as a way of generating particular trace element ratios which then have to be stored in the mantle for long enough to generate the observed radiogenic isotopes (Hofmann & White, 1982; Hart, 1984).

Faced with a myriad of possible mechanisms for fractionating trace element ratios, and sustained in the belief that mass fractionation effects are accounted for in the measurement of Nd-, Sr-, and Pb-isotopes, many authors have sought to describe the observed isotope compositions of basalts as mixtures of discrete components. This was a successful paradigm for earth evolution models, but as the number of inferred components increases (e.g. Zindler & Hart, 1986) the approach arguably becomes less testable, and hence less satisfactory. An additional concern is that as components are defined on the basis of radiogenic isotopes, it gives them an age connotation which may be misleading. Thus, although the depleted mantle reservoir appears to be sufficiently well mixed for its inferred age to represent a reasonable average, mixing cannot be invoked to suggest that a fragment of enriched continental mantle should have the isotope composition of some average value. Averages are important ingredients in bulk earth

S. R. Hart and L. Gülen (eds.), Crust/Mantle Recycling at Convergence Zones, 227–237.
© 1989 by Kluwer Academic Publishers.

evolution models, but they are of limited use in interpreting the isotope ratios in a suite of rocks because average compositions need not actually be present in any of the possible components. An alternative view is that enrichment processes have occurred within the mantle through much of its history, and that they should therefore be discussed in terms of the inferred trace element fractionation patterns (Hawkesworth et al., 1984), particularly because that allows such processes to be of almost any age.

Our approach has been to investigate the age and composition of the continental lithosphere, both because it is a natural laboratory which preserves a unique record of mantle processes that demonstrably fractionate trace element ratios, and because it may be an important sink for incompatible elements which should in turn be incorporated in bulk earth evolution models. Available materials range from the mantle rocks which are present as inclusions in both kimberlites and alkali basalts, to lamproites and kimberlites many of which are now regarded as small volume melts of the sub-continental mantle.

The characteristic features of the sub-continental mantle lithosphere as viewed through xenolith suites, are that it tends to be depleted in major elements and variously enriched in incompatible trace elements (Harte et al., 1987; Erlank et al., 1987). It is therefore buoyant (Jordan, 1979), and it exhibits a range of parent/daughter trace element ratios that with time develop variable Sr-, Nd-, and Pb-isotope compositions. Rock types include dunites, variously depleted, enriched and metasomatised peridotites, and eclogites. The processes involved in the generation of continental mantle lithosphere probably range from komatiite extraction (O'Hara et al., 1975), to magmatic and tectonic accretion above subduction zones (Oxburgh & Parmentier, 1978), to rising diapirs from within the asthenosphere, or even from deep-seated mantle megaliths (Ringwood, 1982). Detailed studies on xenoliths and continental volcanic rocks are required to evaluate both the age and nature of the enrichment and depletion processes in the sub-continental mantle, and several have recently been undertaken on rocks from southern Africa.

SOUTHERN AFRICA

Southern Africa consists of a jig-saw of crustal provinces with ages varying from 3.5Ga to the Phanerozoic. After the last major orogenic event and the subsequent stabilisation of the crust, both crustal and mantle lithosphere have been sampled by the Jurassic continental flood basalt magmatism of the Karoo, as xenoliths in mostly Cretaceous kimberlites, and by a Tertiary suite of nephelinites and melilitites. Thus southern Africa offers a rare opportunity to investigate the age structure at depth within a segment of continental lithosphere.

Dating events in the continental lithosphere is problematic; first, because for the most part its pressures and temperatures are above the blocking conditions for co-existing minerals, and second, because whole rock events can only be dated if they involve both a change in parent/daughter element ratio, and little subsequent alteration of the samples. A common approach is to use model Nd ages, but these can only reasonably be used to date enrichment events that involve a reduction in Sm/Nd. Depleted xenolith whole rocks and clinopyroxenes often have Sm/Nd ratios similar to MORB, and so the isotope data cannot resolve whether their old model Nd ages represent a true lithosphere forming event, or simply the average age of the MORB reservoir (Harte & Hawkesworth, 1988). In our experience it is more often the latter.

Boyd & Gurney (1986) published a preferred cross-section through the south African lithosphere, and this is reproduced in figure 1 as a framework on which to summarise the available age data. The section is orientated NW-SE and includes Pan

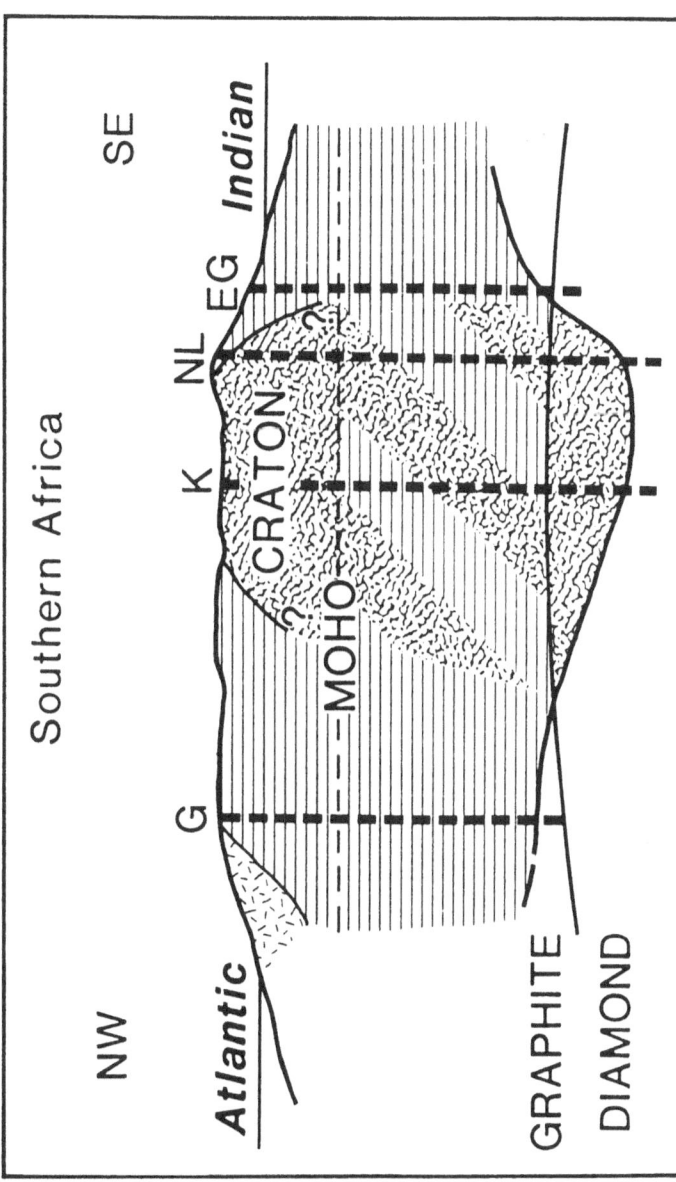

Figure 1 A diagrammatic cross section through southern Africa (modified after Boyd & Gurney, 1986) showing the greater thickness of lithosphere beneath the Archaean terrain compared with that beneath the surrounding Proterozoic and younger mobile belts. Also shown are the relative locations of kimberlite pipes at Gibeon, Namaqualand (G), Northern Lesotho (NL) and Kimberley (K) and East Griqualand (EG) in South Africa. Inferred age zones are illustrated by marble (Archaean), horizontal ruling (Mid- to Late Proterozoic) and granite ornament (Pan African). Mantle beneath the Archaean craton reflects open system behaviour both in the Proterozoic, and during hydrous metasomatism in the Cretaceous (Erlank et al., 1987).

African, mid-Proterzoic and Archaean terrains in the surface geology.

Crustal xenolith suites provide upper and lower crustal material from both the Archaean and Proterozoic areas. Many rocks are similar to those observed on the surface, with the possible exception of a suite of mafic to felsic granulites which appear to be confined to kimberlite pipes in Proterozoic basement, and which record pressures and temperatures of 4.5-20 kb and 550-1000°C (van Calsteren et al., 1986, and references therein). These have been successfully modelled as cumulates from basalts (Rogers & Hawkesworth, 1982) and as they are not observed in the Archaean areas, there appears to have been a significant change in lithosphere-forming processes between the Archaean and the mid-Proterzoic (see also Ellam & Hawkesworth, 1988). Xenolith samples have been analysed from three kimberlite pipes, one in Archaean basement and two in the surrounding mobile belts. In Namaqualand and at Kimberley (fig. 1) the range of model Nd ages in lower crustal material is the same as that observed in upper crustal (amphibolite facies) xenoliths from the same pipe, indicating that there is no significant variation in whole rock ages with depth. In Lesotho, however, the amphibolite facies gneisses yield Archaean, and the granulites yield Proterzoic, model ages. Thus in this area, which is very close to the inferred margin of the Archaean craton, Archaean upper crust appears to be underlain by Proterzoic lower crust. Rb/Sr and Sm/Nd mineral data on these samples are discussed elsewhere (van Calsteren et al., 1986).

The oldest, well constrained ages from mantle materials are the 3.3-3.2 Ga Sm/Nd ages on silicate inclusions in diamond reported by Richardson et al. (1984). They suggest that a thick lithosphere was present in the Archaean, and for argument's sake we've shown that portion of the lithosphere which lies within the diamond stability field to be Archaean in age (fig. 1). Significantly continental lithosphere generated in the Archaean, when the earth was hotter, appears to have been thicker than that generated subsequently. This is consistent with the available major element data on mantle xenoliths which indicate that the Archaean mantle lithosphere is more depleted, and hence more buoyant, than that beneath the younger terrains (Hawkesworth et al., in prep.).

Isotope analyses on separated mantle minerals and whole rock peridotites from Kimberley yield present day ε_{Nd} = +7.8 to -14 and ε_{Sr} = -15 to +100 (Erlank et al., 1987, and references therein). None of these values are sufficiently extreme to date from the Archaean, and the present consensus is that they reflect a mid-Proterzoic enrichment event followed by amphibole-LIMA-phlogopite metasomatism in the Cretaceous (Erlank et al., 1987). Clearly conditions in the lithospheric mantle were such that old ages could only survive in silicates encapsulated in diamond, but the details of the open system evolution are still poorly constrained. Comparison of the data from Kimberley with those from pipes marginal to the craton, for example in Lesotho, shows that the latter are characterised by higher ε_{Nd} (van Calsteren et al., 1987). They don't preserve a record of even the Proterzoic enrichment inferred from Kimberley, and they suggest that mantle material also gets younger away from the Archaean nucleii of southern Africa.

Evidence for the age of mantle segments from magmatic rocks is further hindered by the need to see back through the possible effects of crustal contamination. However, particularly Group II kimberlites and the high-Ti rocks of the northern Karoo are now generally thought to have remobilised lithospheric mantle of mid-Proterzoic age (Hawkesworth et al., 1984; Smith, 1983; Fraser et al., 1986). While the kimberlites represent sufficiently small volume melts to have been scavenged from mantle similar to that sampled in the xenolith suites of Kimberley, it is more difficult to imagine that a magmatic event on the scale of the Karoo could have sampled mantle beneath the

Archaean nucleii without involving some of the Archaean material which is known to have been present from the diamond data, unless such Archaean mantle is residual after komatiite generation, and hence too depleted to be the source for continental tholeiites. Thus we prefer a model in which the mid-Proterzoic mobile belts were underlain by mid-Proterozoic mantle, which was in turn remobilised during the Karoo. It has long been argued that in the context of crustal evolution, it is the youngest crustal material which is most likely to be reworked in the next orogenic event, and so it is interesting to note that the same seems to have occurred in the south African mantle. The Karoo magmatism reworked Proterozoic, but not Archaean mantle, suggesting perhaps that the mantle lithosphere in this area consists of roughly concentric age zones similar to those in the overlying crust.

INTRA-MANTLE FRACTIONATION PROCESSES

Metasomatism implies chemical change associated with the interaction between a fluid or melt and pre-existing host rock. If the net chemical change is simply attributed to the addition of a given volume of fluid with a particular composition, then it is the bulk composition of the fluid rather than any metasomatic interaction which determines the chemical change. Nonetheless studies of mantle xenoliths have documented examples where the bulk chemical changes do reflect the observed mineral reactions, and may therefore be confidently attributed to fluid-rock interaction (Erlank et al., 1987). However, there are comparatively few basalt provinces with trace element features that require some form of metasomatic interaction, rather than simple chemical enrichment, in their source regions (Hawkesworth et al., 1987).

In general, H_2O-metasomatism (as used here) appears to be confined to relatively shallow levels within the continental lithosphere, and is arguably best seen in the phlogopite-K-richterite xenoliths from the Kimberley area. Basalts which have some trace element, and hence with time radiogenic isotope, affinities with such metasomatism are the low-Ti Mesozoic flood basalts of Gondwanaland (Kyle 1980; Mensing et al., 1984), and the potassic orogenic volcanic rocks of Italy and Indonesia (Rogers et al., 1987). CO_2-metasomatism is much more difficult to assess because most of its suggested characteristics are equally consistent with very small degrees of partial melting of four phase lherzolite. Thus the main role of CO_2 may simply be to reduce the viscosity sufficiently for very small volume partial melts to be extracted (McKenzie, 1985).

It follows that the dominant processes for generating trace element enriched segments of mantle is the migration of small volume partial melts. Melts and metasomatic fluids are all characterised by low Sm/Nd, although presumably with significantly different REE abundances. However, Rb/Sr ratios are only noticeably increased when significant H_2O is present, either as a hydrous fluid, or in a small volume partial melt capable of stabilising phlogopite (e.g. lamproites), and again this appears to be confined to continental areas (Hawkesworth et al., 1985).

Generating the observed range of Pb-isotope compositions in oceanic basalts is more controversial, and in particular the implied high U/Pb-low Rb/Sr combination of the HIMU component has traditionally been attributed to distinctive processes such as sulphide separation into the core (Allegre et al., 1980), or the preferential take-up of U into altered ocean crust prior to subduction (Michard & Albarede, 1985). However, a recent study of the Tertiary melilitite-nephelinite suite of southwest Africa (Palacz et al., in prep.) demonstrated a range in measured μ from 20-400, with high μ tending to be associated with low Rb/Sr, and vice versa. Pb- and Sr-isotope results on mantle xenoliths and continental flood basalts (fig. 2) also indicate that inferred U/Pb and Rb/Sr ratios in their source regions are correlated negatively, and there is increasing evidence that this may be a distinctive feature of trace element enriched continental mantle. In general there appears to be a broad negative correlation between measured U/Pb and K/Nb (fig. 3) which further suggests that a potassic phase, apparently phlogopite, strongly influences U/Pb variations in the upper mantle. Partial melting models with liquids in equilibrium with residual phlogopite thus offer an alternative intra-mantle mechanism for generating HIMU material, and significantly the resultant U/Pb ratios might be high enough to generate the observed $^{206}Pb/^{204}Pb$ variation in oceanic basalts within the age of an ocean. At present this represents an extreme view as yet unsupported by U/Pb data on HIMU basalts, but if even partially correct it implies that the larger inferred variations in μ are young, or at least comparatively short-lived. In contrast, the older (1-2Ga) heterogeneities reflect relatively little U/Pb fractionation, and are probably associated with the generation of continental crust.

COMPOSITION OF THE SUB-CONTINENTAL MANTLE

Garnet peridotite xenoliths are largely confined to kimberlite pipes in the older cratonic areas, and they are highly depleted in major elements (average CaO = 1.1%, MgO = 42.5%) relative to common spinel peridotites (CaO = 2.55%, MgO = 40.2%) which tend to occur in younger crustal terrains. The difference corresponds to a 1-1.3% difference in normative densities, recalculated at similar temperatures and pressures, and the preferred interpretation is that it primarily reflects greater degrees of melting in the Archaean. Thus the proposed thicker lithosphere beneath the Archaean (e.g. Boyd & Gurney, 1986, and fig. 1) is due to the more depleted, buoyant nature of the Archaean mantle lithosphere, while younger, more fertile material has a higher average density and may therefore be more readily recycled.

Average minor, trace element and isotope abundances are more difficult to estimate. Xenoliths yield present day average isotopic values of $^{87}Sr/^{86}Sr$ = .7038 ± .0014 and $^{143}Nd/^{144}Nd$ = .51296 ± .00028 for 106 spinel peridotites and $^{87}Sr/^{86}Sr$ = .7065 ± .0028 and $^{143}Nd/^{144}Nd$ = .51260 ± .00023 on 95 garnet peridotites. Most spinel periodite data are on separated clinopyroxenes, whereas most garnet periodite results are on whole rock samples which tend to be displaced to lower $^{87}Sr/^{86}Sr$ and higher $^{143}Nd/^{144}Nd$ by any contamination from the host kimberlites. Lamproites and kimberlites exhibit a much greater range in isotope ratios and although most of them are derived from the mantle lithosphere, they probably sample a rather small portion of such mantle. Many continental flood basalts which appear not to have been contaminated *en route* through the continental crust are characterised by fairly restricted ε_{Nd} = +2 to -10 and ε_{Sr} = -10 to +60 (see also fig. 2). They are derived from isotopically enriched source regions and yield ages in the range 1.8-1.1 Ga (e.g. Hawkesworth et al., 1984, 1986), consistent with spinel peridotite-type source material within the post-Archaean mantle lithosphere. Thus the evidence from both xenoliths and continental mafic rocks is that even enriched continental mantle has ε_{Nd} and ε_{Sr} just slightly lower and higher respectively, than the bulk earth.

Erlank et al. (1987) estimated the proportions of different peridotite types among the mantle xenoliths analysed from the Kimberley area. The weighted means yield an average K_2O = 0.43% for the mantle in this area, but this is at least twice as high as that required to be consistent with the low surface heat flow of Archaean terrains. Similarly the K_2O content in the source of most continental flood basalts is unlikely to be greater than 0.15%, and since that is also the upper limit of K_2O abundances in spinel

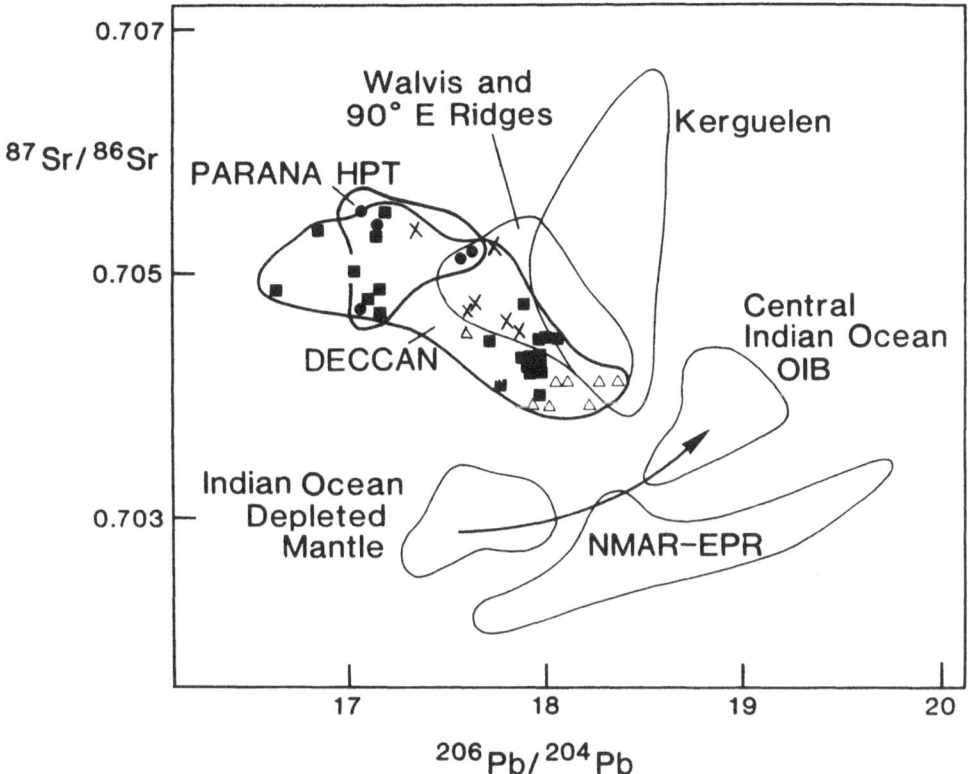

Figure 2 The variation of $^{87}Sr/^{86}Sr$ and $^{206}Pb/^{204}Pb$ in selected oceanic and continental basalts. Basalts from the northern mid-Atlantic ridge and the east Pacific rise (NMAR-EPR) show a positive correlation at relatively low $^{87}Sr/^{86}Sr$. Indian Ocean basalts show a similar covariation but at slightly elevated $^{87}Sr/^{86}Sr$ (Hamelin et al., 1986). Both imply that high time-integrated source Rb/Sr ratios correspond to high μ. In contrast, uncontaminated continental flood basalts from the Deccan and Parana Provinces show a broad negative correlation indicating that high Rb/Sr in the sub-continental lithosphere is associated with low μ (Hawkesworth et al., 1986, Lightfoot & Hawkesworth, 1988).

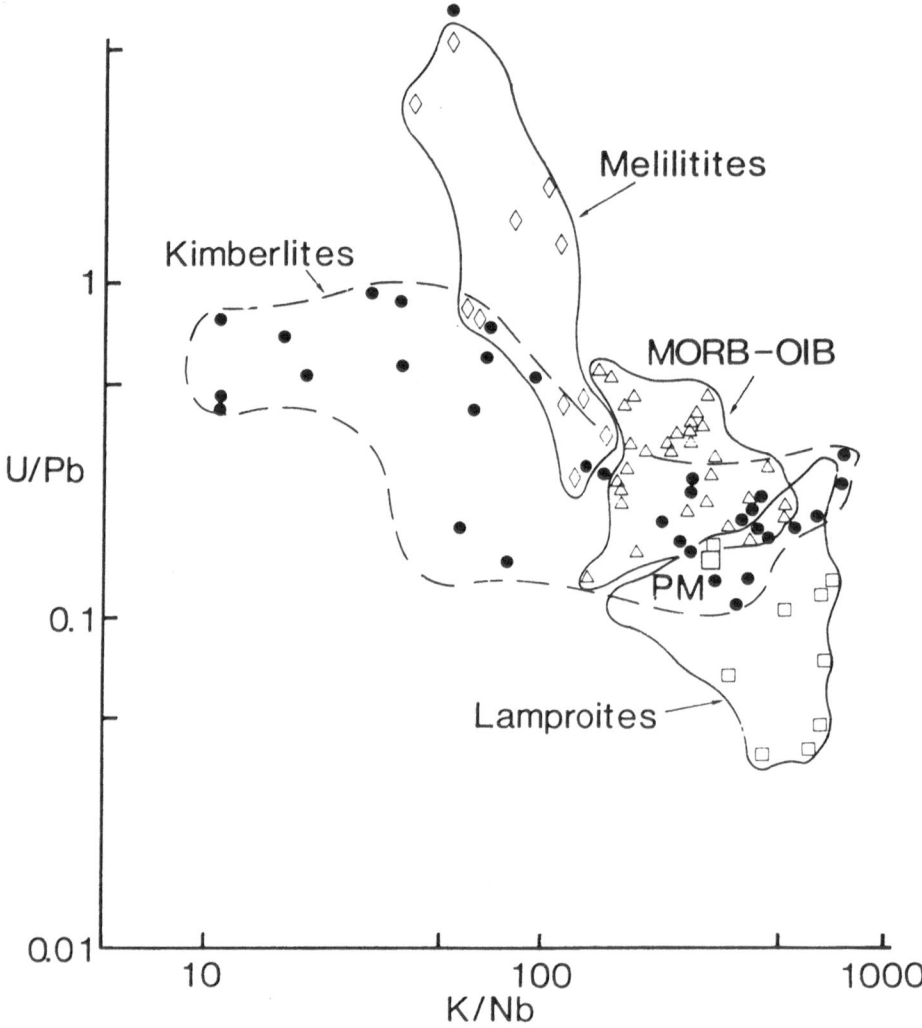

Figure 3 U/Pb vs. K/Nb for selected mantle-derived magmas illustrating the large range in U/Pb (corresponding to μ varying from 20-400) and its covariation with K/Nb. The tendency for high U/Pb rocks to have low K/Nb suggests that U/Pb fractionation in these rocks may be related to the presence or absence of residual phlogopite.

peridotites it is suggested as an upper limit for the K_2O content of the sub-continental lithospheric mantle. Since such mantle is $\sim 2.2\%$ of the silicate portion of the earth, it would on this basis contain just 11% of the potassium in the crust-mantle system.

LITHOSPHERE RECYCLING

As indicated above lithosphere recycling is variously invoked as the source of unusual radiogenic isotope ratios which have aged within the lithosphere, and as an explanation of distinctive trace element ratios which with time could develop the required isotope compositions. As a source of isotope ratios not otherwise observed in oceanic basalts, the lithosphere may have been enriched by processes similar to those inferred for the oceanic mantle, and simply be older, but also by processes such as H_2O-metasomatism which have so far only been documented at shallow levels in the continental lithosphere. The trace element ratio arguments tend to focus on oceanic lithosphere, for there is tremendous scope for fractionating trace elements at crust formation, by interaction with sea water, and on dehydration and partial melting during subduction.

At present it would appear that almost any argument for a contribution from recycled crust in the upper mantle, based solely on radiogenic isotopes, can be countered by one that appeals to intra-mantle fractionation. At issue would be whether the end-member composition can be defined, and if so whether it is deemed too old to have evolved within the convecting upper mantle. By contrast, light elements, such as H, C, and O, are strongly fractionated by low temperature surface processes and thus offer a potentially unique fingerprint of crustal material in mantle derived rocks. Stable isotope studies of hydrogen and carbon have been used to identify a crustal component in both island arc and marginal basin basalts (Mattey et al., 1984; Dobson & O'Neil, 1987), and Kempton et al. (in prep.) argue that oxygen isotope disequilibria between coexisting pyroxenes and olivine in mantle xenoliths is due to the introduction of subduction related fluids. Of greater significance, however, is the observation that trends to low $\delta^{13}C$ values occur in diopsides from mantle xenolith suites from areas which have not neccessarily been associated with recent subduction (Mattey et al., 1986).

Trace element discussions have sought to distinguish material released from subducted crust at comparatively shallow levels, from that taken down to greater depths where it may be remobilised by intraplate magmatism. Some have been tempted to argue that subduction is the source of all hydrous fluids in the upper mantle, but most arguments have been linked to the high LIL/HFS element ratios which appear to be a feature of subduction related magmatism. In southern Africa the Karoo basalts are characterised by relatively low Nb contents (Duncan et al. 1984), with high Ba/Nb and Th/Ta and late Proterozoic model Nd ages (Hawkesworth et al. 1984), while mantle diopsides exhibit a negative trend between $^{87}Sr/^{86}Sr$ and $\delta^{13}C$ (Mattey et al., 1986). In such a cratonic environment it is unlikely that the latter reflects recent mixing with subducted crust, but both the xenolith and the Karoo data are consistent with stabilisation of the continental lithosphere by subduction related processes in the late Proterozoic. Oceanic crust which is subducted to greater depths is presumably characterised by relatively high HFSE contents, which has also fuelled speculation that it may then contribue to plume-type magmatism (Hofmann & White, 1982; Saunders et al., 1987).

Finally, however, there is the question of just how relevant recent subduction related magmatism is to discussions of crust-mantle differentiation. There is increasing evidence that the net flux from mantle to crust at subduction zones is basaltic, and that such environments are therefore unlikely to generate average continental crust as presently understood (Kay and Kay, 1986; Ellam and Hawkesworth, 1988). That can be accomodated by invoking different crust-forming processes in the Archaean and early Proterozoic, when most of the continental crust stabilised, but discussions of lithosphere recycling should then consider material residual after the generation of ancient crust, rather than that subducted recently.

REFERENCES

Allegre, C.J., Brevart, Dupré B & Minster, J.L., 1980. *Phil. Trans. R. Soc. Lond.* **A297**, 447-477.

Boyd, F.R & Gurney, J.J., 1986. *Science,* **232**, 472-477.

Chase, C.G., 1981. *Earth Planet. Sci. Lett.,* **52**, 277-284.

Dodson, P.F & O'Neil, J.R., 1987. *Earth Planet. Sci. Lett.,* **82**, 75-86.

Duncan, A.R., Erlank, A.J. & Marsh, J.S. 1984. *Geol. Soc. S. Afr. Spec. Publ.***13**, 355-388.

Ellam, R.M. & Hawkesworth, C.J., 1988. *Geology* (submitted).

Erlank, A.J., Waters, F.G., Hawkesworth, C.J., Haggerty, S.E., Allsopp, H.L., Rickard, R.S. & Menzies, M.A., 1987. In: *Mantle Metasomatism* (Eds. M.A. Menzies & C.J. Hawkesworth). Academic Press, London, 221-312.

Fraser, K.J., Hawkesworth, C.J., Erlank, A.J., Mitchell, R.H. & Scott-Smith, B.H., 1985. *Earth Planet. Sci. Lett.,* **76**, 1/2, 57-70.

Hamelin, B., Dupré, B. & Allegre, C.J., 1986. *Earth Planet. Sci. Lett.,* **76**, 288-298.

Hart, S.R., 1984. *Nature,* **309**, 753-757.

Harte, B. & Hawkesworth, C.J., 1988. *Spec. Pub. Geol. Soc. Aust.* (in press).

Harte, B., Winterburn, P.A. & Gurney, J.J., 1987. In: *Mantle Metasomatism* (Eds. M.A. Menzies & C.J. Hawkesworth). Academic Press, London, 145-220.

Hawkesworth, C.J., van Calsteren, P.W.C., Rogers, N.W. & Menzies, M.A., 1987. In: *Mantle Metasomatism* (Eds. M.A. Menzies & C.J. Hawkesworth). Academic Press, London, 365-388.

Hawkesworth, C.J., Mantovani, M.S.M., Taylor, P.N. & Palacz, Z., 1986. *Nature ,* **322**, 356-359.

Hawkesworth, C.J., Fraser, K. & Rogers, N.W., 1985. *Trans. Geol. Soc. S. Afr.,* **88**, 439-448.

Hawkesworth, C.J., Rogers, N.W., van Calsteren, P.W.C. & Menzies, M.A., 1984. *Nature,* **311**, 331-335.

Hofmann, A. & White, W.H., 1982. *Earth Planet. Sci. Lett.,* **57**, 421-436.

Jordan,T.H., 1979. In: *The Mantle Sample* (Eds. Boyd, F.R. & Meyer, H.O.A.). AGU, Washington, 1-14.

Kay, R.M. & Kay, S.M., 1986. *Geol. Soc. Lond. Spec. Publ.*, **24**, 147-160.

Kyle, P.R., 1980. *Contrib. Mineral. Petrol.*, **304**, 51-54.

Lightfoot, P.C. & Hawkesworth, C.J., 1988. *Earth Planet. Sci. Lett.* (submitted).

Mattey, D.P., Carr, R.M., Wright, I.P. & Pillinger, C.T. 1984. *Earth Planet. Sci. Lett.*, **70**, 196-206.

Mattey, D.P., Exley, R.A., Porcelli, D.R., Menzies, M.A., Pillinger, C.T., Galer, S. & O'Nions, R.K., in press. Extended Abst., 4th Int. Kimberlite Conf., *Spec. Publ. Geol. Soc. Aust.*, **16**, 276-278.

McKenzie, D.P., 1985. *Earth Planet. Sci. Lett.*, **74**, 81-91.

McKenzie, D.P. & O'Nions, R.K., 1983. *Nature*, **301**, 229-231.

Mensing, T.M., Faure, G., Jones, L.M., Bowman, J.R. & Hoefs, J., 1984. *Contrib. Mineral. Petrol.*, **87**, 101-108.

Michard, A. & Albarede, F., 1985. *Nature*, **317**, 244-246.

O'Hara, M.J., Saunders,M.J. & Mercy, E.L.P., 1975. *Phys. Chem. Earth,* **9**, 571-604.

Oxburgh, E.R. & Parmentier, E.M., 1978. *Phil. Trans. Roy. Soc. Lond.*, **A288**, 415-429.

Richardson, S.H., Gurney, J.J., Erlank, A.J. & Harris, J.W., 1984. *Nature*, **310**, 198-202.

Ringwood, A.E., 1982. *J. Geol.*, **90**, 611-643.

Rogers, N.W. & Hawkesworth, C.J., 1982. *Nature*, **299**, 409-413.

Rogers, N.W., Hawkesworth, C.J., Mattey, D.P. & Harman, R.S. 1987. *Geology*, **15**, 451-453.

Saunders, A.D., Norry, M.J. & Tarney, J., 1987. *Terra Cognita*, **7**, 620.

Smith, C.B., 1983. *Nature*, **304**, 51-54.

Van Calsteren, P.W.C., Harris, N.B.W., Hawkesworth, C.J., Menzies, M.A. & Rogers, N.W., 1986. *Geol. Soc. Spec. Publ..*, **24**, 351-352.

Van Calsteren, P.W.C., Winterburn, P.A., Hawkesworth, C.J. & Harte, B., 1987. *Terra Cognita*, **7**, 395.

Zindler, A. and Hart, S.R., 1986. *Ann. Rev. Earth Planet. Sci.*, **14**, 493-571.

CONVECTION AND GRAVITY CLOSE TO SUBDUCTION IN A TWO LAYER MANTLE: A 3-D MODELISATION

M. Rabinowicz, L. Cserepes* and C. Rosenberg
CNES/CRGS, 18, Avenue Edouard Belin
31055 Toulouse Cedex, France

Geophysical Department, Eotvös University
Kun Bela Ter 2, 1089 Budapest, Hungary

Three dimensional experiments of convection with Rayleigh numbers up to 100 times the critical in a two layer rectangular box have been performed (Cserepes et al., submitted). The thickness of the bottom layer is twice that of the top one. The viscosity ratio of the bottom layer versus the one of the top layer is progressively increased. When this ratio is small (between 1 and 10) dynamical coupling prevails and bimodal or hexagonal cells formed in the lower layer induce similar counter rotating cells in the top thin layer. When this ratio is large (up to 300) thermal coupling prevails: the convective structure of the lower layer is also duplicated in the top one but now the flow in both layers circulates in the same direction. When the viscosity ratio is moderate (~25) rolls with aspect ratio of the order of 1 are set in both the top and bottom layers (see figure 1). The axis of rotation of the top small rolls are perpendicular to the one of the large bottom rolls. This situation is similar to that described by Richter and Parsons (1975) for convective flow with a moving boundary. In our case the shear is produced by the large bottom cell flow along the upper/lower interface. The gravity is well and positively correlated. For the case of moderate viscosity ratio gravity and topography are not well correlated with topography. The main wavelength in the gravity signal reflects convection pattern in the lower layer. The submultiple wavelengths reflect both the structure of the top and bottom flow.

A lateral cooling in the upper layer is respectively imposed along one of the two lateral faces of the box parallel to the axis of the bottom layer cell (figures 2 and 3). This simulates the lateral cooling of the upper mantle due to subduction (Rabinowicz et al., 1980). Convective currents flowing away from this cold slab are generated. The flow structure generated in the direction normal to the cold slab is quite similar to that obtained with two dimensional models (Cserepes and Rabinowicz, 1985): when

S. R. Hart and L. Gülen (eds.), Crust/Mantle Recycling at Convergence Zones, 239–244.
© 1989 by Kluwer Academic Publishers.

FIGURE CAPTION: From left to right and top and bottom: Isotherms along the 4 vertical faces of the computing box. Isoline of the vertical velocity at mid depth of the upper layer, flow directions along the top interface, isoline of the vertical velocity at mid-depth of the lower layer flow, directions of flow along the upper/lower layer interface. Shaded areas are positive values. The Rayleigh number is $3\ 10^6$ and the lower/upper layer viscosity ratio is 25. Free slip boundary conditions are set along the faces of the cavity. Along the top face and bottom faces, temperature is imposed lateral faces are insulated except for cases of figures 2 and 3 where a temperature equal to 0 is imposed in the upper layer along the y left plane (for figure 2) or along the right y plane (for figure 3).

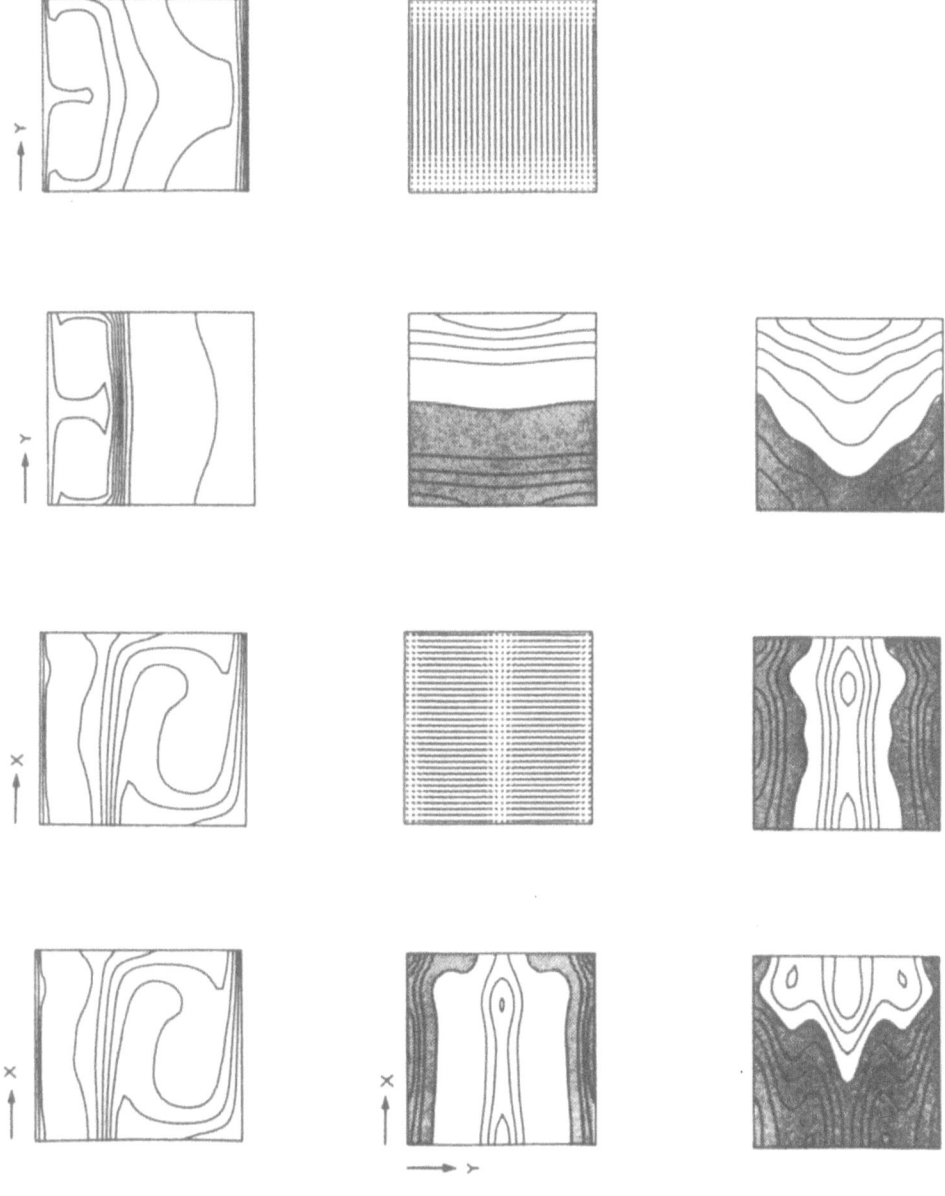

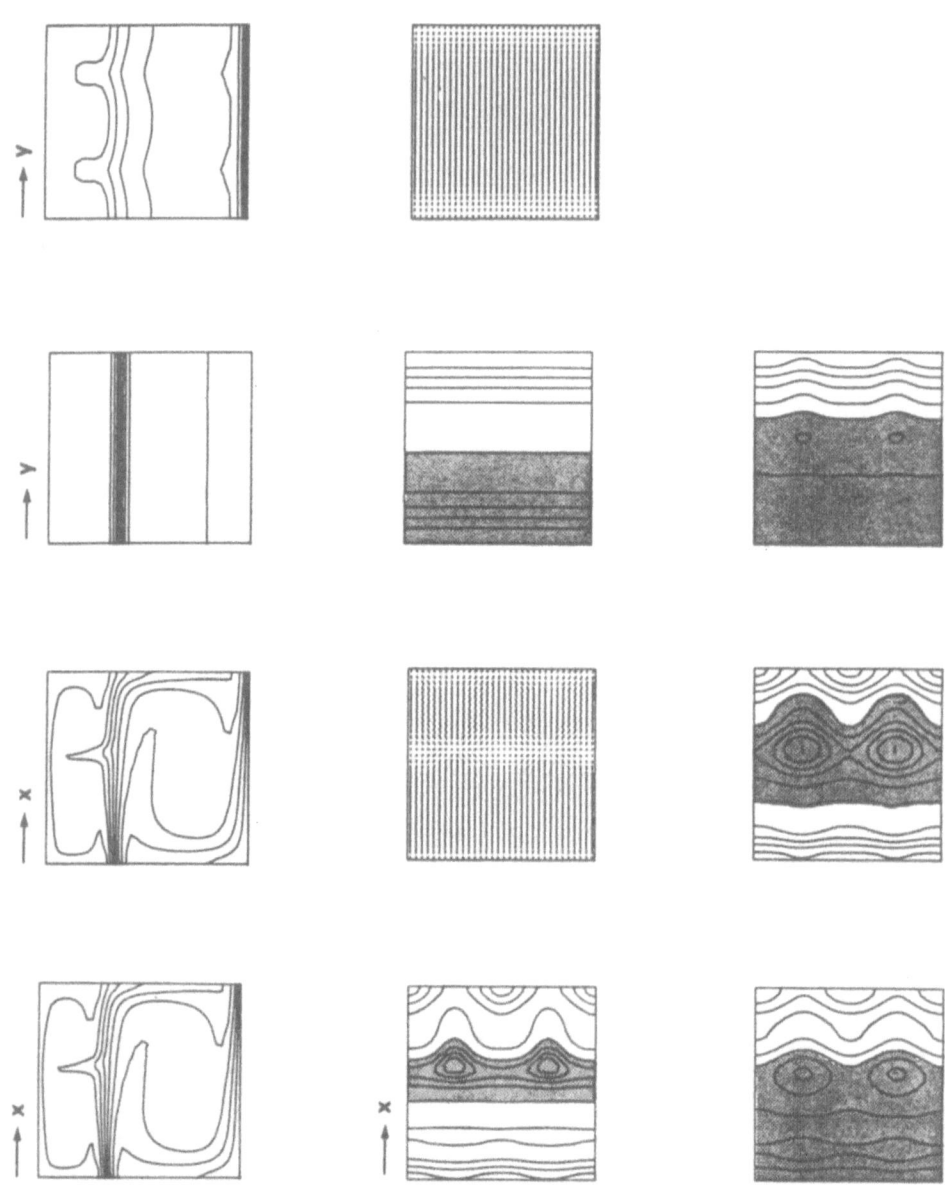

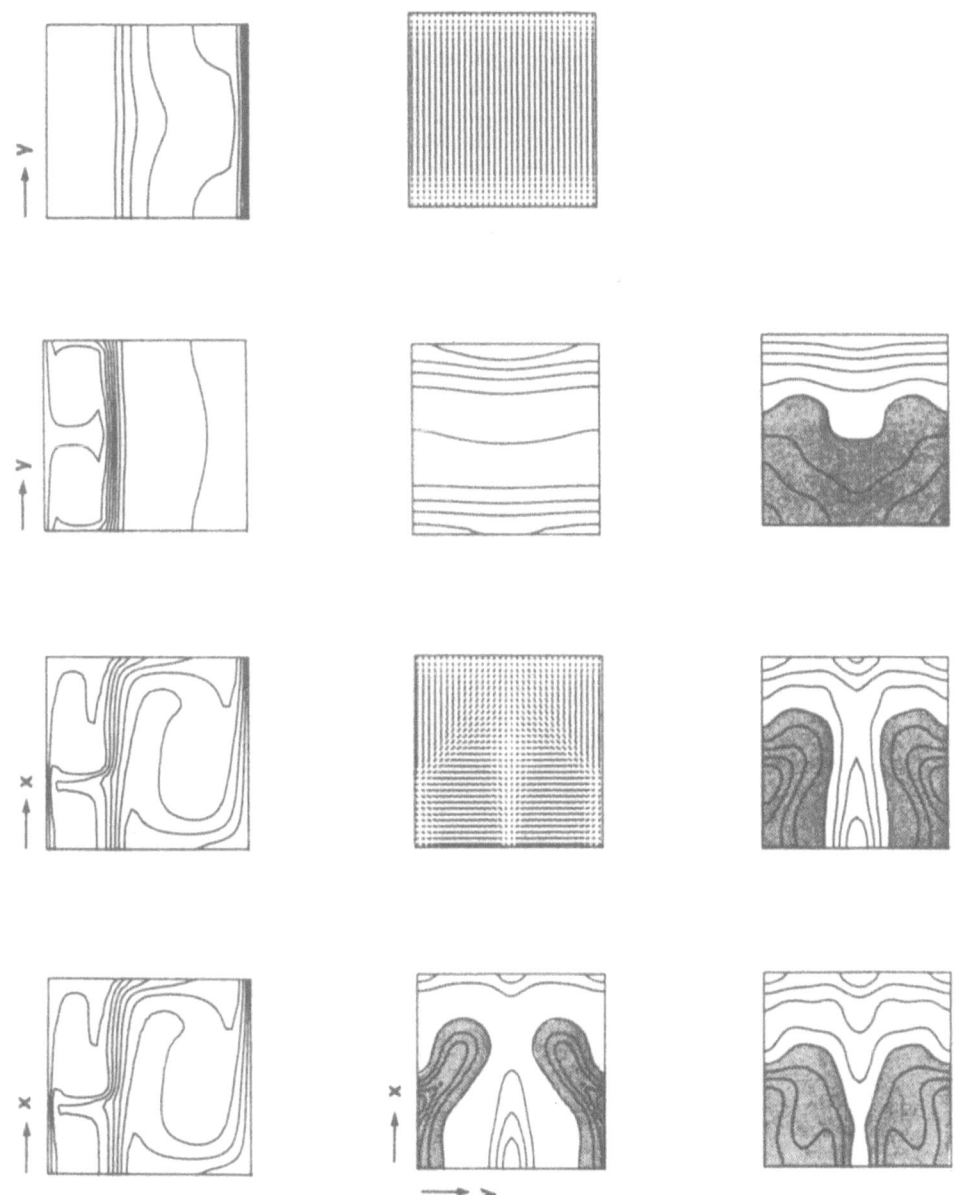

the cold slab is located above the sinking limb of the
bottom layer cell, thermal coupling type of flow prevails;
when it is located above the upwelling sites, dynamical
coupling prevails. In the dynamical coupling case, one
notes that the convective flow in the direction parallel to
the cold slab is inhibited up to some distance of this cold
slab. This distance is known to strongly depend on the
Rayleigh number and on the upper/lower layer thickness ratio
(Cserepes and Rabinowicz, 1985). Further away, upwelling
spouts are generated. When the cold slab is located above
the downwelling currents of the lower layer, the lateral
current driven by this slab is cut by upwelling currents
moving diagonally toward the subduction one. Although these
mathematical exercises are by nature far away from the real
Earth, they put some light on the processes which could
occur close to subduction. The important point, derived
from these models, is the strength of convective circulation
parallel to the subduction zone. This gives way to explain
why subducted zones are broken into pieces of 1000 km.
These models give also some ideas on the nature of
convection at the junction of the subduction sheets.

REFERENCES

Cserepes L., Rabinowicz M., Rosemberg C., 'Three dimensional
convection in a two-layer mantle' (submitted to JGR)

Cserepes L. and Rabinowicz M., 'Gravity and convection in a
two-layer mantle,' Earth and Planet. Sci. Lett. 76, 193-207
(1985).

Richter F. and Parsons B., 'On the interaction of two scales
of convection in the mantle,' J. Geophys Res., 80, 2529-2541
(1975).

Rabinowicz M., Lago B., Froideveaux C., 'Thermal transfer
between the continental asthenosphere and the oceanic
subducting lithosphere: its effect on subcontinental
convection,' J. Geophys. Res. 85, 1839-1953, 1980.

DYNAMICS OF RECYCLING

D.L. Turcotte*
Institut de Physique du Globe de Paris
4, Place Jussieu
75252 Paris Cedex 05
France

ABSTRACT. A relatively simple model is proposed for the rate of
production of continental crust; it is scaled with the rate of heat
loss from the earth's interior. The rate of recycling of continental
crust is scaled with the mass of continental crust present. An
additional constraint is the mean age of separation of the continental
crust from the mantle; this is taken to be 2.1 Gyr based on isotope
studies. We find that the present rate of crustal addition is 1.58
km^3/yr and the present rate of crustal recycling is 1.49 km^3/yr. This
rate of crustal recycling is somewhat high to be explained by the
direct subduction of sediments. We suggest that partial delamination
of the continental crust is a likely mechanism for substantial
recycling of continental crust. We also determine the associated
variations in sea level and argue that they are consistent with
observations.

1. INTRODUCTION

Some recycling of material from the continental crust back into the
mantle must occur. A variety of elements from the continental crust
are dissolved in sea water and some of these elements are deposited in
the oceanic crust during hydrothermal alteration. Sediments deposited
on the sea floor can also be entrained in the oceanic crust. It is
likely that both of these processes lead to the subduction of
continental material. Estimates of the volume of sediments that can
be subducted have been given by Karig and Kay (1981) and Reymer and
Schubert (1984). An important question, however, is whether subducted
sediments can survive the island arc volcanic processes. If the
subducted sediments are simply returned to the continental crust by
these volcanics then there is no net loss of crustal material to the
mantle.

*Permanent address: Snee Hall, Cornell University, Ithaca, NY 14853
USA

S. R. Hart and L. Gülen (eds.), Crust/Mantle Recycling at Convergence Zones, 245–257.

Crustal recycling is directly related to the question of the volume of the continental crust through time. Since new material from the mantle is being added to the continents continually by island arc volcanism and flood basalt volcanism, a near constant volume for the continental crust requires substantial volumes of crustal recycling. Quantitative constraints on the volume of the continental crust come from sea level data. It appears reasonable to hypothesize that the volume of water has remained nearly constant for at least the last 3 Gyr. If the volume of the continents had increased substantially during this time substantial quantities of water would have been displaced and sea level would have increased. If this had occurred the stable cratons would be covered with sediments, this is not the case. In fact, the observational evidence argues that sea level has fallen relative to the stable continents (Ambrose, 1964; Wise, 1974; Watson, 1976; Windley, 1977; Hallam, 1977; Dewey and Windley, 1981; McLennan and Taylor, 1983).

However, the volume of the ocean basins has increased substantially over the last 3 Gyr. Because less heat is being lost from the earth's interior the volume of the ocean ridges has decreased with time. This hypothesis has been successfully used to explain the variations in sea level in the last 500 Kyr (Hays and Pitman, 1973; Pitman 1978). Based on estimates of how the heat flux from the earth's interior has varied with time, estimates of the volume of the ocean basins have been given (Turcotte and Burke, 1978; Abbott, 1984; Reymer and Schubert, 1984). These calculations show that a substantial increase in the volume of the continents can occur without increasing sea level.

Important evidence for the volume of the continents as a function of time comes from isotopic studies. The incompatible elements associated with the U-Th-Pb, Rb-Sr, Sm-Nd, and Lu-Hf isotopic systems are strongly concentrated into the continental crust leaving a residual depleted mantle. The isotopic evolution of the mantle is sensitive to the time of removal of the elements. Studies of the Rb-Sr system indicate that the mean age of the continents relative to the mantle is 2.1 ± 0.3 Gyr (Allègre et al. 1983; Turcotte and Kellogg, 1986). This is the mean time since the material in the continents was removed from the mantle.

Substantial depletion of the mantle has been found from the earliest available rocks. These results have been used to infer that the volume of the continents was a significant fraction of their present size at 3 Gyr before present (Armstrong, 1968, 1981; DePaola, 1983; Nelson and DePaola, 1985). A large continental volume in the Archean and a relatively young mean age of segregation implies very substantial recycling. A number of authors have studied the implications of isotope data for the evolution of the volume of the continents (McCulloch and Wasserburg, 1978; O'Nious et al., 1979; DePaola, 1979, 1980; Allègre, 1982; Hamilton et al., 1983; Allègre and Rousseau, 1984).

Another constraint on recycling is the chemical evolution of the continents with time. There is some evidence that the concentration of silicon relative to magnesium and calcium has increased with time.

Since magnesium and calcium readily dissolve in sea water and can be precipitated either during hydrothermal circulations through the oceanic crust or on the sea floor, mechanisms for the preferential recycling of these elements are available. Uranium may also be preferentially recycled relative to thorium as well as other examples (McLennan and Taylor, 1980, 1982).

In this paper we consider a relatively simple model for recycling of the continental crust; it follows the approach set forth by Allègre and Jaupart (1985) and by Gurnis and Davies (1985, 1986). We also consider alternative mechanisms of recycling.

2. CRUSTAL RECYCLING - A MODEL

An infinite variety of models can be proposed for the evolution of the continental crust. We propose a simple model based on uniformatarianism; basic processes in the past were essentially the same as today. The rate of change of the mass of the continental crust M_c is given by

$$\frac{dM_c}{dt} = F_o \exp\left(\frac{\tau_e - \tau}{\tau_o}\right) - \frac{1}{\tau_r} M_c \tag{1}$$

where F_o is the present rate of addition of mass to the crust. This rate is hypothesized to have been higher in the past because of the more vigorous convection in the mantle due to larger concentrations of heat producing elements. The time τ_e is the time over which crustal growth has occurred; we take $\tau_e = 4$ Gyr. Time t is measured forward from the time of initial crustal growth. We associate τ_o with the time variation of heat production in the mantle and take $\tau_o = 3.6$ Gyr (Turcotte and Schubert, 1982, pp. 139-142). We further hypothesize that the rate of crustal recycling is linearly related to the mass of the crust with the constant τ_r determining the rate. It could be argued that τ_r was smaller in the past but its time dependence would be poorly constrained.

The solution of (1) must satisfy the conditions:

$$M_c = 0 \text{ at } \tau = 0$$
$$M_c = M_{co} \text{ at } \tau = \tau_e$$

where M_{co} is the present mass of the continental crust. The required solution is

$$M = M_o \frac{\exp(-t/\tau_o) - \exp(-t/\tau_r)}{\exp(-\tau_e/\tau_o) - \exp(-\tau_e/\tau_r)} \tag{2}$$

with

$$F_o = \frac{M_o \, (1/\tau_r - 1/\tau_o)}{[1 - \exp \, (\tau_e/\tau_o - \tau_\epsilon/\tau_r)]} \tag{3}$$

We require an additional constraint in order to determine both τ_r and F_o.

We next determine a mean age for the continental crust $\bar{\tau}_c$. This is the age defined as the time in the past that the crust was extracted from the mantle. The required expression is

$$\bar{\tau}_c = \frac{(1/\tau_r - 1/\tau_o) \exp \, (-t/\tau_r)}{\exp \, (-t/\tau_o) - \exp \, (-t/\tau_r)}$$

$$\int_o^{\tau_e} (t - t') \exp \, [(1/\tau_r - 1/\tau_o)t'] dt' \tag{4}$$

and integration gives

$$\bar{\tau}_c = \frac{1}{(1/\tau_r - 1/\tau_o)} + \frac{\tau_e}{[1 - \exp \, (\tau_e/\tau_r - \tau_e/\tau_o)]} \tag{5}$$

The present ratio of crustal recycling to crustal production is given by

$$\frac{M_{co}}{\tau_r F_o} = \frac{1 - \exp \, (\tau_e/\tau_o - \tau_e/\tau_r)}{1 - \tau_r/\tau_o} \tag{6}$$

With the mean continental age specified in (5) the recycling time constant τ_r can be determined.

If no recycling occurs then $\tau_r \to \infty$ and we have from (2), (3), and (5)

$$M_c = M_{co} \left[\frac{1 - \exp \, (- t/\tau_o)}{1 - \exp \, (- \tau_e/\tau_o)} \right] \tag{7}$$

$$F_o = \frac{M_o/\tau_o}{\exp \, (\tau_e/\tau_o) - 1} \tag{8}$$

$$\bar{\tau}_c = \frac{\tau_e}{1 - \exp \, (- \tau_e/\tau_o)} - \tau_o \tag{9}$$

We next consider specific examples.

Taking $\tau_e = 4$ Gyr, $\tau_o = 3.6$ Gyr, and no crustal recycling ($\tau_r \to \infty$) the dependence of M_c/M_{co} on τ (time measured back from the present) is given in Figure 1. The mean age of the continental crust from (9)

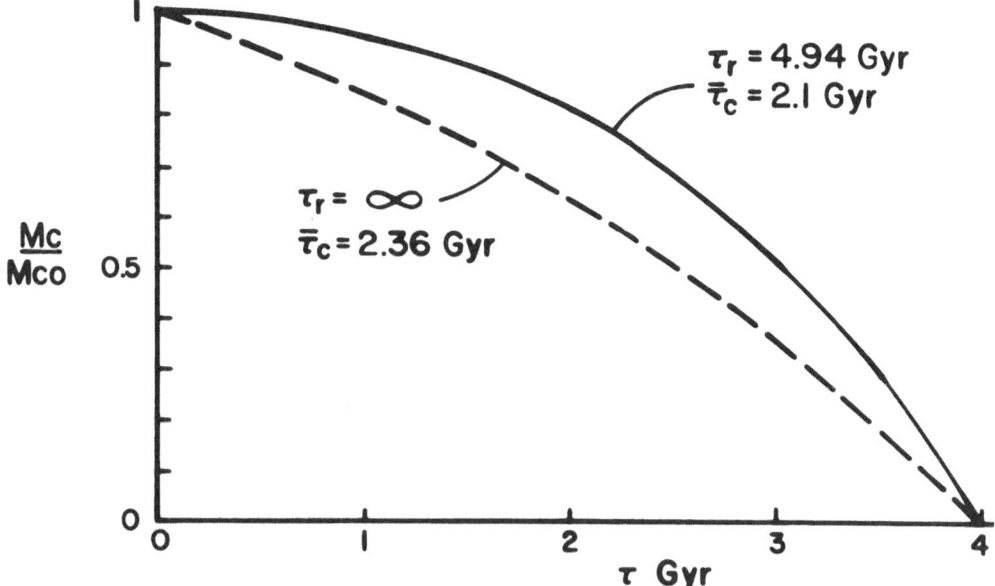

Figure 1. Mass of the continental crust M_c, relative to the present value M_{co}, as a function of time measure backward from the present τ. The dashed line includes no crustal recycling and has a mean segregation age of 2.36 Gyr. The solid line includes crustal recycling with a mean segregation age of 2.1 Gyr.

is $\bar{\tau}_c$ = 2.36 Gyr. Based on a variety of isotopic data it is estimated (Turcotte and Kellogg, 1986) that $\bar{\tau}_c$ = 2.1 Gyr. With this value and the others given above we find from (5) that τ_r = 4.94 Gyr. The corresponding dependence of M_c/M_{co} on τ is given in Figure 1. It is seen that the mass of continental crust is considerable greater in the past with this amount of crustal recycling even though the mean age was only reduced by 0.2 Gyr. With no crustal recycling the present rate of crustal addition from (8) is F_o = 1.01 km^3/yr (with ρ = 2750 kg/m^3). With $\bar{\tau}_c$ = 2.1 Gyr and τ_r = 4.94 Gyr we find from (3) that the present rate of crustal addition is F_o = 1.58 km^3/yr and from (6) the present rate of crustal recycling is M_{co}/τ_r = 1.49 km^3/yr. Thus the net rate of crustal growth is only F_o - M_{co}/τ_r = 0.09 km^3/yr.

Our analysis given above gives a present recycling flux from the continental crust into the mantle of 1.5 km^3/yr. Several mechanisms can result in recycling, we will consider three.

3. MECHANISMS OF RECYCLING

3.1 Subduction of sediments

We first hypothesize that sediment subduction at ocean trenches is
responsible for the recycling of continental crust into the mantle.
The mean rate of subduction of the sea floor averaged over the last
500 Myr is 4.2 km^2/yr (Turcotte and Schubert, 1983, pp. 163-167).
Assuming a sediment density of 2,350 kg/m^3 the subduction of 1.5
km^3/yr of continental crust (density 2,750 kg/m^3) requires a mean
sediment thickness of 420 m to be subducted. It should be emphasized
that for this type of recycling to be effective the sediments cannot
be melting and returned to the continental crust in island arc
volcanism.

Kay (1980) and Karig and Kay (1981) estimate that about 50 m of
sediments must be subducted in order to account for the potassium
budget of the Marinas. Dewey and Windley (1981) estimated the volume
of sediments transported to subduction zones and subtract the volume
accumulating in accretionary prisms and conclude that about 200 m of
sediments are subducted. Thus a substantial fraction of the recycling
flux predicted by our model can be attributed to the direct subduction
of sediments. However, with the volume of sediments available, it is
difficult to attribute the entire recycling flux to direct sediment
subduction.

3.2 Continental crust erosion

A second mechanism for subduction of continental crust is by direct
erosion of the continental crust by the subducting plate (Karig,
1974). If a 35 km thick continental crust is eroded and entrained by
the subducting plate we require that a 90 km depth be eroded each 100
Myrs along the entire 4.8 x 10^4 km length of the subduction zones in
order to provide a volume flux of 1.5 km^3/yr. It is very difficult to
make realistic estimates of the importance of crustal erosion as a
mechanism for the return of continent crust to the mantle.

3.3 Delamination of the continental crust

Continental crust can also be returned to the mantle by the
delamination of the continental lithosphere if continental crust is
included in the delaminated layer. If 15 km of continental crust is
included in the delaminated layer, then the 10^7 km of continental
lithosphere must be delaminated every 100 Myrs in order to provide a
flux of 1.5 km/yr (this corresponds to an area 3,000 km by 3,000 km).

Continental delamination was proposed and studied by Bird (1979)
and Bird and Baumgardner (1981). Direct evidence for continental
delamination does not exist but there are a number of continental
areas in which the mantle lithosphere is absent. The best example is
the western United States. Plateau uplifts such as the Altiplano are
also associated with the absence of mantle lithosphere. Crustal
doubling such as in Tibet has also been attributed to the absence of

mantle lithosphere beneath Asia (Molnar and Topponnier, 1981). Delamination is an efficient mechanism for the separation of the mantle continental lithosphere from the continental crust. It is, of course, quite possible to also delaminate a fraction of the continental crust. Delamination associated with subduction has also been proposed by McKenzie and O'Nious (1983) to explain isotopic anomalies associated with ocean islands.

It is of interest to determine how much continental crust can be delaminated. The net buoyancy mass per unit area of the continental lithosphere below a depth y is given by (Turcotte and Schubert, 1982, pp. 181-183)

$$\delta M = \int_y^\infty [(\rho_{mo} - \rho_{co}) - \rho_{mo} \, \alpha \, (T_m - T)] dy \tag{10}$$

where $\rho_{mo} - \rho_{co}$ is the buoyancy mass of the crust and $\rho_{mo}\alpha \, (T_m - T)$ is the negative buoyancy mass due to the thermal contraction of the continental lithosphere. Following Turcotte and McAdoo (1979) we assume

$$\delta M = \int_y^\infty (\rho_{mo} - \rho_{co}) dy + \frac{\rho_m \alpha \, (T_m - T_o) y_L}{1.16} \int_{1.16 \frac{y}{y_L}}^\infty \mathrm{erfc} \, \eta \, d\eta \tag{11}$$

with $\rho_{ml} = 3300$ kg/m^3, $\rho_{mo} - \rho_{co} = 600$ kg/m^3 for $0 < y < 20$ km and $\rho_{mo} - \rho_{co} = 500$ kg/m^3 for 20 km $< y < 35$ km, $\alpha = 3.2 \times 10^{-5}$ °C^{-1}, $T_m - T_o = 1300$°C, and $y_L = 180$ km. With these values the buoyancy mass per unit area beneath a depth y in normal continental lithosphere is given as a function of y in Figure 2. Beneath a depth of about 16 km the lithosphere has negative buoyancy and can be delaminated. It would be relatively easy to delaminate the continental crust beneath a depth of 20 km. Due to the softness of continental rocks it is likely that a soft zone exists in the continental crust between depths of 15 and 20 km, this has been referred to as an intracrustal asthenosphere (Turcotte et al., 1984).

4.   SEA LEVEL FLUCTUATIONS

The model given is section 2 gives a prediction of the mass of the continental crust as a function of time. We will assume as a reasonable approximation that the thickness of the continental crust has been a constant, thus the mass of the continental crust is directly proportional to the area

$$\frac{A_c}{A_{co}} = \frac{M_c}{M_{co}} \tag{12}$$

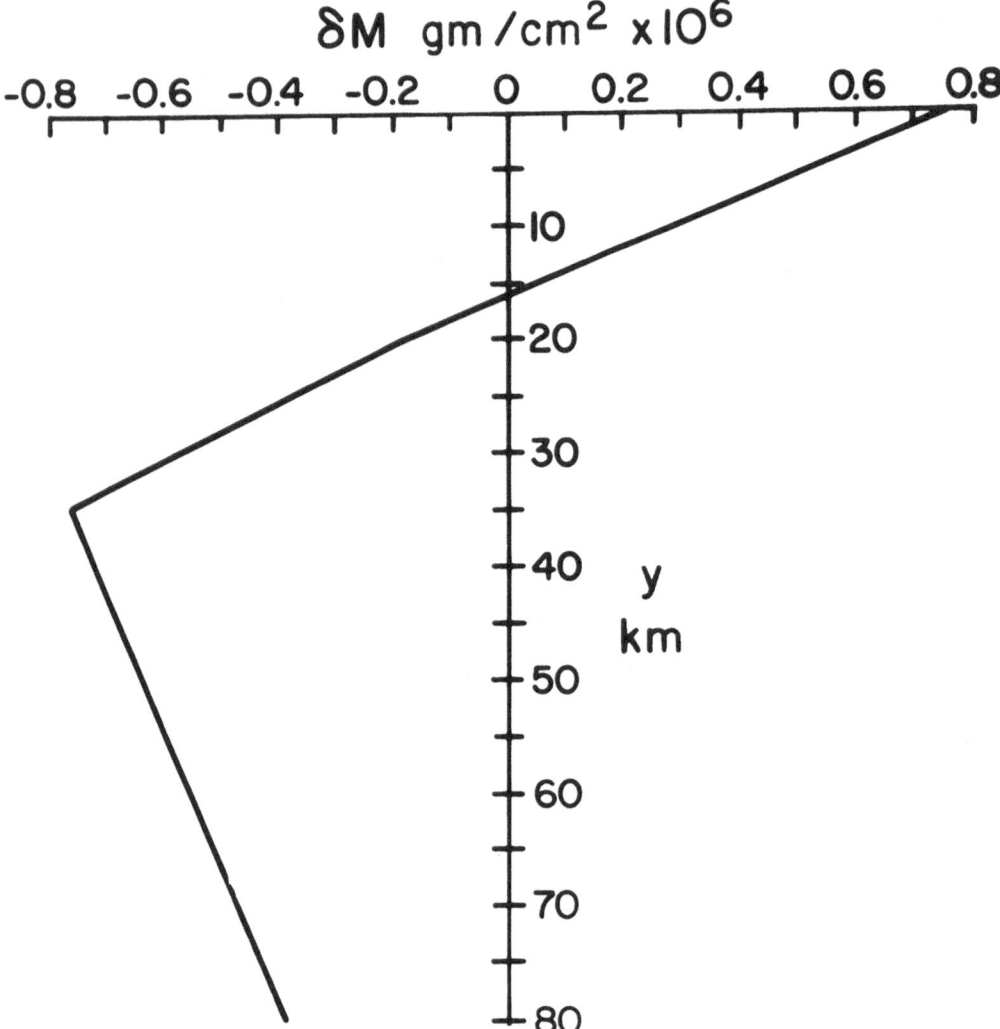

Figure 2. Buoyancy mass per unit area of the continental lithosphere δM below a depth y as a function of that depth.

Assuming the volume of water has been constant we have

$$A_o d_o = A_{oo} d_{oo} \tag{13}$$

where A_o and d_o are the area and mean depth of the oceans. We also have

$$A_o + A_c = A_e \tag{14}$$

where A_e is the total area of the earth. Combining (12) - (14) the variation in sea level due to changes in crustal mass $\delta h_c = d_o - d_{oo}$ is given by

$$\delta h_c = -d_{oo} \left[1 - \frac{A_{oo}}{A_e - A_{co} \dfrac{M_c}{M_{co}}} \right] \qquad (15)$$

The increase in continental mass results in an increase in sea level with time.

It is also necessary to consider the variation in the volume of the ocean basins due to variations in mean heat flux. The change in the reciprocal mean heat flux $\delta(\bar{q}_s^{-1})$ is related to the change in the mean depth of the ocean basins by

$$\delta h_q = - \frac{8}{3\pi} \frac{k \, \rho_m \, \alpha \, (T_m - T_o)^2}{(\rho_m - \rho_w)} \delta \left(\frac{1}{\bar{q}_s} \right) \qquad (16)$$

Following our approach in section 2 the variation in mean heat flux is assumed to be given by

$$\delta \left(\frac{1}{\bar{q}_s} \right) = \frac{1}{\bar{q}_{so}} \left(\frac{\bar{q}_{so}}{\bar{q}_s} - 1 \right) = \frac{1}{\bar{q}_{so}} \left[e^{-\tau/\tau_o} - 1 \right] \qquad (17)$$

Combining (16) and (17) yields

$$\delta h_q = \frac{8}{3\pi} \frac{k \, \rho_m \, \alpha \, (T_m - T_o)^2}{(\rho_m - \rho_w) \, \bar{q}_{so}} \left[1 - e^{-\tau/\tau_o} \right] \qquad (18)$$

Due to shallower oceans the sea level was higher in the past.

Taking values given above along with $k = 3.3$ Wm^{-1} ^{o}K^{-1}, $\bar{q}_{so} = 78.5$ mWm^{-2}, $\rho_m - \rho_w = 2,300$ kg m^{-3}, $d_{oo} = 4.6$ km, $A_{co} = 2 \times 10^{14}$ m^2, and $A_e = 5.1 \times 10^{14}$ m^2 the predicted variation of sea level with time in the past is given in Figure 3. We consider both examples given in Figure 1. With no crustal recycling the sea level was higher in the past with a maximum height of about 200 m at 2 Gyr before present. With the amount of crustal recycling required to reduce the mean age of segregation of the continents to 2.1 Gyr, the sea level was considerable higher in the past. The maximum height was about 600 m at 2 Gyr before present. Fluctuations during the last 500 Myr have been of the order of 300 m, however there is no long-term direct evidence on sea level favoring one or the other of the two models considered.

254

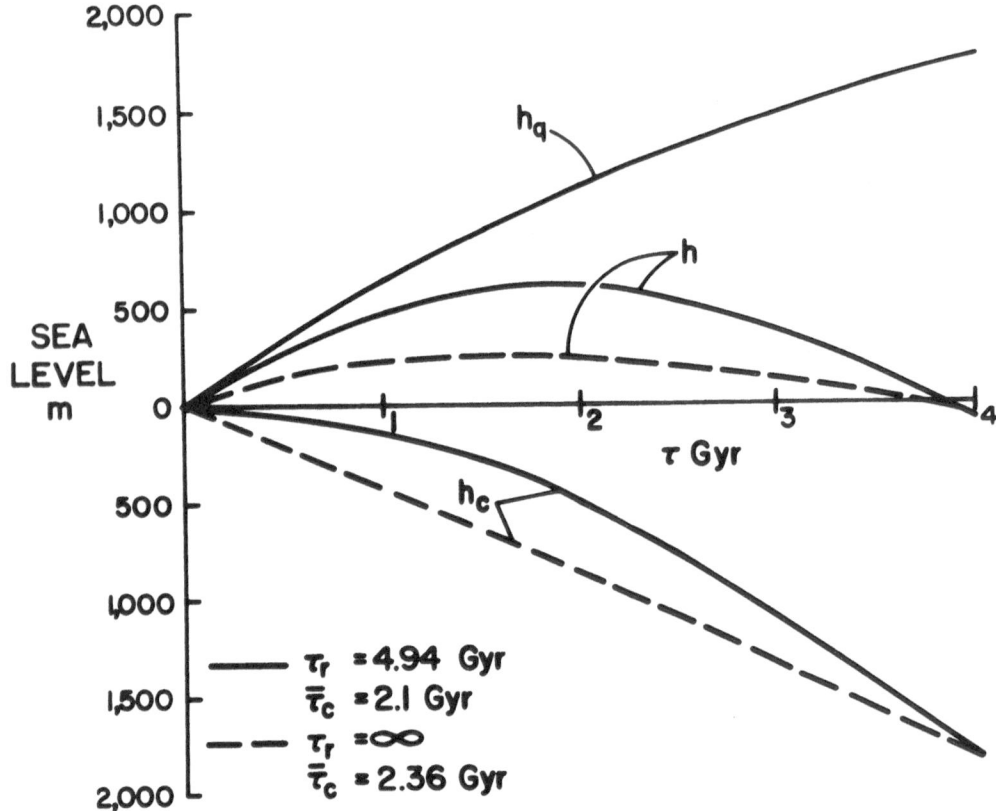

Figure 3. Sea level h as a function of time measured backward from the present τ. Results with crustal recycling (solid lines) and without crustal recycling (dashed line) are given. The contribution due to changes in the volume of the continental crust h_c and the volume of the ocean basins h_q are included.

5. CONCLUSIONS

We have proposed a relatively simple model for the growth of the continents. The rate of crustal addition is scaled with the rate of heat loss from the mantle. Without crustal recycling we find that the mean age of crustal segregation from the mantle is $\tau_c = 2.36$ Gyr. This is somewhat higher than the preferred value of 2.1 Gyr based on rubidium - strontium isotope systematics. In order to reduce the mean age we introduced crustal recycling scaled with the mass of the continental crust. With a recycling time $\tau_r = 4.94$ Gyr the mean age of crustal segregation is reduced to $\bar{\tau}_c = 2.1$ Gyr. The present required rate of crustal addition is 1.58 km^3/yr and the present required rate of crustal recycling is 1.49 km^3/yr.

Three mechanisms can be proposed for the recycling of continental crust into the mantle. The first is the direct subduction of sediments at ocean trenches. In order to recycle 1.49 km^3/yr an average sediment thickness of 420 m would have to be subducted. Estimates based on the volume of sediments transported to subduction zones indicate that only about 25 - 50% of this amount is actually subducted. Continental crust may also be eroded during subduction. In both cases a fraction of the subducted material may be melted and returned to the continental crust in island arc volcanism. Delamination of the continental lithosphere may include a fraction of the continental crust. It is difficult to quantitatively estimate the importance of this process.

Sea level changes place constraints on the volume of the continental crust in the past. There is ample observational evidence that sea level relative to the continents was higher in the past. A smaller volume for the continental crust in the past would have led to a lower sea level. However, larger ocean ridge volumes in the past would have decreased the volumes of the ocean basins in the past. With higher mean heat flows in the past, the estimates of this volume effect indicate that sea level studies do not place tight constraints on alternative models for the evolution of the continents with time.

ACKNOWLEDGMENTS

This research has been supported in part by the Division of Earth Sciences, National Science Foundation under grant EAR-8518019.

REFERENCES

Abbott, D.H., 1984, Archaean plate tectonics revisited. 2. Paleo-sea level changes, continental area, oceanic heat loss and the area-age distribution of the ocean basins, Tectonics, 3, 709-722.

Allègre, C.J., 1982, Chemical geodynamics, Tectonophys., 81, 109-132.

Allègre, C.J., S.R. Hart, and J.F. Minster, 1983, Chemical structure and evolution of the mantle and continents determined by inversion of Nd and Sr isotopic data, II. Numerical experiments and discussion, Earth Planet. Sci. Let., 66, 191-213.

Allègre, C.J. and C. Jaupart, 1985, Continental tectonics and continental kinetics, Earth Planet. Sci. Let., 74, 171-186.

Allègre, C.J. and D. Rousseau, 1984, The growth of the continent through geological time studied by Nd isotope analysis of shales, Earth Planet. Sci. Let., 67, 19-34.

Ambrose, J.W., 1964, Exhumed paleoplains of the Precambrian Shield of North America, Am. J. Sci., 262, 817-857.

Armstrong, R.L., 1968, A model for the evolution of strontium and lead isotopes in a dynamic Earth, Rev. Geophys., 6, 175-200.

Armstrong, R.L., Radiogenic isotopes: the case for crustal recycling on a near-steady-state no-continental-growth Earth, Phil. Trans. Roy. Soc. London, 301A, 443-472.

Bird, P., 1979, Continental delamination and the Colorado Plateau, J. Geophys. Res., 84, 7561-7571.

Bird, P. and J. Baumgardner, 1981, Steady propagation of delamination
    events, J. Geophys. Res., 86, 4891-4903.
DePaolo, D.J., 1979, Implications of correlated Nd and Sr isotopic
    variations for the chemical evolution of the crust and mantle,
    Earth Planet. Sci. Let., 43, 201-211.
DePaolo, D.J., 1980, Crustal growth and mantle evolution: inferences
    from models of element transport and Nd and Sr isotopes, Geochim.
    Cosmochim. Acta, 44, 1185-1196.
DePaolo, D.J., 1983, The mean life of continents: estimates of
    continent recycling rates from Nd and Hf isotopic data and
    implications for mantle structure, Geophys. Res. Let., 10, 705-
    708.
Dewey, J.F. and B.F. Windley, 1981, Growth and differentiation of the
    continental crust, Phi. Trans. Roy. Soc. London, 301A, 189-206.
Gurnis, M. and G.F. Davies, 1985, Simple parametric models of crustal
    growth, J. Geodynamics, 3, 105-135.
Gurnis, M. and G.F. Davies, 1986, Apparent episodic crustal growth
    arising from a smoothly evolving mantle, Geology, 14, 396-399.
Hallam, A., 1977, Secular changes in marine inundation of USSR and
    North America through the Phanerozoic, Nature, 269, 769-772.
Hamilton, P.J., R.K. O'Nions, D. Bridgwater, and A. Nutman, 1983, Sm-
    Nd studies of Archean metasediments and metavolcanics from west
    Greenland and their implications for the earth's early history,
    Earth Planet. Sci. Let., 62, 263-272.
Hays, J.D. and W.C. Pitman, 1973, Lithospheric plate motion, sea level
    changes and climatic and ecological consequences, Nature, 246,
    18-22.
Karig, D.E., 1974, Tectonic erosion at trenches, Earth Planet. Sci.
    Let., 21, 209-212.
Karig, D.E. and R.W. Kay, 1981, Fate of sediments on the descending
    plate at convergent margins, Phil. Trans. Roy. Soc. London, 301A,
    233-251.
Kay, R.W., 1980, Volcanic arc magmas: Implications of a melting-
    mixing model for element recycling in the crust-upper mantle
    system, J. Geology, 88, 497-522.
McCulloch, M.T. and G.J. Wasserburg, 1978, Sm-Nd and Rb-Sr chronology
    of continental crust formation, Science, 200, 1003-1011.
McKenzie, D. and R.K. O'Nions, 1983, Mantle reservoirs and ocean
    island basalts, Nature, 301, 229-231.
McLennan, S.M. and S.R. Taylor, 1980, Th and U in sedimentary rocks:
    crustal evolution and sedimentary recycling, Nature, 285, 621-
    624.
McLennan, S.M. and S.R. Taylor, 1982, Geochemical constraints on the
    growth of the continental crust, J. Geol., 90, 347-361.
McLennan, S.M. and S.R. Taylor, 1983, Continental freeboard,
    sedimentation rates and growth of continental crust, Nature, 306,
    169-172.
Molnar, P. and P. Topponnier, 1981, A possible dependence of tectonic
    strength on the age of the crust in Asia, Earth Planet. Sci.
    Let., 52, 107-114.

Nelson, B.K. and D.J. DePaolo, 1985, Rapid production of continental crust 1.7 to 1.9 b.y. ago: Nd isotopic evidence from the basement of the North American mid-continent, Geol. Soc. Am. Bull., 96, 746-754.

O'Nions, R.K., N.M. Evensen, and P.J. Hamilton, 1979, Geochemical modeling of mantle differentiation and crustal growth, J. Geophys. Res., 84, 6091-6101.

Pitman, W.C., 1978, Relationship between eustacy and stratigraphic sequences of passive margins, Geol. Soc. Am. Bull., 89, 1389-1403.

Reymer, A. and G. Schubert, 1984, Phanerozoic addition rates to the continental crust and crustal growth, Tectonics, 3, 63-77.

Schubert, G. and A.P.S. Reymer, 1985, Continental volume and freeboard through geological time, Nature, 316, 336-339.

Turcotte, D.L. and K. Burke, 1978, Global sea-level changes and the thermal structure of the earth, Earth Planet. Sco. Let., 41, 341-346.

Turcotte, D.L. and L.H. Kellog, 1986, Isotopic modeling of the evolution of the mantle and crust, Rev. Geophys., 24, 311-328.

Turcotte, D.L., J.Y. Liu, and F.H. Kulhawy, 1984, The role of an intracrustal asthenosphere on the behavior of major strike-slip faults, J. Geophys. Res., 89, 5801-5816.

Turcotte, D.L. and D.C. McAdoo, 1979, Geoid anomalies and the thickness of the lithosphere, J. Geophys. Res., 84, 2381-2387.

Turcotte, D.L. and G. Schubert, 1982, Geodynamics, (John Wiley and Sons, New York), p. 450.

Watson, J.V., 1976, Vertical movements in Proterozoic structural provinces, Phil. Trans. Roy. Soc. London, A280, 629-640.

Windley, B.F., 1977, Timing of continental growth and emergence, Nature, 270, 426-428.

Wise, D.U., 1974, Continental margins, freeboard and the volumes of continents and oceans through time, in The Geology of Continental Margins, C.A. Burk and C.L. Drake, eds., pp. 45-56, Springer-Verlag, New York.

SOME SPECULATIONS ON CONTINENTAL EVOLUTION

Thomas H. Jordan
Department of Earth, Atmospheric and Planetary Sciences
Massachusetts Institute of Technology
Cambridge, MA   02139 USA

ABSTRACT. There appears to exist a thick (>300 km) thermal boundary layer (TBL), or tectosphere, beneath the ancient cratonic nuclei that is stabilized against convective disruption by a buoyant, viscous, chemical boundary layer (CBL) depleted in basaltic constituents and enriched in large-ion lithophile (LIL) elements relative to the source mantle of mid-ocean ridge vulcanism. This paper discusses some implications of this tectosphere model for continental formation and evolution. The total volume of basalt extracted from the sub-lithospheric CBL is estimated to be at least 3×10^9 km^3, about 40% of the present-day volume of continental crust. A multi-stage evolutionary scenario is required to decouple the process of basalt depletion from the process of CBL thickening. A plausible mechanism for the formation of the continental tectosphere is by the advective thickening of a basalt-depleted, LIL-charged CBL during major episodes of compressive orogenesis, particularly those accompanying the assembly of supercontinents. The seismic and heat flow data indicate that most of the cratons having thick TBLs are of Archæan and early Proterozoic age. Since these crustal ages mark the stabilization dates of the subcrustal tectospheric columns, it is inferred that the depletion and advective-thickening processes which produced the extant cratons were operating more efficiently prior to mid-Proterozoic time. The cold, refractory CBL of the continental tectosphere constitutes a large and particularly stable reservoir for LIL elements; crude estimates based on nodule data suggest that the subcrustal CBL's total inventory of light rare earths and other incompatibles could be comparable to the crust's. Both the geochemical and thermometric constraints indicate high heat production within the CBL and low heat flow through its base. It is inferred that very little of the internal heat being convectively transported out of the earth is escaping through the continental cratons. The stability of the continental tectosphere and the discrepancy in the heat fluxing from the deep mantle beneath continents and oceans are evidence that the surficial CBL is strongly coupled to large-scale convective flow in the mantle. A model of the earth is proposed consisting of four major dynamical systems: the two convecting shells of the mantle and core, and the two CBLs at the free surface and core-mantle interface. Strong interactions among the low-wavenumber states of these four systems offer new possibilities for explaining the earth's large-scale, long-term behavior.

1. INTRODUCTION

Seismic data show that the ancient continental cratons are underlain by an extensive layer of anomalous mantle material, geographically variable and up to several hundred kilometers thick, which translates with the continents during plate motions (Jordan,

259

S. R. Hart and L. Gülen (eds.), Crust/Mantle Recycling at Convergence Zones, 259–276.
© 1989 by Kluwer Academic Publishers.

1975a,b; Woodhouse and Dziewonski, 1984; Grand and Helmberger, 1984a,b; Lerner-Lam and Jordan, 1987; Grand, 1987; Jordan et al., 1988). It has been argued that the typical thickness of the thermal boundary layer (TBL), or tectosphere, in stable continental regions exceeds the thickness of the mechanical boundary layer (MBL), or lithosphere, by an amount inconsistent with the boundary-layer cooling models (Jordan, 1978, 1981a, 1988).

The model of the continental tectosphere advocated by the author is based on the notion that the mass excesses associated with cold, thick TBLs are locally compensated by variations in the composition of an augmented chemical boundary layer (CBL). This dynamical effect can be quantified in terms of a normative density $\hat{\rho}$; *i.e.*, the mass per unit volume of rock normalized to a standard temperature and pressure (Jordan, 1979a). According to the chemical compensation model, $\hat{\rho}$ of the subcontinental upper mantle is less than the oceans, and the perturbations in the radii of isopycnic (equal-density) surfaces are much smaller than predicted by a constant-composition, plate-cooling model. A particular profile that satisfies the appropriate stability condition is one where the positive buoyancy due to composition exactly cancels the negative buoyancy due to lower temperatures at every level, and isopycnic surfaces are thereby horizontal. This idealization is here called the *isopycnic hypothesis* and is expressed by the equation

$$\delta\hat{\rho}(z) = \hat{\rho}\,\alpha\,\delta T(z) \tag{1}$$

where δT is the continent-ocean temperature difference (Fig. 1).

The existence of a thick CBL beneath the cratons is suggested by petrological data, as well as by the geophysical constraints. The primary mechanism for chemical compensation of the continental TBL appears to be the extraction of a basalt-like component from mantle peridotites; i.e., the depletion of Fe and Al relative to Mg. The subtraction of basalt from a garnet lherzolite leaves a residue less dense than the parental material (O'Hara, 1975; Boyd & McCallister, 1976; Green & Liebermann, 1976; Jordan, 1979a). Peridotite nodules from kimberlite pipes in cratonic areas are observed to have major-element compositions depleted in basaltic constituents relative to compositions derived for the oceanic upper mantle (Ringwood, 1966; Nixon *et al.*, 1973; Boyd & McCallister, 1976; Boyd & Mertzman, 1987; Nixon, 1987).

The differences in normative density calculated from these compositions are sufficient to stabilize the superadiabatic thermal gradients inferred by geophysical methods. For the range of garnet lherzolite compositions observed in the earth's mantle, the normative densities computed by Jordan (1979a) are approximately related to the whole-rock molar magnesium number $N_{Mg} \equiv 100\,X_{Mg}/(X_{Mg} + X_{Fe})$ by the linear equation

$$\hat{\rho} = (5.093 - .0191\,N_{Mg})\,\text{Mg/m}^3$$

Therefore, to the extent that compositional variations can be parameterized by the magnesium number (see Jordan (1981a) for a discussion of the effect of an independent variation in alumina content), the isopycnic eqn. (1) can be written as a proportionality between the continent-ocean difference in N_{Mg} and the continent-ocean temperature contrast δT,

$$\delta N_{Mg} = -\delta T\,/\,187°C \tag{2}$$

In other words, increasing N_{Mg} by a little as one unit offsets the negative buoyancy caused by a temperature decrease of nearly 200°C.

An average continental garnet lherzolite (ACGL) constructed from 78 whole-rock analyses of kimberlite xenoliths (Jordan, 1978) has a magnesium number of 91.1, which is 1.5 to 2.3 units larger than typical pyrolite compositions (Ringwood, 1975) and comparable to the residuum obtained by fractionating 10-20% of "primitive" basalt from such parental rocks. The granular peridotite nodules that dominate the ACGL appear to have equilibrated at depths of 150-200 km and temperatures of 1000-1200°C (Boyd, 1973; Finnerty & Boyd, 1987). Oceanic temperatures at this depth are inferred to be about 200-400°C higher (Jordan, 1975a; Pollack & Chapman, 1977; Sclater *et al.*, 1980), which is in good agreement with the 280-430°C calculated from eqn. (2). Moreover, the xenoliths display a decrease in N_{Mg} with an increase in their depth of equilibration (Boyd & Nixon, 1975), providing direct evidence for compositional stratification of the continental CBL (Jordan, 1978).

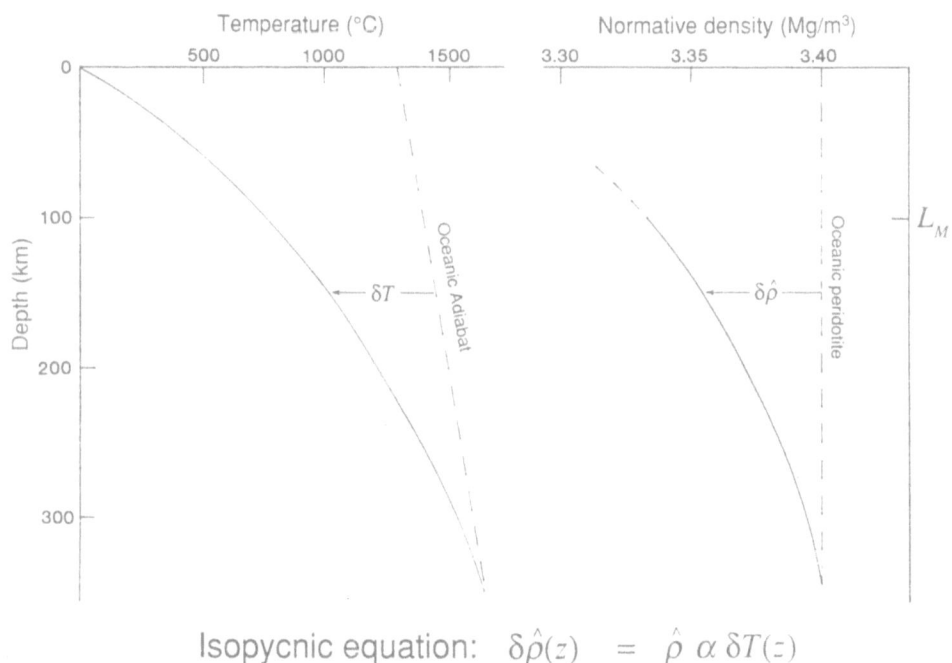

Isopycnic equation: $\delta\hat{\rho}(z) = \hat{\rho} \, \alpha \, \delta T(z)$

FIGURE 1. An upper-mantle structure for a continental craton conforming to the isopycnic hypothesis, which states that, below the base of the MBL at depth L_M, the temperature drop within the cratonic CBL is chemically compensated by a depletion in Fe and Al relative to Mg. *Left panel:* Temperature difference between Pollack & Chapman's (1977) 40-mW/m^2 continental geotherm and an average oceanic adiabat having a potential temperature of 1300°C. *Right panel:* Normative density of the continental upper mantle calculated from the isopycnic equation assuming $\alpha = 3 \times 10^{-5}$°C and an oceanic reference density of 3.40 Mg/m^3.

The model of the continents as a compound chemical boundary layer, comprising both crustal and mantle components, has a number of ramifications pertinent to the problems of continental evolution. For example, it implies that the process of "cratonization" or "chelogenesis" involves the compositional modification of the mantle, not just the crust. It has been speculated that the fundamental transition from the permobile behavior of the early Archæan crust to the linear geosynclinal and mobile-belt tectonics of the Proterozoic (e.g., Windley, 1977) marks the time when the area occupied by a stable, thickened CBL became significant (Jordan, 1978). The Archæan cratonic nuclei would then be those areas where the CBL built up at an early date, whereas active mobile belts and orogenic zones would be regions where a thick CBL was never formed or was destroyed by convective activity. This mechanism appears to offer more hope of explaining why the present-day distinction between cratons and mobile belts has, in many cases, persisted since the Proterozoic and why continents tend to break apart near old sutures than does a model based solely on thermal variations in the uppermost mantle.

Of course, given the paucity of the constraints on CBL structure and the dynamical complexities associated large-scale chemical heterogeneities, any ruminations on the fundamental issues of continental evolution are necessarily speculative. There are, however, some difficult problems raised by the existence of continental deep structure that need to be addressed in the context of an evolutionary model. The discussion in this paper focuses on several of these problems with intention of stimulating inquiry rather than establishing firm conclusions.

2. VOLUME OF BASALT EXTRACTED

The isopycnic hypothesis can be used to estimate the amount of basalt that has been removed from the part of the CBL below the MBL. For the purposes of a rough calculation, the oceanic geotherm is assumed to be an adiabat with a potential temperature of 1300°C and a slope of 0.5°C/km. The temperature profile for Precambrian shields and platforms (Region S in Jordan's (1981b) tectonic regionalization) is taken to be Pollack & Chapman's (1977) 40-mW/m^2 geotherm. For $\bar{\rho} = 3.4$ Mg/m^3 and $\alpha = 3 \times 10^{-5}$°C, integrating eqn. (1) from the base of the MBL at 100 km depth to the intersection of oceanic and shield geotherms at ~400 km yields an areal density of 8.2×10^6 kg/m^2. The total mass excess obtained by multiplying by the area A of Region S (34×10^6 km^2) is $\Delta M = 2.8 \times 10^{20}$ kg.

The volume of basalt that must be extracted to compensate this mass excess is $\Delta V \approx \Delta M/f\Delta\bar{\rho}$, where $\Delta\bar{\rho}$ is the normative density difference between parental and residual mantle and f is the residuum/basalt ratio. The product of the last two quantities depends only weakly on the particular melting model. A numerical value of 210 kg/m^3 (Jordan, 1979a) gives $\Delta V \approx 1.4 \times 10^9$ km^3 for Region S, which is about 30% greater than its crustal volume. The Phanerozoic platforms (Region P, $A = 53 \times 10^6$ km^2) have an area about 50% larger than Region S, but the heat-flow and seismic data indicate a significantly thinner TBL, so less chemical compensation is needed. The integrated volume of basalt extracted from this region should thus be comparable to Region S. In the Phanerozoic orogenic zones (Region Q, $A = 111 \times 10^6$ km^2), where the tectosphere is relatively thin and generally behaves similar to the oceanic TBL, there is no evidence for chemical compensation below the MBL.

In round numbers then, the total volume of basalt extracted from the sub-lithospheric CBL is predicted by the isopycnic hypothesis to be ~3×10^9 km^3, about 40% of the present-day volume of continental crust. Added to this estimate will be the basalt corresponding to any net depletion of the continental MBL itself, which could as much as double the total. To reduce the estimate substantially, we must appeal to an implausibly low oceanic adiabat or unrealistically steep continental geotherm. Hence, if the isopycnic hypothesis is correct, the volume of basalt extracted from the sub-cratonic mantle must be very large indeed.

The total amount of chemical differentiation presents no particular problem; sea-floor spreading is now generating new crust at a rate of 2×10^7 km^3/Ma, so that the present-day plate system fluxes 3×10^9 km^3 of basalt between the mantle and oceanic crust in only 150 Ma. However, we do not know what part of the basalt extracted from the sub-lithospheric CBL can be found in the present-day continental crust, what part has been trapped in the sub-crustal MBL, and how much has been recycled back into the mantle. The answers to these questions are central to any evolutionary model.

3. ADVECTIVE THICKENING AS A MECHANISM FOR TECTOSPHERIC STABILIZATION

Several lines of evidence lead to the conclusion that a thick continental tectosphere was not formed by simple conductive cooling (Jordan, 1988): (1) the growth of a thick TBL by conductive decay requires a monotonic subsidence of the cratons over an extended period since their stabilization, so that they should lie in deep water or be covered by thick sediments, which is not observed; (2) the geoid anomalies for thick TBLs are substantial, so that they should be prominent in the long-wavelength geoid, which they are not; and (3) stability calculations based on a local Rayleigh number indicate that TBLs of sufficient thickness to satisfy the seismic data are likely to be unstable with respect to small-scale convective disruption. It is more likely that the continental tectosphere was produced by dynamic processes prior to crustal stabilization.

If the CBL controls the thickness of the continental TBL, continued growth of the tectosphere after crustal stabilization requires a mechanism that either steadily depletes the continental upper mantle in its basaltic constituents or adds depleted material below the MBL. Impressive outpourings of continental flood basalts do occasionally occur, but the volume depleted by these episodes is trivial compared to the amount of mantle involved in the CBL, only on the order of 10^6 km^3 in a major episode such as the Miocene flooding of the Columbia River Basin (Carlson & Hart, 1987). While the production of flood basalts might be locally important in establishing the geochemical character of the sub-continental mantle in places like the Karoo province of southern Africa (Hawkesworth *et al.*, 1983), the isopycnic hypothesis implies that the amount of basalt missing from the subcontinental upper mantle is orders of magnitude greater than the integrated volume of basalt erupted in continental regions since their stabilization in late-Archæan, early-Proterozoic time.

Oxburgh & Parmentier (1978) have suggested that the CBL beneath the cratons increases progressively by the addition of depleted mantle rising diapirically from subducted oceanic plates. It is certainly possible (perhaps necessary) that a significant amount of peridotite depleted by ridge-crest or oceanic hot-spot vulcanism has been incorporated into the continents over the course of geological history, especially by the lateral transfer of a thickened oceanic CBL (*e.g.*, oceanic plateaus and aseismic ridges) across subduction zones (Vogt *et al.*, 1976; Jordan, 1979a). However, for the progressive underplating of the cratons to be a viable model for tectospheric growth requires that the processes of thermal decay and chemical accretion operate on nearly identical spatial and temporal scales; otherwise, the balance between chemistry and temperature implied by the isopycnic hypothesis would be difficult to establish and maintain. This seems implausible.

In fact, any one-stage process suffers from this regulation problem. Multi-stage formation is needed to decouple the process of basalt depletion from the process of CBL thickening. One attractive mechanism for the latter is the advective thickening of the tectosphere during major episodes of orogenic compression, such as those which accompany the formation of Pangea-like supercontinents. The advective-thickening hypothesis is the basis for the multi-stage model proposed by Jordan (1978, 1981*a*). Vulcanism along the active continental margins, and to a much lesser extent in intracontinental rifts, generates buoyant, refractory mantle resistant to convective recycling. Augmenting this material is any depleted peridotite transferred from the oceanic CBL by lateral accretion across subduction zones or by diapirism from descending slabs. Dispersed regions of buoyant mantle and their superjacent (but not chemically complementary) crustal columns are amalgamated during the course of plate motions and eventually swept up and accreted to the major continental masses. Large-scale compressive events at convergent plate boundaries, particularly the violent episodes of continent-continent collision, further consolidate and thicken both the CBL and the TBL by downward advection.

Compressive thickening of the continental crust by a factor of two or more has recently occurred in the Tibetean Plateau and is thought to be the primary mechanism for generating ancient terrains of high metamorphic grade observed elsewhere (Dewey & Burke, 1973; Windley, 1977; Park, 1982). During such an event, a similar amount of thickening takes place within the sub-crustal tectosphere. Because it occurs at plate-tectonic speeds of centimeters per year, this thickening is essentially instantaneous relative to the time scale for thermal decay. If there is no chemical compensation, the advection of the geotherms leads to an instability; the thickened TBL is swept into a downgoing plume and a thin-plate regime is quickly reestablished (Houseman *et al.*, 1981). On the other hand, if a nearly isopycnic situation is enforced prior to thickening, the potential energy available to drive convection is not substantially increased by the advection, and a large-scale instability is not initiated. Of course, in such a violent episode, some portions of the tectosphere are likely to be recycled back into the convecting mantle, as advocated by McKenzie & O'Nions (1983), but most of the basalt-depleted, low-density peridotite presumably remains as part of the CBL. Therefore, in a highly idealized, uniform compression by two-dimensional flow with a thickening factor f (McKenzie, 1978; Houseman *et al.*, 1981), a neutrally stable volume satisfying eqn. (1) at depth z will be advected along an adiabat to a depth of f times z without perturbing the (horizontal) isopycnic surfaces.

In more realistic situations, these constructive processes will be competing against the destructive, double-diffusive instabilities associated with thermally uncompensated chemical heterogeneities and differential heating. Instabilities of the double-diffusive type are intrinsic to convection in chemically heterogeneous systems, such as thermohaline circulation in the oceans (Turner, 1973). Although the theory of double-diffusive convection is beginning to be applied to some geological systems, such as mixing in magma chambers (*e.g.*, Sparks *et al.*, 1984; Hansen & Yuen, 1987; Clark *et al.*, 1987), suprisingly little work has been done on the large-scale dynamics of the earth's chemical boundary layers. An exception is a linear stability analysis presented in abstract form by Stevenson (1979), who argues that the growth time of small-scale, double-diffusive instabilities along cratonic margins is so rapid (~200 Ma) that a thick, chemically compensated cratonic tectosphere would be hard to maintain for the thousands of million years envisaged by Jordan (1978). A more elaborate treatment of this important dynamical problem is needed, however, since Stevenson's analysis was done for a constant viscosity fluid, and in the case of a cold, mature craton, the sorts of instabilities he considers will tend to be damped by the strong temperature-dependence of viscosity.

On the other hand, small-scale, double-diffusive instabilities are likely to play a crucial role in setting up and regulating the compositional stratification implied by the isopycnic hypothesis. To attain stability, the tectosphere must not only be compositionally stratified, its thermal structure must have evolved to a nearly steady state. In regions where it has not, or where it is perturbed by convective upwellings from the deep mantle, double-diffusive instabilities can lead to disruption and extensional thinning of the CBL. They presumably play a major role in continental rifting and vulcanism and may even contribute to the initiation Atlantic-type drift (Jordan, 1978). Perhaps only after repeated cycles of differentiation, assembly, thickening, and disruption of the continental CBL would the interaction of large-scale compression and small-scale, double-diffusive instabilities generate a stable cratonic tectosphere.

Given the dynamical complexity of a heterogeneous system of thermal, mechanical and chemical boundary layers, a numerical simulation of tectospheric evolution is by no means a straightforward task. In addition to contributing to the buoyancy driving the system, chemistry plays a role through the differential heating by local concentrations of radiogenic elements within the CBL; furthermore, melt migration provides a path for the differential transport of major and minor elements. Testing the feasibility of hypotheses for the formation and stabilization of the continental tectosphere is nevertheless a key problem for dynamical modeling.

4.   AGE OF THE CONTINENTAL TECTOSPHERE

Whereas the oceanic tectosphere can be thought of as a cooling igneous assemblage with a single age of formation, the continental tectosphere is best characterized as a metamorphic complex resulting from a long series of tectonic events. The history of the sub-crustal tectosphere is probably just as messy as the continental crust itself, if not more so. In this context, the occasional samples of these metamorphic rocks transported to the surface as xenoliths by deep-seated vulcanism (Nixon, 1987) are very precious, because they provide the few data pertaining to this history.

In a multi-stage model involving stabilization by advective thickening, the primary episodes of basalt extraction and extended intervals of conductive cooling are required to precede cratonization; i.e., the overall TBL density structure of an ancient craton must date from the time of its crustal stabilization. Therefore, the stabilization ages of sub-crustal tectospheric columns can be inferred from the ages of their superjacent crustal assemblages. Jordan (1981a) emphasized that this is a nontrivial and testable consequence of the evolutionary model, pointing out that the data from South Africa, in particular the lead ages obtained by Kramers (1979) on sulphide inclusions in diamonds, are consistent with the model's predictions.

In a recent, more definitive study of southern African diamond inclusions, Richardson et al. (1984) have obtained Sm-Nd and Rb-Sr model ages of 3200-3300 Ma for sub-calcic garnets encapsulated by diamonds. Peridotitic inclusions in diamonds from the Kaapvaal craton (Gurney et al., 1979) show residual major-element compositions characteristic of the entire kimberlite nodule suite (Mathais et al., 1970), providing strong evidence that a basalt-depleted CBL extends below the minimum depth of diamond stability. Although Richardson et al. (1984) did not comment on the tectospheric model, their age data are consistent with diamond formation during an episode of advective thickening of the CBL prior to the late-Archæan (Hunter, 1974) stabilization of Kaapvaal craton (see also Shimizu and Richardson, 1987).

If a thick, basalt-depleted CBL was formed by this time, then so must have been a thick TBL, since the low-density CBL would not otherwise be stable. This consequence of the isopycnic hypothesis is corroborated by the application of geothermometry to monomineralic silicate inclusions in South African diamonds, which show equilibration temperatures of 900-1200°C (Boyd *et al.*, 1985), similar to the range observed in the granular garnet lherzolite xenoliths (Finnerty and Boyd, 1987). The former were frozen in at the time of diamond formation, presumably in the Archæan, whereas the latter were equilibrated in the Cretaceous (Boyd & Gurney, 1986). The diamond-inclusion data are consistent with a variety of other evidence, summarized by Burke & Kidd (1978), Davies (1979), and Richter (1985), that the geotherm beneath the cratons was not much different than it is today. Estimates of subcontinental temperatures in the Archæan calculated from convection models invoking simple TBL cooling (oceanic-type tectosphere) invariably give much higher temperatures and imply significant melting, inconsistent with these observations (Davies, 1979; Richter, 1985).

Most of the regions with thick (> 200 km) TBLs deduced from seismic and heat-flow data occur within the ancient cratons having Archæan and early Proterozoic ages. If we accept the inference that these tectonic ages reflect the stabilization dates of the sub-crustal tectospheric column, then we are led to conclude that the depletion and advective-thickening processes which produced the extant cratons were operating most efficiently prior to mid-Proterozoic time, and that the average thickness of the continental TBL created since then has been decreasing.

5. FORMATION OF ARCHÆAN CRATONS

The hypothetical scenario for the development of the continental tectosphere by repeated cycles of differentiation and thickening illustrates that a thick CBL is a plausible consequence of the present-day Wilson cycle (Jordan, 1978). However, the mechanisms responsible for forming thick Archæan-type cratons have been much less effective since the mid-Proterozoic, and it is appropriate to question this uniformitarian principle. For example, Boyd (1987) has pointed out that the South African kimberlite xenoliths with low equilibration temperatures (< 1100°C) and high values of N_{Mg} (> 91.5) may be residua (and perhaps cumulates) associated with the generation of komatiitic, rather than basaltic, liquids. If so, the rocks with the low normative densities needed to satisfy the isopycnic hypothesis at shallower depths (< 150 km) within the continental tectosphere may have been generated by Archæan magmatic processes no longer operating in modern tectonic regimes (Bickle, 1986).

A more fundamental problem with an actualistic model concerns the greenstone belts that make up a significant fraction of the Archæan cratons (Condie, 1981). Whereas the high-grade gneiss-granulite terrains show abundant evidence for the compression and crustal thickening that should accompany the advective thickening of the CBL, the granite-greenstone terrains are characterized by relatively low metamorphic temperatures and pressures. In plate-tectonic models of the sort propounded by Tarney *et al.* (1976), Burke *et al.* (1976), and Windley & Smith (1976), the greenstones are viewed as back-arc basins that have been obducted onto Andean-type continental margins. The collisional tectonics associated with this one-stage process does not appear to be sufficient to advectively thicken the CBL to cratonic values, at least in modern subduction environments.

Perhaps the advective thickening responsible for stabilizing the upper mantle beneath the granite-greenstone terrains took place without much involvement of the upper crust; say, by the stacking of cold, depleted peridotitic layers along lower- or sub-crustal décollement faults. It should be noted, however, that some greenstone belts have been recently reinterpreted as thin thrust sheets obducted and deformed during episodes involving much higher sub-horizontal strains than in previous schemes for their development. Jackson *et al.* (1987) discuss the Archæan terrain around the Barberton greenstone belt from this point of view, and Park (1982) outlines a general model for Archæan crustal tectonics that invokes extensive crustal shortening by piggyback underthrusting. Park (1982) explicitly states that "the excess exhausted mantle material not dense enough to sink may itself be piled up in successive layers below the thickened continent forming a thick continental lithospheric 'root'." Whether or not the data on Archæan crustal development are consistent with a multi-stage history of CBL formation implied the advective thickening hypothesis is an interesting and, at this point, open question.

6. THE CONTINENTAL CBL AS A GEOCHEMICAL RESERVOIR

The cold, refractory CBL of the continental tectosphere constitutes a large and particularly stable reservoir for large-ion lithophile (LIL) elements: the removal of easily fusible components during primary differentiation of basaltic (or komatiitic) magmas leaves behind peridotites that are resistant to large-scale melting and, like a dry sponge, can soak up any volatiles and highly differentiated liquids fluxing from below (Jordan, 1978, 1981a). Abundant evidence that the CBL has been impregnated by LIL-rich fluids is now available from a number of petrological and geochemical studies on kimberlite xenoliths (e.g., Harte et al., 1975; Menzies & Murthy, 1980; for a comprehensive review, see Menzies & Hawkesworth, 1987). In some cases, the alteration is petrographically evident owing to replacement textures and the development of hydrous phases within the rocks (the "patent metasomatism" of Dawson (1982) or "modal metasomatism" of Harte (1983)); in others, trace-element enrichment occurs apparently unaccompanied by mineralogical changes (Dawson's "cryptic metasomatism"). There is considerable debate about the nature of these enrichment events; whether, for example, they result from "true" metasomatism by low-density fluids or alteration by the infiltration of partial melts (Menzies & Hawkesworth, 1987).

When and where the enrichment events occur are the key questions for an evolutionary model. The ubiquitous and diachronous emplacement of kimberlites in cratonic regions indicates that some LIL-rich material has fluxed from the sub-tectospheric mantle, progressively modifying the CBL after it was stabilized by advective thickening. Erlank et al. (1987) argue that the modal metasomatism which introduced phlogopite and K-richterite into peridotite nodules from the Kimberley pipes was caused by the upward migration of potassium-charged, H_2O-rich fluids during the 100 Ma preceding kimberlite emplacement, perhaps in an episode related to Karoo volcanicity. On the other hand, the isotopic data demonstrate that the minor-element compositions of high-N_{Mg} peridotite nodules have been altered by cryptic metasomatism that took place much earlier (Menzies & Murthy, 1980; Kramers et al., 1983; Richardson et al., 1985). The isotopic and trace-element compositions of diamond inclusions suggest that significant LIL enrichment occurred beneath South Africa in the interval between basalt depletion and diamond crystallization (Richardson et al., 1984; Shimizu & Richardson, 1987).

In the context of the actualistic model discussed by Jordan (1978), the most likely site for this latter type of alteration is the mantle wedge above the descending slab, the metasomatic agent being the fluids released from the slab itself. This suggests a scenario where subduction drives the initial differentiation of the CBL as well as its subsequent LIL enrichment prior to advective thickening; continued LIL enrichment may then proceed, although presumably at a much lower average rate. (Again, the question must be raised as to whether or not the subduction processes were fundamentally different during the Archæan.)

As an example of a rock that might typify the sub-crustal CBL beneath a craton, consider the granular garnet peridotite AJE25 analyzed by Richardson *et al.* (1985) and Erlank *et al.* (1987). This xenolith shows no modal metasomatism, but an ancient (> 1 Ga) enrichment event is indicated high strontium and low neodymium ratios in unaltered garnet ($^{87}Sr/^{86}Sr$ = 0.7055, $^{143}Nd/^{144}Nd$ = 0.5126). Its whole-rock Nd and Sm contents are 2.37 and 0.394 ppm; 15% and 11%, respectively, of Taylor & McLennan's (1985) bulk crustal values. Peridotites showing evidence of modal metasomatism (GPP, PP and PKP suites) have light rare earth abundances that are typically several times greater than AJE25 (Erlank *et al.*, 1987, Fig. 20). The geophysical arguments indicate that the volume of the mantle included in the continental tectosphere probably exceeds the crustal volume by a factor of three to five, so that the subcrustal CBL's total inventory of light rare earths (and presumably other incompatibles) could easily be comparable to the crust's.

Taking this enrichment into account significantly increases the LIL inventory and heat production of the tectosphere over models which consider the crust to be the only significant component of the continental CBL (*e.g.,* Sclater *et al.,* 1980). Given the rather large consequences for mantle models, there has been surprisingly little effort by the geochemical and geophysical communities to include the subcrustal CBL into quantitative models of mantle dynamics. It has been popular, for example, to estimate the volumetric fraction of the mantle that acts as a source for mid-ocean ridge basalt (MORB) from Nd-Sr systematics by assuming that the MORB reservoir has been depleted in LILs by the extraction of the continental crust (Jacobsen & Wasserberg, 1979; O'Nions *et al.,* 1979; Allègre *et al.,* 1983). Although the uncertainties are large (Allègre *et al.,* 1983), the volume of the MORB reservoir thus obtained is about equal to the upper-mantle volume, and the results have been used to argue for convective stratification at the 650-km discontinuity (*e.g.,* Turcotte & Kellog, 1986). Although there is now general recognition that the sub-crustal CBL exists and may be geochemically peculiar, none of the global mass-balance calculations known to the author have attempted to quantify it as a long-term LIL reservoir isolated from mantle convection like the continental crust; doing so should raise the estimates of the MORB source volume to be near that of the entire convecting mantle.

7. IMPLICATIONS FOR MANTLE HEAT FLOW

From a geophysical perspective, the most interesting incompatible elements in the continental upper mantle are the radioactive isotopes of U, Th and K. The introduction of these heat-producing elements to the continental CBL by mantle metasomatism can contribute significantly to the surface heat flux. Because they tend to be concentrated in minor phases, along grain boundaries, and in other unstable sites susceptible to contamination by the host kimberlite and enrichment events precursory to eruption, the content of U, Th and K inherited from tectospheric formation is difficult to infer from the xenolith data. In an attempt to get a crude bound on the contribution of the sub-crustal CBL to the surface heat flux, the author used the whole-rock potassium content of an average continental garnet lherzolite (ACGL) and an assumed K/U ratio to estimate the heat production (Jordan, 1981*a*). It was found that the integral over the column could be as much as 15 mW/m^2, over half of the heat flux at the base of the crust in stable continental environments.

Fig. 2 is a graph of heat production per unit volume as a function of K and K/U for Th/U = 3. Plotted on this graph are six clinopyroxenes from southern African peridotite xenoliths whose potassium and uranium contents were measured by Kramers (1979) and Kramers *et al.* (1983); K varies from 181 to 1169 ppm and K/U from 4.5×10^3 to 2.4×10^4, respectively, yielding heat productions in the range .01-.05 μW/m^3. The ACGL has a relatively high potassium content, K = 913 ± 90 ppm (Jordan, 1978), presumably owing to the high K_2O associated with the minor but ubiquitous mica and amphibole phases found in most metasomatized nodules (Menzies & Hawkesworth, 1987). Allowing K/U to vary over the full range displayed by the cpx data gives a heat production of .03-.13 $\mu W/m^3$, corresponding to an integrated flux of 3-13 mW/m^2 for a nominal 100-km-thick layer. Jordan's (1981a) estimate is somewhat greater than the upper bound obtained here because the former was based on a K/U ratio of 3000, which lies below the lowest cpx point.

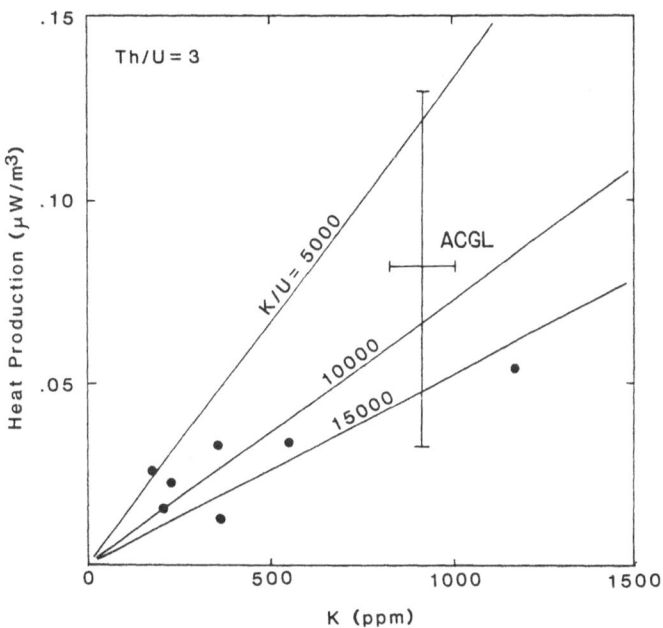

FIGURE 2. Heat production versus potassium content for a thorium/uranium ratio of 3 and various uranium/potassium ratios. Point labeled ACGL is the estimated value for Jordan's (1978) average continental garnet lherzolite (K = 913 ± 90 ppm). Solid circles correspond to measurements on clinopyroxenes from granular kimberlite xenoliths by Kramers (1979) and Kramers et al. (1983). Vertical error bar on ACGL includes the range of K/U observed in Kramer's data ($4.5 \times 10^3 \le$ K/U $\le 2.4 \times 10^4$). The high heat production of the sub-crustal CBL inferred from the xenolith data indicates that the heat flux into the base of the cratonic CBL is small, providing evidence that the dynamics of the continental tectosphere is strongly coupled to the convecting mantle.

Of course, the validity of this calculation is easily criticized: all of the assumptions are subject to large potential errors, and more sophisticated approaches (and better data!) are sorely needed. However, what makes a high heat production plausible to the author is the corroborating evidence derived from the low equilibration temperatures of the granular xenoliths. This additional evidence comes from a variation on a calculation discussed by Jordan (1975a). For a conductive geotherm $T(z)$ in a layer of thickness $\Delta z = z_1 - z_0$ with constant conductivity k and heat production A, the steady-state temperature equation can be solved to give the expression

$$A = 2Q_0/\Delta z - 2k \, \Delta T / \Delta z^2 \qquad (3)$$

where $\Delta T = T_1 - T_0$ is the temperature drop across the layer and Q_0 is the heat flux through its upper boundary. Reasonable crustal models give $T_0 = 430 \pm 50°C$ and $Q_0 = 25 \pm 5$ mW/m^2 for an M discontinuity at $z_0 = 35$ km (Pollack & Chapman, 1977; Sclater et al., 1980). The application of thermobarometry to granular garnet lherzolite nodules yields a temperature of $1000 \pm 100°C$ at a depth of 150 km beneath the Kaapvaal and other continental cratons (Finnerty & Boyd, 1987), consistent with the large thickness of the TBL inferred from seismology and the analysis of the flexural data discussed previously. Hence, for $\Delta z = 115$ km, we obtain $\Delta T = 550 \pm 111°C$. Taking $k = 3.4$ W/m/°C (Sclater et al., 1980) yields $A = .15 \pm .10 \, \mu$W/m^3, which is in the same ballpark as the heat production estimated from the potassium content of the ACGL (Fig. 2).

Looked at another way, eqn. (3) says that if the subcrustal heat production is zero, then the geotherm is linear and extrapolates to $T_1 = T_0 + Q_0 \Delta z / k = 1276°C$, which is too hot to satisfy the xenolith data. Some curvature is needed to bend the geotherm to lower temperatures, and curvature is proportional to heat production.

A high heat production in the subcrustal CBL implies that the heat flow into the base of the tectosphere is significantly lower than the flux across the crust-mantle boundary. Since $Q_1 = Q_0 - A\Delta z$, eqn. (3) requires

$$Q_1 = 2k \, \Delta T/\Delta z - Q_0 = 8 \pm 8 \text{ mW/m}^2 \qquad (4)$$

Subtracted from this value will be any heat production in the CBL below 150 km. Jordan (1975a) used slightly different numerical values but reached essentially the same conclusion: very little of the internal heat being convectively transported out of the earth is escaping through the continental cratons. Knowing why (4) is so small is fundamental to understanding mantle dynamics.

8. CONTINENTS AND MANTLE DYNAMICS

The discrepancy in the heat coming from the deep mantle below the continents and oceans has been recognized as a major dynamical problem since the first observations of the approximate equality in continental and oceanic surface heat fluxes (Bullard, 1952). Sclater et al. (1980, 1981) attempted to minimize this discrepancy by constructing continental models with thin TBLs and negligible heat production below the upper crust. The problem, of course, is that such end-member models are inconsistent with a wide variety of seismic and other data. The thick TBL observed beneath old continents requires lower geothermal gradients and, as we have seen, a greater concentration of U, Th and K in the CBL, which increases the heat-flow discrepancy. If the heat production of the continental CBL is as high as the arguments in the last section suggest, then the heat flux at the base of the TBL beneath cratons cannot be much greater than 15-20 mW/m^2, less than half the oceanic value. In fact, it may be nearly zero (eqn. 4).

Heat is brought out of the deep interior almost entirely by advection; therefore, the heat-flow discrepancy implies that the continental CBL is dynamically coupled to large-scale convective flow in the earth's mantle. In a high Rayleigh number system like the mantle, we expect the upward advection of heat to be dominated by narrow plumes of low-viscosity rising material (Morgan, 1971; McKenzie *et al.*, 1974; Loper, 1985). If the planform distribution of the deep-mantle plumes is essentially random with respect to the locations of the continents, then their flow must be diverted away from the continents in some way that preserves the low basal heat flux inferred for the cratonic tectosphere. To do so may require the vertical scale lengths associated with these diverting flows to be fairly large, say, 700 km or more. The subduction of oceanic lithosphere around the periphery of the continents could be the mechanism responsible for this pattern, although its efficacy in this regard seems questionable. Alternatively, the distribution of upwelling may be anticorrelated with the positions of the continents through a coupling between the large-scale structure of the upper and lower boundary layers to the plate-tectonic convection system.

Many of the details remain to be worked out, but recent seismological studies indicate that the aspherical structure of the earth's solid shell is dominated by the boundary layers formed at the free surface and the base of the mantle (Dziewonski & Woodhouse, 1987; Jordan *et al.*, 1988). The role of the 650-km discontinuity in regulating mantle convection remains controversial, but the documentation of significant downward flux through this boundary by slab penetration into the lower mantle suggests that the plate-tectonic return flow occurs at considerable depth (Jordan, 1975*b*, 1977; Creager & Jordan, 1984, 1986*a*). Moreover, the locations of major mantle hotspots display a large-scale organization that correlates with the long-wavelength, lower-mantle heterogeneity evident in the geoid and seismic tomographic images, suggesting a deep source for mantle plumes (Crough & Jurdy, 1980; Richards *et al.*, 1988). Circulation patterns having large vertical scale lengths are favored by these studies. Deep-mantle circulation is also more compatible with the existence of a thick continental tectosphere than models postulating convective stratification into separate upper-mantle and lower-mantle systems (Jordan, 1975*b*).

Furthermore, the density contrast between the mantle and core, like the density jump at the free surface, is likely to act as a trap for chemically differentiated material. Tomographic images derived from core-penetrating and core-reflecting seismic phases have demonstrated the existence of large-amplitude, large-wavelength heterogeneity in the vicinity of the core-mantle interface (Creager & Jordan, 1986*b*; Morelli & Dziewonski, 1987). As in the case of the free surface, it is difficult to account for the magnitude of the asphericity without postulating the existence of a laterally heterogeneous CBL (Jordan, 1979*b*; Creager & Jordan, 1986*b*). Although not much is known about its composition, properties and dynamics, this boundary-layer structure must interact with convection in the mantle and play a crucial role in coupling this flow to convection in the core (Bloxham & Gubbins, 1987).

We envisage, therefore, an earth comprising four major dynamical systems: the two convecting shells of the mantle and outer core, and the two CBLs at the free surface and core-mantle interface. Strong interactions among the low-wavenumber states of these four systems would appear to offer new possibilities for explaining the earth's large-scale, long-term behavior that have not been considered in previous models of mantle dynamics (Jordan *et al.*, 1988).

An example may include the nature of the forces that drive the continents together. The continents are gregarious; that is, they tend to congregate close to one another to form supercontinents. The hemispheric dichotomy of Atlantic (continental) and Pacific (oceanic) domains appears to be a long-standing aspect of earth history, and continental dispersals of the present magnitude are perhaps exceptional (Piper, 1976; McElhinny & McWilliams, 1977; Windley, 1977). Our evolutionary model is consistent with this observation (Jordan, 1981*a*); indeed, it is founded on the ability of large-scale compression to advectively thicken the continents despite their tendency to be dispersed by small-scale, double-diffusive instabilities. Of course, at the most basic level, the force that drives the continents together must be related to thermal buoyancy: a configuration in which the continental CBL is spread out across the earth's surface in a uniform, thin layer much less efficient in removing heat from the deep interior than one with the CBL bunched up in thick patches of more limited surface area on one side of the earth. Quantifying this effect is an important problem for dynamical modeling that strikes at the heart of the heat-flow discrepancy and should lead to a better understanding of the CBLs at both the top and bottom of the mantle.

ACKNOWLEDGEMENTS

This research was sponsored by the Defense Advanced Research Projects Agency and the Air Force Geophysical Laboratory under contract F19628-87-K-0040.

REFERENCES

Allègre, C. J., Hart, S. R., & Minster, J.-F., 1983. 'Chemical structure and evolution of the mantle and continents determined by inversion of Nd and Sr isotopic data, II. Numerical experiments and discussion.' *Earth Planet. Sci. Lett.,* **66**, 191-213.

Bickle, M. J., 1986. 'Implications of melting for stabilization of the lithosphere and heat loss in the Archaean.' *Earth Planet. Sci. Lett.,* **80**, 314-24.

Bloxham, J., & Gubbins, D., 1987. 'Thermal core-mantle interactions.' *Nature,* **325**, 511-13.

Boyd, F. R., 1973. 'A pyroxene geotherm.' *Geochim. Cosmochim. Acta,* **37**, 2533-46.

Boyd, F. R., 1987. 'High- and low-temperature garnet peridotite xenoliths and their possible realtion to the lithosphere-asthenosphere boundary beneath southern Africa.' In Nixon, P. H. (ed.), 1987. *Mantle Xenoliths,* Chichester, Wiley-Interscience, 403-12.

Boyd, F. R., & Gurney, J. J., 1986. 'Diamonds and the African lithosphere.' *Science,* **232**, 472-7.

Boyd, F. R., Gurney, J. J., & Richardson, S. H., 1985. 'Evidence for a 150-200-km thick Archaean lithosphere from diamond inclusion thermobarometry.' *Nature,* **315**, 387-9.

Boyd, F. R., McCallister, R. H., 1976. 'Densities of fertile and sterile garnet peridotites.' *Geophys. Res. Lett.,* **3**, 509-12.

Boyd, F. R., & Mertzman, S. A., 1987. 'Composition and structure of the Kaapvaal lithosphere southern Africa.' In press.

Boyd, F. R., & Nixon, P. H., 1975. 'Origins of the ultramafic nodules from some kimberlites of northern Lesotho and the Monastery mine, South Africa.' *Phys. Chem. Earth,* **9**, 431-54.

Bullard, E. C., 1952. 'Discussion of a paper by Revelle and Maxwell.' *Nature,* **170**, 200.

Burke, K., Dewey, J. F., & Kidd W. S. F., 1976. 'Dominance of horizontal movements, arc and microcontinental collision during the later permobile regime.' In: Windley, B. F. (ed.), *The Earth History of the Earth,* London, Wiley-Interscience, 113-29.

Burke, K., & Kidd, W. S. F., 1978. 'Were Archean continental geothermal gradients much steeper than today?' *Nature,* **272**, 240-41.

Carlson, R. W., & Hart, W. K., 1987. 'Flood basalt volcanism in the northwestern United States.' In: Macdougall, J.D. (ed.), *Continental Flood Basalts,* Reidel, in press.

Clark, S., Spera, F. J., & Yuen, D. A., 1987. 'Steady state double-diffusive convection in magma chambers heated from below.' In: Mysen, B. O. (ed.), *Magmatic Processes: Physiochemical Principles*, Special Pub. 1, Geochemical Society, 289-305.

Creager, K. C., & Jordan, T. H., 1984. 'Slab penetration into the lower mantle.' *J. Geophys. Res.*, **89**, 3031-49.

Creager, K. C., & Jordan, T. H., 1986*a*. 'Slab penetration into the lower mantle beneath the Mariana and other Island Arcs of the Northwest Pacific.' *J. Geophys. Res.*, **91**, 3573-89.

Creager, K. C., & Jordan, T.H., 1986*b*. 'Aspherical structure of the core-mantle boundary from *PKP* travel times.' *Geophys. Res. Lett.*, **13**, 1497-1500.

Condie, K. C., 1981. *Archean Greenstone Belts, Developments in Precambrian Geology*, 3, Amsterdam, Elsevier, 434 pp.

Crough, S. T., & Jurdy, D., 1980. 'Subducted lithosphere, hotspots, and the geoid.' *Earth Planet. Sci. Lett.*, **48**, 15-22.

Dawson, J. B., 1982. 'Contrasting types of upper-mantle metasomatism.' In Kornprobst, J. (ed.) *Kimberlites II: The Mantle and Crust-Mantle Relationships*, Holland, Elsevier, 289-294.

Davies, G.F., 1979. 'Thickness and thermal history of continental crust and root zones.' *Earth Planet. Sci. Lett.*, **44**, 231-38.

Dewey, J. F., & Burke, K. C. A., 1973. 'Tibetan, Variscan, and Precambrian basement reactivation: products of continent collision.' *J. Geol.*, **81**, 683-92.

Dziewonski, A. M., & Woodhouse, J. H., 1987. 'Global images of the earth's interior.' *Science*, **236**, 37-48.

Erlank, A. J., Waters, F. G., Hawkesworth, C. J., Haggerty, S. E., Allsopp, H. L., Rickard, R. S., & Menzies, M., 1987. 'Evidence for mantle metasomatism in peridotite nodules from the Kimberley pipes, South Africa.' In Menzies, M. A., & C. J. Hawkesworth (eds.), *Mantle Metasomatism*. London, Academic, 221-311.

Finnerty, A. A., & Boyd, F. R., 1987. 'Thermobarometry for garnet peridotites: basis for the determination of thermal and compositional structure of the upper mantle.' In Nixon, P. H. (ed.), *Mantle Xenoliths*, Chichester, Wiley-Interscience, 381-402.

Grand, S. P., 1987. 'A tomographic inversion for shear velocity beneath the North American plate.' *J. Geophys. Res.*, **92**, 14065-90.

Grand, S. P., & Helmberger, D. V., 1984*a*. 'Upper mantle shear structure of North America.' *Geophys. J. R. Astr. Soc.*, **76**, 399-438.

Grand, S. P., & Helmberger, D. V., 1984*b*. 'Upper mantle shear structure beneath the northwest Atlantic Ocean.' *J. Geophys. Res.*, **89**, 11465-75.

Green, D. H., & Liebermann, R. C., 1976. 'Phase equilibria and elastic properties of a pyrolite model for the oceanic upper mantle.' *Tectonopyhsics*, **32**, 61-92.

Gurney, J. J., Harris, J. W., & Rickard, R. S., 1979. 'Silicate and oxide inclusions in diamonds from the Finsch kimberlite pipe.' In Boyd F. R. & Meyer, H. O. A. (eds.) *Kimberlites, Diatremes, and Diamonds: Their Geology, Petrology, and Geochemistry*. Washington, D.C., American Geophysical Union, 1-15.

Hansen, U., & Yuen, D. A., 1987. 'Evolutionary structures in double-diffusive magma chambers.' *Geophys. Res. Lett.*, **14**, 1099-1102.

Harte, B., 1983. 'Mantle peridotites and processes - the kimberlite sample.' In Hawkesworth, C. J., & Norry, M. J. (eds.) *Continental Basalts and Mantle Xenoliths*, Cheshire, Shiva, 46-91.

Harte, B., Cox, K. G., & Gurney, J. J., 1975. 'Petrography and geological history of upper mantle xenoliths from the Matsoku kimberlite pipe.' *Phys. Chem. Earth*, **9**,

477-506.

Hawkesworth, C. J., Erlank, A. J., Marsh, J. S., Menzies, M. A., & van Calsteren, P., 1983. 'Evolution of the continental lithosphere: evidence from volcanics and xenoliths in southern Africa.' In Hawkesworth, C. J. , & Norry, M. J. (eds.) *Continental Basalts and Mantle Xenoliths,* Cheshire, Shiva, 111-38.

Houseman, G. A., McKenzie, D. P., & Molnar, P., 1981. 'Convective instability of a thickened boundary layer and its relevance for the thermal evolution of continental convergent belts.' *J. Geophys. Res.,* **86,** 6115-32.

Hunter, D. R., 1974. 'Crustal development in the Kaapvaal Craton.' *Precambr. Res.,* **1,** 259-326.

Jackson, M. P. A., Eriksson, K. A., & Harris, C. W., 1987. 'Early Archean foredeep sedimentation related to crustal shortening: a reinterpretation of the Barberton Sequence, Southern Africa.' *Tectonophysics,* **136,** 197-221.

Jacobsen, S. B., & Wasserburg, G. J., 1979. 'The mean age of mantle and crustal reservoirs.' *J. Geophys. Res.,* **84,** 7411-27.

Jordan, T. H., 1975a. 'The continental tectosphere.' *Rev. Geophys. Space Phys.,* **13,** 1-12.

Jordan, T. H., 1975b. 'Lateral heterogeneity and mantle dynamics.' *Nature,* **257,** 745-50.

Jordan, T. H., 1977. 'Lithospheric slab penetration into the lower mantle beneath the Sea of Okhotsk.' *J. Geophys.,* **43,** 473-96.

Jordan, T. H., 1978. 'Composition and development of the continental tectosphere.' *Nature,* **274,** 544-48.

Jordan, T. H., 1979a. 'Mineralogies, densities, and seismic velocities of garnet lherzolites and their geophysical implications.' In Boyd , F. R. & Meyer, H. O. A. (eds.) *The Mantle Sample: Inclusions in Kimberlites and Other Volcanics.* Washington, D. C., American Geophysical Union, 1-14.

Jordan, T. H., 1979b. 'Structural geology of the Earth's interior.' *Proc. Natl. Acad. Sci. U.S.A.,* **76,** 4192-4200.

Jordan, T. H., 1981a. 'Continents as a chemical boundary layer.' *Phil. Trans. R. Soc. Lond.,* **A301,** 359-73.

Jordan, T. H., 1981b. 'Global tectonic regionalization for seismological data analysis.' *Bull. Seismol. Soc. Am.,* **71,** 1131-41.

Jordan, T. H., 1988. 'Structure and formation of the continental tectosphere.' *J. Petrol.,* in press.

Jordan, T. H., Lerner-Lam, A. L., & Creager, K. C., 1988. 'Seismic imaging of boundary layers and deep mantle convection.' In Peltier, W. R., (ed.) *Mantle Convection,* in press.

Kramers, J.D., 1979. 'Lead, uranium, strontium, potassium, and rubidium in inclusion-bearing diamonds and mantle-derived xenoliths from Southern Africa.' *Earth Planet. Sci. Lett.,* **42,** 58-70.

Kramers, J. D., Roddick, J. C. M., & Dawson, J. B., 1983. 'Trace element and isotopic studies on veined, metosomatic and "MARID" xenoliths from Bultofontein, South Africa.' *Earth Planet. Sci. Lett.,* **65,** 90-106.

Lerner-Lam, A. L., & Jordan, T. H., 1987. 'How thick are the continents?' *J. Geophys. Res.,* **92,** 14007-26.

Loper, D. E., 1985. 'A simple model of whole-mantle convection.' *J. Geophys. Res.,* **90,** 1809-36.

Mathais, M., Siebert, J. C. & Rickwood, P. C., 1970. 'Some aspects of the mineralogy and petrology of ultramafic zenoliths in kimberlite.' *Contr. Mineral. Petrol.,* **26,** 75-123.

McElhinney, M. W., & McWilliams, M. O., 1977. 'Precambrian geodynamics - a palaeomagnetic view.' *Tectonophysics,* **40,** 137-59.

McKenzie, D., 1978. 'Some remarks on the development of sedimentary basins' *Earth Planet. Sci. Lett.,* **40,** 25-32.

McKenzie, D., & O'Nions, R. K., 1983. 'Mantle reservoirs and ocean island basalts.' *Nature,* **301,** 229-31.

McKenzie, D. D., Roberts, J. M., & Weiss, N. O., 1974. 'Convection in the earth's mantle: towards a numerical simulation.' *J. Fluid Mech.,* **62,** 465-538.

Menzies, M. A., & C. J. Hawkesworth (eds.), 1987. *Mantle Metasomatism.* London, Academic, 472 pp.

Menzies, M., & Murthy, R. V., 1980. 'Enriched mantle: Nd and Sr isotopes in diopsides from kimberlite nodules.' *Nature,* **283,** 634-36.

Morelli, A., & Dziewonski, A. M., 1987. 'Topography of the core-mantle boundary and lateral homogeneity of the liquid core.' *Nature,* **325,** 678-83.

Morgan, J., 1971. 'Covection plumes in the lower mantle.' *Nature,* **230,** 42-3.

Nixon, P. H. (ed.), 1987. *Mantle Xenoliths,* Chichester, Wiley-Interscience, 844 pp.

Nixon, P.H., Boyd, F.R., & Boullier, A.M., 1973. 'The evidence of kimberlite and its inclusion on the constitution of the outer part of the earth.' In Nixon, P.H. (ed.) *Lesotho Kimberlites.* Lesotho, Lesotho National Development Corp., 312-18.

O'Hara, M. J., 1975. 'Is there an Icelandic mantle plume?' *Nature,* **253,** 708-10.

O'Nions, R. K., Eversen, N. M., & Hamilton, P. J., 1979. 'Geochemical modeling of mantle differentiation and crustal growth.' *J. Geophys. Res.,* **84,** 6091-101.

Oxburgh, E. R., & Parmentier, E. M., 1978. 'Thermal processes in the formation of continental lithosphere.' *Phil. Trans. R. Soc. Lond.,* **A288,** 415-29.

Park, R. G., 1982. 'Archaean tectonics.' *Geol. Rundsch.,* **71,** 22-37.

Piper, J. D. A., 1976. 'Palaeomagnetic evidence for a Proterozoic supercontinent.' *Phil. Trans. R. Soc. Lond.,* **A280,** 469-90.

Pollack, H. N., & Chapman, D. S., 1977. 'On the regional variation of heat flow, geotherms, and lithospheric thickness.' *Tectonophysics,* **38,** 279-96.

Richards, M. A., Hager, B. H., & Sleep, N. H., 1988. 'Dynamically supported geoid highs over hotspots: observations and theory.' *J. Geophys. Res.,* in press.

Richardson, S. H., Gurney, J. J., Erlank, A. J., & Harris, J. W., 1984. 'Origins of diamonds in old enriched mantle.' *Nature,* **310,** 198-202.

Richardson, S. H., Erlank, A. J., & Hart, S. R., 1985. 'Kimberlite-borne garnet peridotite xenoliths from old enriched subcontinental lithosphere.' *Earth Planet. Sci. Lett.,* **75,** 116-28.

Richter, F. M., 1985. 'Models for the Archean thermal regime.' *Earth Planet. Sci. Lett.,* **73,** 350-60.

Ringwood, A. E., 1966. 'The chemical composition and origin of the earth.' In Hurley, P. (ed.) *Advances in Earth Composition.* Cambridge, Mass., M.I.T. Press, 287-356.

Ringwood, A. E., 1975. *Composition and Petrology of the Earth's Mantle.* New York, McGraw-Hill, 619 pp.

Sclater, J. G., Jaupart, C., & Galson, D., 1980. 'The heat flow through oceanic and continental crust and the heat loss of the Earth.' *Rev. Geophys. Space Phys.,* **18,** 269-311.

Sclater, J. G., Parsons, B., & Jaupart, C., 1981. 'Oceans and continents: similarities and differences in the mechanisms of heat loss.' *J. Geophys. Res.,* **86,** 11535-52.

Shimizu, N., & Richardson, S. H., 1987. 'Trace element abundance patterns of garnet inclusions in peridotite-suite diamonds.' *Geochem. Cosmochem. Acta,* **51,** 755-8.

Sparks, R. S., Huppert, H. E., & Turner, J. S., 1984. 'The fluid dynamics of evolving magma chambers.' *Phil. Trans. Roy. Soc. Lond.,* **A310,** 511-534.

Stevenson, D., 1979. 'Double diffusive instabilities in the mantle.' (abstract) *I.U.G.G. 17th Gen. Assembly, I.A.S.P.E.I. Abstr.* Canberra, Australian National University, 72.

Taylor, S. R., & McLennan, S. M., 1985. *The Continental Crust: its Composition and Evolution.* Oxford, Blackwell, 312 pp.

Tarney, J., Dalziel, I. W. D., & de Wit, M. J., 1976. 'Marginal basin "Rocas Verdes" complex from S. Chile: a model for Archaean greenstone belt formation.' In Windley, B. F., (ed.) *The Earth History of the Earth.* London, Interscience, 131-46.

Turcotte, D. L., & Kellogg, L. H., 1986. 'Isotopic modeling of the evolution of the mantle and crust.' *Rev. Geophys. Space Phys.,* **24,** 311-28.

Turner, J. S., 1973. *Buoyancy Effects in Fluids,* Cambridge, Cambridge, 368 pp.

Vogt, P. R., Lowrie, A., Bracey, D. R., & Hey, R. N., 1976. 'Subduction of aseismic oceanic ridges: effects on shape, seismicity, and other characteristics of consuming plate boundaries.' *Geol. Soc. Amer. Spec. Paper,* **172,** 59 pp.

Windley, B. F., 1977. *The Evolving Continents.* Chichester, John Wiley & Sons, 385pp.

Windley, B. F., Smith, J. V., 1976. 'Archaean high-grade complexes and modern continental margins.' *Nature,* **260,** 671-75.

Woodhouse, J. H., & Dziewonski, A. M., 1984. 'Mapping the upper mantle: three-dimensional modeling of earth structure by inversion of seismic waveforms.' *J. Geophys. Res.,* **89,** 5953-86.

SUBJECT INDEX